PERSONAL DATA PROTECTION LAWS IN INDIA

Recent Developments

Editors

Dr. Harish Kumar Verma
Ms. Priti Chaudhari
Dr. Bhumika Sharma

Table of Contents

Foreword

The collection of papers presented here provides a comprehensive overview of the key issues related to data protection. The papers cover a wide range of topics, from the conceptual framework of privacy to the specific provisions of the Digital Personal Data Protection Law at national and international levels.

One of the key themes that emerges from the papers is the importance of data protection in the digital age, especially for the Global South countries. As our lives become increasingly digitized, our personal data is being collected and stored by a wide range of entities. This data can be used to track our movements, monitor our online activity, and even predict our future behavior. This raises serious concerns about privacy and data security. We cannot ignore the role of digital colonialism per se, and the 'platformisation' of data, especially by FAAMG companies in the United States and BAT companies in China.

Now, we are aware of the fact that India has a large and diverse population, and it is difficult to ensure that all data controllers and processors are complying with the law. Additionally, the Indian legal system is complex and can be slow to respond to new challenges. In addition, the tendencies of policy paralyses sometimes could lead to regulatory subterfuge as well.

Despite the challenges, India is committed to protecting personal data, which would inspire better standards for democratic Asia against the interventionist exploitation of personal data by major powers and entities, especially those based out of the United States and China. Data scraping, for example, is one of those pertinent and immediate issues to be addressed. The Digital Personal Data Protection Laws is a significant step forward in this regard. The Act provides a comprehensive framework for regulating the collection, use, and disclosure of personal data. It also gives individuals a number of rights, including the right to access their data, the right to correct their data, and the right to be forgotten.

Yet, implementing a Data Protection Law is just the tip of the iceberg. The next set of technology law issues concerning our social and economic lives, will be deeply connected to AI systems used to collect and process data at an unprecedented scale. It is important to ensure that AI systems are used in a way that respects privacy and data protection rights. Generative AI systems and their use cases already reflect a pattern. Unless their use cases become sustainable to survive six-twelve months, regulating AI on specific industrial aspects will remain a sweet spot for India and the Global South countries. Hence, the misuse and exploitation of personal data in such turbulent and interesting times can for surely be dealt with India's Digital Personal Protection law. There is optimism and anticipation for the potential of the Data Protection Board of India and the Competition Commission of India to shape global consensus and dilemmas on technology sovereignty and the regulation of artificial intelligence technologies.

Abhivardhan
Managing Partner
Indic Pacific Legal Research
Chairperson and Managing Trustee
Indian Society of Artificial Intelligence and Law

Preface

Justice Louis Brandeis and Samuel Warren as early as 1890 published an article in Harvard Law Review titled "The Right to Privacy" sowing the seeds for this intrinsic human right. Over the past two centuries, the society has witnessed tremendous transformation regarding the concept, nature, and ambit of the right to privacy. Telecommunication advances have led to recognition of naive right of data protection. As per the *United Nations Conference on Trade and Development, 71%* countries of the world (approximately 137 out of 194 countries) have put in place an exclusive legislation to assure the protection of data and privacy.

As far as India is concerned, last decade is crucial to trace the evolution of data protection regime. Information Technology (Reasonable Security Practices and Procedures and Sensitive Personal Data or Information) Rules (Privacy Rules), 2011 were notified under the Information Technology Act, 2000. In August 2011, India's Ministry of Communications and Information issued a 'Press Note' Technology (Clarification on the Privacy Rules). In November 2022, the Government of India released the long-anticipated draft the Digital Personal Data Protection Bill. Prior to this, in August 2022, the Government withdrew earlier draft Bill- the Personal Data Protection Bill, 2019. India is in the process of refurbishing its personal data protection regime.

The book is an attempt to deliberate the global and national developments regarding both the right to privacy and the right to data protection. The constitutional and emerging legal frameworks have been deeply touched by the contributors. It will definitely help to understand the emerging regime of data protection in Indian context. The Chapters reflect the need for introducing 'Habeus Data' Writ in our country.

Chapter 1

Privacy On Social Media: Do's and Don'ts

Vidhi Maheshwari
B.B.A.L.L.B. (Hons.) Student
SSOL, Sharda University
Greater Noida, UP, India

Dr. Nituja Singh
Assistant Professor
SSOL, Sharda University
Greater Noida, UP, India

Abstract

Social media platforms are designed for sharing various forms of information, be it photos, articles, or personal reflections. Typically, these platforms employ standardized contracts, leaving users with limited room to negotiate or modify the terms. Consequently, users are obliged to accept the Terms and Conditions or Privacy Policy to access and use the particular social media service or website. Major social media players like WhatsApp, Instagram, Facebook, Telegram, and others impose such mandatory acceptance of their Terms and Conditions. However, from a broader perspective, these Terms and Conditions and Privacy Policies can potentially infringe upon users' fundamental Right to Privacy, putting their personal information at risk of unauthorized access or exploitation. Active users of social media are statistically more likely, around 30%, to fall victim to identity fraud. Notably, individuals with accounts on platforms like Snapchat, Facebook, and Instagram face a higher risk, with a 46% increased likelihood of such incidents. While social media encourages sharing, it's crucial to exercise caution to safeguard personal safety and financial records. Implementing basic practices, such as customizing privacy settings, refraining from revealing current locations, and establishing robust passwords, can significantly mitigate these risks. Ensuring the security of one's identity on social media is a vital step toward protecting both financial well-being and personal safety. Ultimately, the internet can be a safe space for those who are mindful of its potential risks and are proactive in taking measures to shield themselves.

Keywords - Privacy, risk, Social Media, setting, duty.

I. Introduction

Social Media is not just a source of generating data but also a source of communication between the sources and the end users of the data. The most used social media networks include Google, Instagram, Snapchat, YouTube, Facebook and many more such applications which are used by every age group. It has now become a part of one's daily life. Keeping our social media handles up-to-date and keeping ourselves updated with the news, ongoing activities and trends has now become a part of our lifetime without which one feels boredom. Not just by sharing their own pictures, texts, videos of their whereabouts but even by uploading images of assignments, question papers etc. along with sharing music, jokes, dance videos etc. also. No matter which field it is, social media is now used actively everywhere and for all purposes.

Its excessive use (i.e., increase in the per application usage time-period) is now becoming a bane for individuals as well as for people at large for the breach of privacy of their users. Along with it, the crime is being increasing due to the share of users' current location which allows the share of their preferences and choices. Therefore, it is required to bring legal advancements and to effectively execute them.

II. Types of Social Networking Sites

The common classification of Social Networking Sites is as:

(i) **Social Networking Sites:** It includes various applications which provide people to share about themselves, their lives current events and allow them to share it with others. It also allows users to make friends, send messages, and to share different types of content with them. It includes Facebook, Twitter, LinkedIn, Google and many more such websites and applications.

(ii) **Media Sharing Sites:** It includes all the applications wherein the photos, videos, current updates of an individual's life using applications like YouTube, Instagram, Snapchat, MySpace etc. which allow the users to share their videos and images online. Though these pictures are shared in a zone which is safe but at times they are used to with an ill-will which create problematic situations for the users.

(iii) **Location based Networks:** Applications like Foursquare, Gowalla etc. have entered via the usage of smart phones. These applications have a feature to choose to share the current locations to their friends in their sites.

In today's highly advanced era of information technology and globalized business, it's impossible to remain unaffected by the information and technologies that carry significant transitional and legal implications. This extends to the implementation of control measures and the establishment of regulations tailored for emerging technologies.

III. Privacy and Security Issues

Privacy can be understood differently by different individuals but can be put together in few words as a right to be in personal space without any interference. Though it can be invaded for 'social good.' While upholding privacy is a significant concern, it should be weighed against other collective interests for the greater good.

Ensuring data privacy is vital to safeguard user information from falling into the wrong hands, where it could be exploited for personal gain. This prevents unauthorized and malicious access, which could lead to the unauthorized disclosure, alteration, or harm to data stored or previously shared online by the user or end user. Though, with the increase in the usage of Online communication, communicating has become an easy task now but it has also created a risk of accessibility of sensitive data which leads to the disclosure of the user's privacy.

(i) Privacy concerns in Social Media Sharing Services

Social Media Sharing Services allows the users to generate their own content using different types of content including YouTube, Instagram etc. which by using images and videos of other's content create their own content. The rising volume of personal content shared online gives rise to escalating privacy apprehensions. Even though the Privacy Policy states to protect the privacy of an individual, still there have been end number of cases wherein youngsters and teenagers have been harassed by pedophiles online which ultimately led to suicides just to save their own dignity.

(ii) Location based Social Networks and Privacy Issues

The key feature of several applications is that one can share the live location of themselves which causes trouble at times. The GPS (Global Positioning System) knows where the person is. In fact, the Google also tracks the person's location to not just find its physical location but also involves sharing details like common interests, behavior, tastes, and interest in what kind of activities according to which the recommendation list shows up. Such applications clearly depict that there is an invasion to the privacy to an individual.

IV. Threats to Privacy

There have been threats to Privacy in the Classic way which have now increased and have advanced themselves into modern ways as well.

(i) **Classic Threats:** These are the attempts which have been made to invade the privacy by various means in order to either harm others by doing spams, malware, phishing etc. which led to data breach, data loss and several other harms.

a. Malware

Malware, short for Malicious Software, is a broad term encompassing harmful software designed to infiltrate and compromise a computer system. Its primary purpose is to gain unauthorized access to private data, information, or content on someone else's computer. Social networks are particularly susceptible to malware attacks due to their application structure and user interactions. One of the most severe forms of malware involves stealing users' login credentials to impersonate them and send deceptive messages to their contacts. Engaging in fraud and distributing malware are illegal activities that may involve users accessing a URL and executing a harmful code on the computer of an Online Service Network user.

b. Phishing Attacks

Phishing is a well-known form of fraudulent activity where an attacker acquires a user's personal information by pretending to be a reliable third party, often using a fabricated or stolen identity. Likewise, in various instances, phishers have utilized social media platforms to impersonate others, occasionally receiving compensation for this risky endeavor.

c. Spam Attacks

Unwanted messages, commonly referred to as spam messages, pose a significant issue. Within Online Service Networks, spam can manifest as posts or instant messages, which can be more hazardous than conventional email spam due to users' extended time spent on these platforms. Typically, they contain advertisements or links that may lead to phishing or malware-infested websites. Essentially, they serve as traps for potential threats. These messages often originate from compromised accounts and automated spamming bots.

d. Web Bugs

They are alternatively referred to as web beacons, which are file elements strategically positioned within a web page or email message to surreptitiously track user activities, akin to spyware. In the internet advertising sphere, they are often euphemistically labeled as 'clear GIFs,' 'invisible GIFs,' or 'Beacon GIFs.' Typically, these elements remain imperceptible to users due to their transparency and color-matching with the background, occupying only minimal space. Such practices represent a significant breach of privacy, especially in an era where a computer serves as a repository for an individual's most valuable information and personal data.

(ii) **Modern Threats:** The Modern Threats to the Privacy include Click Jacking, Fake Profiles, so on.

a. Click jacking

This type of attack, also recognized as a user-interface manipulation assault, employs deceptive methods to prompt online users to click on something different from their intended target. Through a clickjacking attack, perpetrators may even exploit the hardware of users' computers.

b. De-anonymization Attacks

De-anonymization employs data-mining techniques to match anonymous information with public, identifiable data sources, effectively revealing the identity of an individual within the previously anonymous dataset. Current online services often utilize pseudonyms to protect data privacy while still allowing it to be publicly accessible.

c. Fake Profiles

In many social networks, a prevalent form of attack is the creation of fake profiles. In this type of assault, the perpetrator establishes an account on a social network using fabricated information and proceeds to communicate with genuine users. Once these users respond and establish connections, the attacker then inundates them with spam. Typically, these fake profiles are either automated or semi-automated, designed to emulate human behavior. Generating artificial followers and retweets has become a substantial industry in the IT sector, facilitated by these fake profiles, albeit providing misleading information to onlookers.

d. Identity Clone Attacks

Profile cloning is an act where an attacker, by stealing credentials from an existing profile, generates a new counterfeit profile while incorporating the pilfered private data. These stolen credentials may then be deployed within the same network or across various networks.

e. Inference Attacks

In social networks, inference attacks are utilized to anticipate sensitive and private details about a user that they might prefer not to reveal, such as age, gender, as well as religious and political affiliations.

f. Information Leakage

Social media platforms revolve around the open exchange of information among friends. Some users willingly disclose personal details, including health-related data. Regrettably, a small portion of them might divulge excessive private information regarding products, projects, organizations, or other sensitive data. Such sharing of delicate content could potentially lead to adverse consequences for the users.

g. User Profiling

User profiling involves the examination of regular user actions within an OSN platform, often employing diverse machine-learning methods. This practice can potentially result in the exposure of private information, as user profiles often contain personal details. Consequently, safeguarding against user profiling is imperative to address privacy concerns within an OSN setting.

h. Surveillance

Social media surveillance represents a novel form of observation, distinct from assessing a person's sociability and societal roles in areas like politics, the economy, and civil society. It involves the systematic tracking of users' diverse activities across their various social roles, relying on their profiles and interactions with others. This form of surveillance employs technology to monitor human activities on social media platforms.

V. Do's and Don'ts of keeping Privacy

Safeguards that should be keep in mind are discussed as:

(i) Do's for keeping the Privacy

Privacy on social media which is, though of paramount importance has become an issue in the today's digital era. It is crucial to protect your personal information and maintain control over who can access it. Here are some do's to consider:

a. **Adjust Privacy Settings:** Familiarize yourself with the privacy settings of the platforms used. Each platform (e.g., Facebook, Twitter, Instagram) has different options to control who can see your content. Set your profiles to private or restricted access, allowing only approved followers or friends to view your content.

b. **Keep Personal Information Minimal:** Avoid sharing sensitive personal details like your home address, phone number, or financial information. Be cautious about oversharing. Consider the potential consequences of sharing location-specific posts in real-time.

c. **Use Limited Profile Information:** Share only necessary information on your profile. Avoid displaying sensitive data such as your full birthday, hometown, or workplace.

d. **Use Strong, Unique Passwords:** Create strong passwords with a combination of upper and lower-case letters, numbers, and special characters. Avoid easily guessable passwords like "password123" and information like your name, birthday, or common phrases.

e. **Enable Two-Factor Authentication (2FA):** Implement 2FA whenever possible. This provides an additional level of security, necessitating a code from your phone in addition to your password.

f. **Be Cautious with Friend/Follower Requests:** Only accept requests from people you know or trust. Be skeptical and cautious of requests from unknowns, strangers or suspicious accounts.

g. **Review and Adjust App Permissions:** Regularly check which third-party apps have access to your social media accounts and revoke access from ones you no longer use or trust.

h. **Avoid Location Tagging in Real-Time:** Be careful about sharing your location in real-time. Wait until you've left a location before posting about it to avoid revealing your current whereabouts.

i. **Think Before You Post:** Consider the potential consequences of your posts, both in terms of privacy and reputation. Once something is on the internet, it can be difficult to fully remove.

j. **Be Mindful of Direct Messages:** Avoid sharing personal information in direct messages, especially with people you don't know well or trust.

k. **Regularly Review and Update Settings:** Social media platforms may update their privacy policies and settings. Periodically review and adjust them to match your preferences.

l. **Educate Yourself:** Stay informed about the latest privacy features and best practices on the platforms you use. This knowledge will help you make informed decisions.

m. **Report Suspicious Activity:** If you encounter any suspicious activity or believe your privacy has been compromised, report it to the platform and take appropriate action to secure your account.

Remember, the online presence is an extension of your real-life identity. Taking these steps to protect your privacy on social media will go a long way in safeguarding your personal information from unwanted exposure.

(ii)   Don'ts

Some key facts and activities that usually harm on social media are as:

a. **Over-Share Personal Information:** Avoid oversharing personal information that could be used maliciously.

b. **Accept All Friend/Follower Requests:** Don't accept requests from unknown or suspicious accounts. Verify their identity before accepting.

c. **Use of Weak or Common Passwords:** Avoid using easily guessable passwords like "password" or "123456" or personal details like birthday etc. These are a causation to risk and can easily be used to breach privacy.

d. **Ignore Privacy Policy Updates:** Don't disregard notifications about updated privacy policies. Take the time to understand the changes and adjust your settings accordingly.

e. **Engage in Public Arguments or Controversial Discussions:** Avoid heated debates or arguments in public posts. It can attract unwanted attention and potentially harm your reputation.

f. **Post Vacation Plans in Real-Time:** Avoid sharing details about vacations in real-time, as it can signal that your home is empty and vulnerable to theft.

g. **Trust Unknown Sources:** Be cautious when clicking on links or interacting with messages from unfamiliar or unverified sources.

h. **Assume Everything Is Private:** Even with strong privacy settings, remember that nothing is completely private on the internet. Be cautious about what you share.

By following these do's and don'ts, you can significantly enhance your privacy on social media platforms and reduce the risk of unauthorized access to your personal information.

VI. Need for Critical Analysis

Analyzing privacy on social media from a legal perspective reveals a complex and evolving landscape. Laws and regulations governing online privacy vary from country to country, but there are some common themes and challenges:

(i) **Data Protection Laws:** Many countries have enacted data protection laws to regulate how organizations, including social media platforms, collect, process, and store personal data. For instance, the European Union's General Data Protection Regulation (GDPR) has had a global impact by requiring companies to obtain explicit consent for data collection and empower users with increased authority over their data.

(ii) **Terms of Service and Privacy Policies:** Social media platforms typically have terms of service and privacy policies that users agree to when creating accounts. These documents outline how user data will be used and shared. However, they are often lengthy and filled with legalese, making it challenging for users to fully understand their rights and the implications of their consent.

(iii) **Data Sharing and Third Parties:** Social media platforms often share user data with third-party companies for advertising and analytics purposes. This practice has raised concerns about the extent to which users' data is shared and how it is used. Legal frameworks may require platforms to disclose such practices, but the effectiveness of these disclosures can vary.

(iv) **User Consent:** Obtaining meaningful consent for data collection and processing is a significant challenge. Users often grant permissions without fully understanding the implications, and obtaining informed consent can be challenging due to the complexity of privacy policies.

(v) **Children's Online Privacy:** Protecting the privacy of children online is a particular concern. Laws like the Children's Online Privacy Protection Act (COPPA) in the United States require parental consent for data collection from children under 13. However, enforcing these regulations can be challenging.

(vi) **Cross-Border Data Flows:** Social media platforms operate globally, making it difficult to enforce privacy laws consistently across borders. The international nature of these platforms often leads to conflicts regarding jurisdiction and data transfer.

(vii) **Data Breaches and Security:** Legal frameworks also address data breaches and the obligations of companies to notify users when their data is compromised. However, the effectiveness of these laws in preventing breaches and protecting user data remains a subject of debate.

(viii) **Evolving Technology:** The rapid evolution of technology, including artificial intelligence and data analytics, poses challenges for existing privacy laws. These technologies can be used to process vast amounts of user data, potentially raising new ethical and legal concerns.

VII. Conclusions and Suggestions

While legal frameworks exist to address privacy on social media, they often struggle to keep pace with the rapidly evolving digital landscape. Social media privacy emerges as a complex and vital concern that necessitates thoughtful deliberation from individuals, social media entities, and policymakers alike. In the realm of the digital age, achieving equilibrium between the advantages of connectivity and self-expression and the safeguarding of personal data becomes paramount. In the ever-evolving digital landscape, the discourse surrounding social media privacy is poised to undergo continual transformations. Striking a delicate balance between fostering innovation and ensuring protection represents an enduring challenge, necessitating cooperative efforts among users, social media platforms, and regulatory bodies. As technological advancements and societal norms continue to evolve, upholding a steadfast commitment to privacy stands as a fundamental pillar in upholding trust and the integrity of online interactions.

Achieving meaningful privacy protection requires a combination of robust legislation, user education, and proactive enforcement efforts. Users should be vigilant about understanding their rights and the implications of sharing personal information on social media platforms, and governments and regulatory bodies must adapt their approaches to address emerging privacy challenges in the digital age. The guidelines for maintaining social media privacy underscore the significance of proactive and informed user behavior. Key practices encompass vigilant management of privacy settings, the adoption of robust passwords, and the responsible sharing of content. Moreover, exercising prudence in terms of oversharing and adopting a discerning approach to online information assumes pivotal roles.

From a legal standpoint, data protection regulations, terms of service agreements, and privacy policies endeavor to provide a foundation for preserving user data. However, these legal frameworks are not devoid of their inherent complexities, including the challenges related to securing genuine consent and addressing issues concerning the international transfer of data.

Ultimately, it is vital to equip social media users with the requisite knowledge and tools to effectively shield their privacy. This includes remaining well-informed about the policies of the platforms they engage with, comprehending the consequences of data disclosure, and advocating for enhanced privacy safeguards when warranted.

References

1   Kaplan, A.M.; Haenlein, M. Users of the world, unite! The challenges and opportunities of Social Media. Bus. Horiz. 2010.

2   Sakshi Rewaria, "Data Privacy in Social Media Platform: Issues and Challenges" , available at https://papers.ssrn.com/sol3/papers.cfm?abstract_id=3793386.

3   Social media privacy issues , available at https://www.slideshare.net/NousheenArshad1/social-media-privacy-issues.

4   There Isn't Enough Privacy On Social Media And That Is A Real Problem, available at https://www.forbes.com/sites/petersuciu/2020/06/26/there-isnt-enough-privacy-on-social-media-and-that-is-a-real-problem/?sh=7b67dbe744f1.

5   On Privacy and Security in Social Media – A Comprehensive Study, available at https://www.sciencedirect.com/science/article/pii/S1877050916000211.

6   Data Privacy in Social Media Platform: Issues and Challenges, available at https://papers.ssrn.com/sol3/papers.cfm?abstract_id=3793386.

7   Tips for protecting your social media privacy, available at https://us.norton.com/blog/privacy/protecting-privacy-social-media.

8   https://www.technology.pitt.edu/security/best-practices-safe-social-networking

9   https://www.uvic.ca/news/topics/2023+social-media-precautions+notice.

10 https://ico.org.uk/for-the-public/online/social-networking/.

Chapter 2

Data Protection Law: Journey Till Now and the Road Ahead

Somesh Tiwari
Managing Partner
Vardharma Chambers
New Delhi, India

Abstract

Technology is the most imperative tool in the 21st century, it is everywhere. It has made humans efficient like never before. The benefits of using technology are abundant but the less talked and the left-out question in this world where technology is so pierced into human life. Is our data safe in the hands of people who are controlling technology? Data privacy and its protection is one of the most important challenges that we face today, yet it is recently that we have started to talk about it. Part I of the present chapter deals with the genesis of privacy as a concept and how Indian judiciary has been the torch bearer on this topic, it also gives a chronological order on how the privacy has evolved in India. As privacy is an imperative matter, The Government of India have come with a legislation called the 'Personal Data Protection Bill, 2019' which tried to encapsulate the major problems that the public at large faces, but the Bill has not yet been passed by the parliament. Part II of the chapter explains this aspect. In this paper, an attempt has been made to explain the variety of challenges that India is facing into making data protection law a reality. Also, a view has been presented on how other countries apart from India have been tackling this problem, and what the future holds for us in the realm of data protection.

Keywords: Data Protection, Privacy, Genesis, Technology, Judiciary.

I. Introduction

Technology is the most imperative tool in the 21st century. It has made humans efficient like never before, the benefits of using technology are abundant but the less talked and the left-out question in this world where technology is so pierced into human life is – Is our data safe in the hands of people who are controlling technology. In an era of liberalization, privatization and globalization, the role of technology has become indispensable to achieve planned economic goals of the country. However, it is also true that it has become a challenging task to the law enforcement agencies to counter technology driven crimes that are emerging now. Data privacy and its protection is one of the most important challenges that we face today, yet it is recently that we have started to talk about it.

The chapter discusses about the genesis of privacy as a concept and how Indian judiciary has been the torch bearer on this topic, it also gives a chronological order on how the privacy has evolved in India. As privacy is an imperative matter, The Government of India have come with a legislation called the 'Personal Data Protection Bill, 2019' which tried to encapsulate the major problems that the public at large faces, but the Bill has not yet been passed by the parliament. In this chapter an attempt has been made to explain the variety of challenges that India is facing into making data protection law a reality. Also, a view has been presented on how other countries apart from India have been tackling this problem, and what the future holds for us in the realm of data protection.

II. Genesis of the concept of Privacy Concept and Indian Judiciary

The Indian Constitution leaves room for judicial interpretation because the right to privacy was not explicitly mentioned in the constitution, the right to privacy is included in the scope of the fundamental rights according to the legal interpretation of those rights. Although the Indian Constitution does not expressly guarantee the right to privacy, this area of law has developed since it was interpreted in accordance with Article 21.

(i) Role of Indian Judiciary

Indian judiciary, particularly Supreme Court of India has interpreted Article 21 i.e. right to life and personal liberty in a liberal way to include right to privacy within it. For example, the Apex Court in **M. P. Sharma and Others vs Satish Chandra case**[1] considered whether or not the "right to privacy" is a basic right in this case for the first time. It was argued that the search and seizure warrant issued in accordance with Sections 94 and 96(1) of the Code of Criminal Procedure violated the individual's right to privacy. The Hon'ble Supreme Court ruled that no constitutional clause is violated by the ability to search and seize property. According to a report, the state's legal right to search and seize property is mandated by the need to maintain social security. It was claimed in this judgement that the constitution's makers did not specify or foresee the right to privacy as a basic right, analogous to the 4th Amendment in the United States.

Also, in **Kharak Singh v. State of Uttar Pradesh and Ors.**[2] the issue raised was "whether the right to privacy was inclusive of Article 21." The question posed was whether Article 21 of the Indian Constitution was violated by a domiciliary visit for night surveillance against the accused. The Hon'ble Supreme Court ruled that such a visit violated Article 21.

However, the majority of justices believed that because Article 21 does not include any provisions for privacy, the right to privacy cannot be regarded as a fundamental right. Additionally, they claimed that the right to privacy is not a guaranteed right and that such surveillance does not infringe Article 19(1)(d). Therefore, monitoring the suspect's movements does not contravene any basic rights guaranteed by Part III of the constitution. Although Hon. Mr. Justice Subbarao held the minority position that privacy is a crucial component of human liberty.

Recently, the Supreme court in **Justice K.S. Puttaswamy (Retd.) and Anr. v. Union of India and Ors. Case**[3] rendered a landmark decision in this matter, concluding that Articles 14, 19, and 21 of the Indian Constitution guarantees and/or enshrine the right to privacy. The Kharak Singh and M.P. Sharma ruling was overturned by this judgment.

In this case, the "Aadhar Card Scheme" was contested because it was arbitrarily collecting and it was alleged that it was exploiting people' biometric data for other reasons, which violated their right to privacy. The petitioner stated that Article 21 of the Indian Constitution should include the right to privacy because it is a basic right. In response, the respondents argued that the Constitution merely recognises personal liberty and that it has partially granted persons the right to privacy.

The nine-judge Constitutional bench was set up and it decided this case unanimously. It was held by the Hon'ble Supreme Court that the right to privacy is intrinsic of the right to life and personal liberty enshrined under Article 21. It also is a part of rights guaranteed under Part III of the Constitution of India. It was also said that it is an obligatory duty of the State to protect the privacy of its citizens. Hence, the 'Aadhaar Card Scheme' was held liable for violating the right to privacy of the citizens.

This decision of the Supreme Court empowers the citizens to seek judicial relief in case their data privacy rights are breached. Also, this judgment affects the rules and regulations set by tech companies in India. The Supreme Court in its Puttaswamy judgment, 2017 declared privacy a fundamental right. This set the government in motion to take steps to bring a new data protection legislation for the country. The report has emphasized that interests of the citizens and the responsibilities of the state have to be protected, but not at the cost of trade and industry. The Committee was constituted by the union government in July 2017, to deliberate on a data protection framework. The Committee has also proposed a draft Personal Data Protection Bill[4].

III. Journey so Far: The Proposed Legislation on Data Protection in India

The absence of data protection law in India has posed several challenges relating how to protect privacy of the individual on the one hand and interest of the data processor on the other hand. Before enacting a comprehensive law on protection of personal data, the government took following initiatives.

(i) B.N. Srikrishna Report

The report of BN. Srikrishna ensured extensive deliberation on a data protection framework in India. This committee was constituted by Government of India in July 2017 and headed by retired Supreme Court judge BN. Srikrishna. This was a significant step in the direction of framework an important and required law on data protection.

(ii) Provisions of Data Protection Bill

Following are the significant provisions of the Data Protection Bill -

a. Individual Consent:[5]

The proposed Bill significances individual's consent as the cornerstone of data sharing, grants users' rights, and places duties on data fiduciaries (all those entities, including the State, which determine purpose and means of data processing).

The legal basis for processing personal data shall be consent. Although the law will use a different framework for consent, the data fiduciary will still be responsible for any harm done to the data principal.

The Data Protection Bill specifies terminology like data principal, data processor, data breach and sensitive data, etc. and asks for privacy by design on the part of data processors.

b. Data Protection Authority[6]

A Data Protection Authority (DPA), an independent regulatory organisation tasked with the effective execution of the legislation, would be established as a result of the data protection law.

The key responsibilities of the DPA are as follows: legal matters, policymaking, and standard setting, monitoring and enforcement, research and awareness inquiry, grievance handling and adjudication.

c. Personal data[7]

The law will apply to both public and private organizations handling personal data. If personal data has been used, shared, revealed, gathered, or otherwise processed in India, the law will have jurisdiction over such processing. The Bill has provisions relating to handling of personal data. The Bill covers several categories of information within the purview of sensitive personal data such as Passwords, financial information, health information, official identifiers, sex life, sexual orientation, biometric and genetic information, and information revealing a person's caste, tribe, religion, or political affiliations. The DPA will, however, be given the residuary authority to notify additional categories in compliance with the standards established by law.

Additionally, personal data collected, used, shared, disclosed, or otherwise processed by companies incorporated under Indian law will be covered, irrespective of where it is actually processed in India. However, the data protection law may empower the Central Government to exempt such companies which only process the personal data of foreign nationals do not present in India.

d. Right to be forgotten[8]

The right to be forgotten refers to a person's capacity to restrict, delink, erase, or rectify the exposure of inaccurate, humiliating, irrelevant, or out-of-date personal information online.

e. Cross border data transfer[9]

Cross border data transfers of personal data, other than critical personal data, will be through model contract clauses containing key obligations with the transferor being liable for harms caused to the principal due to any violations committed by the transferee.

Personal data determined to be critical will be subject to the requirement to process only in India (there will be a prohibition against cross border transfer for such data).

f. Data of minors[10]

The Committee has made specific reference to the need for distinct and stricter standards for safeguarding children's data, and it has recommended that businesses be prohibited from processing data in ways like behavioral tracking, targeted advertising, and any other processing that is not in the child's best interest.

The fact that youngsters are unable to completely comprehend the repercussions of their behavior provides the foundation for such discriminatory treatment. In the digital era, where data collection and processing are usually opaque and bogged down in complicated consent forms, this is only made worse.

The committee suggests giving the Data Protection Authority the authority to identify websites or online services that process a significant number of children's personal data as guardian data fiduciaries.

The committee noted that this approach, of placing the onus of properly processing the data of a child on the company, is preferable to the existing regulatory approach which is based solely on a system of parental consent.

The parental concern is prone to circumvention. It risks encouraging children to lie about their age, without achieving the intended purpose of protection.

g. Exceptions[11]

In cases of public welfare, law and order, emergencies when the person cannot provide consent, employment, and justifiable purposes, the state may handle data without the user's consent. Some requirements of the proposed data privacy law may not apply to the processing of data for particular purposes, such as the security of the State, judicial processes, research, and journalistic purposes. To prevent potential abuse, the law must, however, include sufficient security measures.

h. Data Storage[12]

The Bill includes provisions requiring a copy of personal data to be kept on file in India. The current world of commerce and trade depends on cross-border data flows. Investments and transactions (like outsourcing) might be hampered by restrictions or uncertainties on these flows (in data businesses). India must strike a balance between its demand for commercial facilitation and its desire to maintain its sovereignty as a global hub for both.

i. Appellate Tribunal[13]

To hear and decide any appeals against a DPA order, the Central Government must either create an appellate tribunal or give authority to an existing tribunal.

j. Penalties[14]

Infractions of the data protection legislation may result in penalties. If the amount of the fine exceeds the maximum limit that has been set, it will be increased by a percentage of the previous financial year's total worldwide turnover. It is worthy to mention here that the above stated committee has suggested a penalty of Rs. 15 crore or 4% of the total worldwide turnover of any data collection/processing entity, for violating provisions. Failure to take prompt action on a data security breach can attract up to Rs. 5 crore or 2% of turnover as a penalty. Further, the penalties paid by violating entities in this case will be deposited to a Data Protection Fund, which will, among other purposes, finance the functioning of the Data Protection Authority.

k. Impact on Allied laws[15]

The committee has observed that the Aadhaar Act is silent on the authority of the Unique Identification Authority of India (UIDAI) to impose sanctions against violating businesses operating within its ecosystem. To strengthen data protection, the Aadhaar Act has to be changed. The Report also suggests changes to the RTI Act, citing the possibility of private harm from the release of public authorities' information.

I. Obligations on Fiduciaries and rights to principles are the two underlying themes of the Bill[16]

The obligations would include "purpose limitation," which states that data will only be used for explicit, legal purposes, and "collection limitation," which states that only the data required for the purpose will be collected and will only be kept for as long as it is reasonably required for the purpose.

A data protection impact assessment is started before new technologies are implemented, data policies are verified by a data auditor, and they have data protection officers. The Bill lays out requirements for fiduciaries to guarantee no harm to the user, with transparency and security protections. The Act would apply to data processors who are not physically present in India as well as those who conduct business there or engage in other activities like profiling that might endanger the privacy of data principals in India.

IV. Personal Data Protection Bill, 2019

The PDP Bill proposes a significant shift in India's data collection and processing procedures. So far, both the private sector and the government have functioned in a largely uncontrolled environment, free of checks and balances and processes to protect citizens' privacy interests. according to Section 3(28) of the PDP Bill means "personal data" means data about or relating to a natural person who is directly or indirectly identifiable, having regard to any characteristic, trait, attribute or any other feature of the identity of such natural person, whether online or offline, or any combination of such features with any other information, and shall include any inference drawn from such data for the purpose of profiling. The key notion behind deciding what personal data is the identifiability of a natural person.

(i) Features of the Bill

The PDP Bill aims to strengthen data processing and privacy in a manner similar to the GDPR of the European Union.[17] The PDP Bill proposes the establishment of a Data Protection Authority (DPA)[18], similar to those found among European Union countries, and specifies the types of sensitive personal data that must be protected. The PDP Bill establishes the term "data fiduciary" and imposes specific requirements on them about how they collect, deal with/process, and retain personal data. It holds them responsible for upholding their commitments in relation to personal data processing performed by them or on their behalf.

If the PDP Bill becomes law, businesses will be required to inform customers about their data gathering activities and obtain their consent. They would have to gather and store proof that such notice was given, and approval was given. As a result of the PDP Bill giving customers the ability to withdraw their consent, firms would have to devise ways to allow consumers to do so. After their data has been processed for the purpose for which it was intended, consumers have the right to access, correct, and erase it under the PDP Bill. As a result, businesses would have to devise means for customers to do so.

The PDP Bill allows consumers to transmit their data to other firms, including any inferences formed by businesses based on that data. All businesses would have to devise mechanisms to enable customers to do so. The PDP Bill mandates that all firms adopt organizational changes to better secure data. This includes privacy-by-design principles (a business model in which privacy is a primary issue), security precautions, and so on. The PDP Bill mandates data localization, mandating companies to keep specific types of data on Indian servers solely. In this way, it establishes the following three-tiered structure:[19]

(ii) Major Provisions of the Bill

The major provisions of the Bill are as follows -

a. **Personal data**: Personal data that is not designated "sensitive" or "critical" is not subject to localization or data transfer limitations. No transfer limitations would apply if this type of personal data was stored wholly outside of India.

b. **Sensitive personal data**: "sensitive personal data" may be sent outside of India, but it must still be stored there. "Special categories of personal data" include health, religion, sex life, political convictions, biometric, genetic, and financial information, among other things. Passwords, for example, have been eliminated from the definition.

c. **Critical Personal Data:** The Bill allows the government to designate some types of personal data as "critical personal data" that cannot be transferred outside of India. The Bill, however, allows transfers to nations or organizations that are assessed to provide a sufficient level of protection (and will not jeopardize the State's security or strategic interests).

d. **Miscellaneous Provisions of the Bill:** The PDP Bill establishes a right to be forgotten, allowing a data principal to limit or prevent the continued disclosure of his personal data by a data fiduciary where such disclosure has served the purpose for which it was collected or is no longer necessary for that purpose; was made with the data principal's consent and such consent has since been withdrawn; or was made in violation of the provisions of this Act or any other law in force at the time.

The PDP Bill mandates that data fiduciaries notify the DPA of any breach of personal data they process if the breach is likely to cause harm to any data principal. "Significant data fiduciaries" will have additional responsibilities under the Bill, such as conducting data audits and hiring data protection officers. The Bill allows the Central Government to declare any social media intermediary (such as Facebook) with subscribers over a certain threshold as a significant data fiduciary:

The PDP Bill offers the Central Government vast powers to exempt any government agency from the Act's application. The government can access any anonymized or non-personal data from data fiduciaries and processors under Section 91[20] of the PDP Bill. The PDP Bill gives the DPA the authority to punish any business that does not follow the Bill or the DPA's or the Government's rules. The maximum penalty that can be levied is 150 million Indian rupees (about $2.1 million), or 4% of the company's global turnover in the previous financial year.[21]

V. Joint Parliamentary Committee's Report and Suggestions on 2019, Bill

There are certain shortcomings with this proposed Bill of 2019 which were highlighted by the report released by the Joint Parliamentary Committee[22]. The proposed suggestions are as follows:

(i) Difficulties in discerning Personal and Non-personal data

It is difficult to discern between personal and non-personal data, according to the JPC, because the Data Protection Authority (DPA) will manage various types of data at various security levels. According to the committee, the PDP Bill should cover both categories of data until a separate framework to distinguish between personal and non-personal data is implemented.

The report stated, *"As soon as the provisions to regulate non-personal data are finalised, there may be a separate regulation on non-personal data in the Data Protection Act to be regulated by the Data Protection Authority,"*[23]

(ii) A mirror copy of sensitive and critical data should be localized

The JPC recommends that, in addition to the provisions under Clauses 33 and 34 for cross- border data transfer, the government take concrete steps to ensure that a mirror copy of sensitive and critical personal data already in the possession of foreign entities be brought to India in a timely manner[24].

(iii) Data Localization

The central government has been given orders to take precautions and secure sensitive data held by foreign firms. According to the JPC, a copy of such data must be brought to India in a timely way. The central government has also been required to verify that all local and foreign entities meet the new data localization laws. In the coming days, the government has been requested to create and issue a complete data localization strategy.[25] The report suggest the following for data localization.

(iv) Data Localization Norms

"Normally, the imposition of data localization norms can be attributed to mitigation of certain risks in cross-border flow of data and address strategic objectives. These include:

a. **National security and law enforcement**: In a hostile country, the data can be a tool for surveillance and manipulation of consumer behavior/opinion. Timely access to personal information by law enforcement agencies is also one major requirement.

b. **Privacy**: Better informational privacy can be ensured with appropriate data protection regulations within the country.

c. **Employment generation**: Data localization can provide a great boost to the data economy in the domestic market with the emergence of the data centers and other associated industries, which have the potential to create significant employment opportunities.

d. **Bargaining power**: With strong presence on the internet and mammoth generation of consumer data, India can be in better position to bargain with other countries for encouraging data-based innovation for providing digital services and impetus to the digital economy."

(v)   Data Breaches

Clause 25(3) should include a 72-hour reporting timeframe for data breaches, according to the committee. The committee requested that the DPA follow particular guiding principles while creating data breach legislation. It has been suggested that the authorities ensure the privacy of the data principals. The data fiduciary will bear the burden of proving that the delay was reasonable if the data principle has experienced inconsequential or material harm as a result of delayed reporting, according to the study.

The data fiduciary is liable for any harm caused to the data principal as a result of a late complaint. According to the study, fiduciaries would also be required to keep track of all data breaches. A data fiduciary must not keep personal data for longer than is necessary and must erase it once processing is completed. Because the government will also become a data fiduciary, the report suggests that in the event of a breach or a crime, the head of the department in question conduct an internal inquiry first. This procedure is designed to ascertain who was responsible for the specific crime, after which liability can be determined.

(vi)   Social Media and Intermediary Liability

With the revelations of the Facebook Files, social media platforms have become a prominent focus of discussion in technology policy. The JPC

Report has been impacted by this, with various new insertions in the Draft Data Protection Bill, 2021. To begin, we'd like to emphasize that, while data protection laws are crucial tools for regulating data flows to giant digital platforms, they are distinct from broader social media regulation, which requires its own statute (e.g., the UK's Online Harms Whitepaper). "Social media platforms have been designated as intermediaries in the IT Act, and the Act has not been able to regulate social media platforms adequately," according to the JPC Report and that "the current Bill is about personal data protection, and social media regulation is an entirely different aspect that requires detailed deliberation". Despite these observations, the report goes on to suggest major changes to social media regulation.

The JPC Report has identified and remarked on a number of issues with social media, including "the predominance of fraudulent accounts" and "instigated individuals all over the world to plan, organize, and carry out revolutions, protests, riots, and spread violence." The JPC Report suggests the following modifications in response to these concerns:

Publishers: The Committee's main worry was that the IT Act had labeled social media sites as 'intermediaries'. In this regard, the Committee believes that social media platforms should not be identified as such because, in essence, they act as content producers, with the capacity to choose who receives the content and control access to any content placed on their platform.

Verification: A method might be established whereby social media platforms that do not function as intermediaries are held liable for content posted by unverified accounts on their networks. The social media intermediary must verify the account once application for verification is submitted with all required papers.

Local office: "Moreover, the Committee also recommends that no social media platform should be allowed to function in India unless the technology's parent business has an office there.

Social Media Regulatory Body: "Further, the Committee recommend[s] (sic) that a statutory media regulatory authority, on the lines of Press Council of India, may be setup for the regulation of the contents on all such media platforms irrespective of the platform where their content is published, whether online, print or otherwise."[26]

VI. Miscellaneous Legislations relating to Data Privacy

Information Technology Act, 2000 and the relevant amendments empowers individuals to sue a company for damages caused by its failure to develop and maintain "reasonable security measures and procedures" to protect sensitive personal data or information under Section 43- A, (IT Act, 2000). The IT Act and the Information Technology (Reasonable Security Practices and Procedures and Sensitive Personal Data or Information) Rules 2011 (IT Act Rules) define sensitive personal data or information as personal information relating to: "Passwords, Financial information, such as bank account or credit card details or other payment details, Physical, physiological, and mental health condition, Sexual orientation, Medical records and history and Biometric information." [27]

Section 43-A provides compensation in the event that a body corporate that owns, controls, or operates a computer resource and possesses, deals, or handles any sensitive personal data or information is negligent in implementing and maintaining reasonable security practices and procedures, resulting in wrongful loss or gain to any person. Non-digital data is not mentioned in this section. Data saved on any electronic device or in a relevant file system (such as a salesperson's diary) should be covered by data protection legislation.[28]

Data kept on non-electronic media is not protected by this provision. Furthermore, while the part mentions sensitive personal information, it does not compare it to personal information, which is on a whole different level. In essence, there appears to be no distinction between what is traditionally deemed personal information and sensitive personal information under this clause.[29]

Information disclosed in violation of a legitimate contract is punishable under Section 72-A. Anyone who acquires access to any material containing personal information while providing services under a legitimate contract and reveals that material to anyone else without consent or in violation of the contract will be penalised. The problem with this clause is that it lacks a definition of personal information, which, when combined with the provisions of Section 43- A, which refer to personal sensitive information, causes some inherent ambiguity between different sections of the IT Act. While the clause makes it illegal to break a person's confidentiality, it does not provide any compensation to the victims of such a breach. That is arguably the most essential solution in the context of violation of privacy. The section is very specific in that it only applies to personal information gathered within the terms of a service contract. Personal information can be gained in a variety of ways, and all such personal information must be protected.[30]

While these improvements give some protection against privacy intrusions, they are by no means comprehensive. It is critical to clarify terminology such as "personal information" and "sensitive personal information." In terms of collection, use, and disclosure, sensitive personal information must be handled with greater caution. It's also critical to make sure that data saved on non-electronic media is covered and secured. More crucially, while the new provisions in the IT Act 2008 offer a foundation for data protection in the country where none previously existed, a full-fledged data protection legislation must contain regulations on data collection, control, use, and destruction. To have an effective data protection regime in India, several fundamental aspects must be addressed.

VII. Changes Incorporated into Draft Data Protection Bill, 2021

Following suggestions have been offered by the JPC:

(i) Incorporation of Non-Personal Data[31]

One of the most significant revisions recommended by the JPC was to expand the Bill's scope to cover all types of non-personal data, as per Clause 3(28), which states that it includes "data other than personal data." The committee advised this because, in their opinion, it is highly challenging to discern between personal and non-personal data, which might even impact privacy. Additionally, a separate statute will need to be introduced in order to govern non-personal data if it is not included by the data protection Bill.

(ii) Social media terminology change

The word "social media intermediaries" has been changed in the Draft Bill, 2021 to "social media platforms," which is defined under clause 3(44) of the Draft Data Protection Bill, 2021, as any platform that aims to facilitate online interaction between various users. This change was made to make these platforms accountable for any content they host.

However, including these social media companies within the ambit of "platforms" will still not provide a viable solution to the issue of data privacy. Classifying these social media platforms strictly as an "intermediary" or strictly as a "platform" will not be as effective because a middle ground has to be found wherein as per the varying circumstances, these companies can be legally exempted or be held liable for the content posted by their users.

(iii) Defining 'damage'

The JPC report also discusses the issue pertaining to the breach of data and its disclosure and reporting mechanisms. It puts forth a number of rules, one of which specifies a time limit of 72 hours for reporting a violation. In accordance with this, organizations will also need to properly justify any delays in disclosing security breaches. The Data Protection Authority (DPA) only required to be notified by the data fiduciaries in the event that any "damage" was caused, according to the 2019 Bill. The most recent iteration of this definition of "damage" has been updated to include psychological manipulation in addition to the loss of reputation or any type of financial loss. Additionally, this proposal suggests making it essential to notify all data breaches, whether or not any harm was done.

(iv) Obtaining consent:

The express consent requirement and giving consumers the choice of supplying or not disclosing their personal data were two more changes included in the proposed Legislation. According to the JPC committee's suggestion, the individual will be permitted to exercise their right to privacy even if they opt not to divulge their personal information. However, as all rights are subject to a few reasonable limitations, efforts have been made to define the conditions that permit the non-consensual processing of personal data. Non-consensual use is permitted where" such processing is required," according to an earlier version of the 2019 Bill that deals with data protection. However, the 2021 draft of the Bill fails to incorporate the procedural safeguards provided in the landmark judgment of **Justice K.S. Puttaswamy v. Union of India (2017)**[32] which requires the presence of "proportionality" and "legitimate purpose". The 2021 Draft Bill provides that non-consensual usage can be allowed whenever it "can be reasonably expected" thereby failing to incorporate the procedural safeguards laid down in the aforementioned case.

(v)   Storing of data within the boundaries of the country

The JPC report recognised the significance of data localization and asserted that all sensitive information pertaining to national security, personal information, economic activity, etc. must be kept inside the boundaries of the country. The committee even made recommendations for the actions that should be performed to transfer all of the sensitive data that has been held abroad and bring it back within the country's boundaries. However, there have been some discrepancies over the legal bases for the transmission of this data. In the 2019 draft of the Bill, it was provided that the transfer of sensitive data can be restricted if it is against the public or the state policy. But there is no information available as to what all can be included within the definition of public and state policy, and this could lead to the arbitrary use of powers that have been assigned to the DPA.

(vi)   Mandatory appointment of the data protection officer

The JPC changed the draft and mandated the appointment of a Data Protection Officer for all notable data fiduciaries. The committee said that the role of a data protection officer should only be filled by someone who has a senior or significant managerial position inside the organisation. This is done to make sure that whoever is chosen to serve as the DPO is familiar with how the business operates.

(vii) Hardware and software testing and certification

The Committee also took into account the consequences of data breaches brought on by the technology and software used for data gathering and storage. It opined that certain basic minimum criteria need to be set that need to be fulfilled before the approval is granted for the hardware or the software. The JPC advised that the DPA create a framework for monitoring and make sure that frequent testing is done to maintain the data secure.

VIII. Roadblocks and Challenges

There exists a number of challenges regarding data protection in India:

(i) Question of government controlling and accessing public data

While some national security issues may need the state's access to and maintenance of public data records, it is critical that the Bill does not place citizens' right to privacy ahead of the country's security interests. The government will have the power to gather and exploit public data as it sees fit, thanks to an intentional elevation of national security over the right to privacy.

(ii) Effective restraints on large corporations

In 2018, the Central Bureau of Investigation filed a case against the British data analytics firm Cambridge Analytica for allegedly obtaining and exploiting the personal data of 6.1 million Indian Facebook users[33]. The business allegedly utilised the information to sway Indian elections. As a result, imposing effective limits on huge firms' data mining tactics should be a major legislative priority.

(iii) The importance of people's right to privacy

Any data privacy and protection law must be based on the Supreme Court's 2017 decision on the right to privacy. In August 2017, a nine-judge Constitution Bench of India's Supreme Court declared that the right to privacy is "intrinsic to life and liberty" and hence fundamentally guaranteed by the Indian Constitution's many fundamental freedoms. The government must strike a compromise between the protection of personal data and the nation's national security concerns.

(iv) Cyber-attacks and surveillance prevention

According to government data presented in Parliament, India reported 1.16 million cyber security cases in 2020, which is three times more than in 2019. Several nations have recently been accused of using Pegasus spyware[34] for spying purposes. It is unknown whether the spyware was deployed by the Centre or if it was employed by a third party in India. Data security and privacy issues posed by current surveillance technology must be considered in the Bill.

(v) Data leaks

The past few years have also been victims of several major data leaks. In 2020, IRCTC's data leak revealed the personal information of millions of Indian citizens on the dark web. The information included their full names, mobile numbers, e-mail IDs, dates of birth, marital statuses, and cities of residence. Similarly, the data of 45 lakh passengers was leaked on Air India's passenger system service provider SITA, including information about the passport and credit card details. Outlook India reported data localization issues, given SITA is based out of Geneva, an aspect PDP Bill would raise. An 8.2 TB data leak also revealed such sensitive information at Mobi Kwik, with the KYC documents, Aadhar card and passport details of 10 Crore people on sale on the dark web. In an even worse turn, Domino's India's data leak led to the information being on the surface web, accessible to anyone with a search engine.[35]

IX. Global Legislations Relating to Data Privacy

This Part shall specifically deal with legislations at the global level:

(i)   U. S. Fourth Amendment

Stated in the U.S Constitution, the Fourth Amendment provides that "The right of the people to be secure in their persons, houses, papers, and effects, against unreasonable searches and seizures, shall not be violated, and no Warrants shall issue, but upon probable cause, supported by Oath or affirmation, and particularly describing the place to be searched, and the persons or things to be seized." The secretary of State Thomas Jefferson on 1st March 1792 declared this Fourth Amendment.

The focal aim of this provision is to guarantee people's right to privacy and secure them from unreasonable interference by the State. Not all search and seizures are protected by the law but only those which are initiated by the government or an unwarranted search.

With a few exceptions, a search or seizure without a warrant is generally unjustified and illegal. A law enforcement official must show probable cause that a search or seizure is warranted in order to obtain a search warrant or an arrest warrant. The totality of circumstances will be considered by a court-authority, generally a magistrate, in deciding whether or not to issue the warrant.

If an officer has probable cause and obtaining a warrant would be impractical in the given context, the warrant requirement may be waived. For example, an Ohio court concluded in *State v. Helmbright*[36], that a warrantless search of a probationer's person or residence is not a violation of the Fourth Amendment provided the officer conducting the search has "reasonable reasons." to believe that the probationer has failed to comply with the terms of his probation.

The Fourth Amendment's applicability in electronic searches and seizures[37] has gotten a lot of attention in recent years from the courts. With the rise in use of computers and the internet, there has been an increase in the quantity of crime committed electronically. As a result, computers, hard drives, and other electronic devices are frequently used to store evidence of such crimes. Electronic devices are subject to search and seizure under the Fourth Amendment.

In many electronic search situations, the question is whether law enforcement can search a company-owned computer that an employee uses for work. Despite the fact that the law is split, the majority of cases rule that employees do not have a genuine expectation of privacy when it comes to information saved on company-owned computers. The Supreme Court expanded this lack of expectation of privacy to text messages sent and received on an employer-owned pager in the 2010 decision of *City of Ontario v. Quan*[38].

(ii)   European Union- General Data Protection Regulation[39]

The General Data Protection Regulation (GDPR) is the world's most stringent privacy and security law. Despite the fact that it was designed and passed by the European Union (EU), it imposes duties on organizations anywhere that target or collect data about EU citizens. On May 25, 2018, the regulation went into effect. Those who break the GDPR's privacy and security regulations will face severe fines, with penalties ranging in the tens of millions of euros.

The GDPR signals Europe's hard stance on data privacy and security at a time when more individuals are committing their personal data to cloud services and data breaches are becoming more common. The regulation itself is large, far-reaching, and fairly light on specifics, making GDPR compliance a daunting prospect, particularly for small and medium- sized enterprises (SMEs).

"Everyone has the right to respect for his private and family life, his home, and his communications," says the European Convention on Human Rights, which was adopted in 1950. The European Union has worked to assure the protection of this right by legislation based on this foundation.

The EU recognized the necessity for updated protections as technology advanced and the Internet was developed. As a result, in 1995, it passed the European Data Protection Directive, which established minimum data privacy and security criteria on which each member state's implementing legislation was based. However, the Internet was already transforming into the data Hoover it is today. The first banner ad emerged on the internet in 1994. The majority of financial institutions offered online banking in the year 2000. A Google customer filed a lawsuit in 2011 after the corporation scanned her emails. Two months later, the European Commission's data protection authority determined that the EU required "a comprehensive approach to personal data protection," and work on updating the 1995 directive started.

After passing via the European Parliament in 2016, the GDPR became effective in 2016, and all enterprises were obliged to comply by May 25, 2018.

X. Data protection principles

The main Data Protection Principles are summarized -

(i) Transparency, fairness, and lawfulness – Processing must be legal, fair, and transparent to the data subject.

(ii) Purpose limitation: Data processing shall be limited to the legitimate purposes notified to the data subject when the data was obtained.

(iii) Minimize data collection and processing – Only gather and process as much data as is absolutely necessary for the objectives stated.

(iv) Accuracy — Personal data must be accurate and up to date.

 (v) Limitation on storage – You may only keep personally identifying information for as long as it is required for the intended purpose.

(vi) Integrity and secrecy – Processing must be carried out with the utmost security, integrity, and confidentiality (e.g., by using encryption).

(vii) Accountability — It is the responsibility of the data controller to show GDPR compliance with all of these criteria.

XI. Conclusions and Road Ahead

The changing digital world has forever changed how personal data is acquired, kept, and used for a variety of purposes, both known and unknown. Personal data protection is more voluntary than required in the absence of organized and comprehensive data privacy regulations. With the introduction of the Data Protection Bill and its several versions, India has adopted a careful approach to establishing itself as a "nation intent on constructing a solid data privacy framework" that will help enhance global trust and make India a desirable investment destination.

The Bill intends to promote fair and transparent personal data processing through informed permission, making ethical data usage a key component of data management. Furthermore, with the passage of the Bill, there is an urgent need to educate the next generation of India's digital users from all walks of life on their privacy rights and the redressal mechanisms that companies are required to implement. This growing knowledge will be crucial in promoting trust and openness among the general public.

On the one hand, the demand for data localization would increase compliance costs by compelling multinational firms to establish up personal data duplicates in India. On the other hand, it will progress technology and provide future investment and job prospects, transforming India into a real data sovereign nation.

In light of the Bill, we anticipate the development of a number of regulatory policies, including a framework for non-personal data, data-localization, and certification for digital and IoT (Internet of things) devices, all of which emphasize the need for a flexible data privacy framework to meet future requirements in their overarching data privacy programmes.

Endnotes

1   AIR 1954 SC 300.

2   AIR 1963 SC 1295.

3   AIR 2017 SC 4161

4   The Personal Data Protection Bill, 2019(Bill No. 373 of 2019)

5   Justice B.N. SriKrishna Committee, A Free and Fair Digital Economy: Protecting Privacy, Empowering Indians, page no. 24, (2018), Ministry of Electronics & Information Technology, Government of India.

6   *Id* at page no. 151

7   *Id* at page no.170

8   *Id* at page no.68

9   *Id* at page no.82

10   *Id* at page no. 42

11   *Id* at page no.120

12   *Id* at page no. 83

13   *Id* at page no.159

14   *Id* at page no. 163

15   *Id* at page no. 98

16   *Id* at page no.146

17   General Data Protection Regulation – Regulation (EU) 2016/679 of the European Parliament and of the on-data protection and privacy in the European Union and the European Economic Area.

18   The Personal Data Protection Bill, 2019(Bill No. 373 of 2019), s. 41(1)

19   The Personal Data Protection Bill, 2019(Bill No. 373 of 2019), s.3

20   Section 91 of the PDP, Bill 2019: 91. "Act to promote framing of policies for digital economy, etc.", available at:http://164.100.47.4/BillsTexts/LSBillTexts/Asintroduced/373_2019_LS_Eng.pdf

21   The Personal Data Protection Bill, 2019(Bill No. 373 of 2019), s.57 (2) (d)

22   Report of the Joint Committee on the Personal Data Protection Bill, 2019, available at: http://164.100.47.193/lsscommittee/Joint%20Committee%20on%20the%20Personal%20Data%20Protection%20Bill,%202019/17_Joint_Committee_on_the_Personal_Data_Protection_Bill_2019_1.pdf'

23   Mohit Sharma, "Joint committee report on Data Protection Bill tabled in both Houses of Parliament | Details", available at: https://www.indiatoday.in/india/story/joint-committee-report-data-protection-Bill-tabled-houses-parliament- details-1888747-2021-12-17

24   Gulshan Rai, "Personal Data Protection Bill, 2019 - Key Highlights Of Reports Of Joint Parliament Committee (JPC)", available at:https://www.mondaq.com/india/privacy-protection/1161678/personal-data-protection-Bill-2019--key-highlights-of-reports-of-joint-parliament-committee-jpc#:~:text=Salient%20features%20of%20the%20%22Bill,on%2011th%20Dec%2C%202019&text=The%20proposed%20legal%20framework%20would,Non%20Personal%20Data%20(NPD)

25   Supra Note 8

26   Key Takeaways: The JPC Report and the Data Protection Bill, 2021 #SaveOurPrivacy, available at: https://internetfreedom.in/key-takeaways-the-jpc-report-and-the-data-protection-Bill-2021-saveourprivacy-2/

27   Supratim Chakraborty, Privacy in India: Overview, available at: https://www.khaitanco.com/sites/default/files/202101/Privacy%20in%20India%20Overview.pdf

28   Approach Paper for legislation of Privacy, available at: https://documents.doptcirculars.nic.in/D2/D02rti/aproach_paper.pdf

29   Id.

30   Id.

31   Siddarth raj choudhary," Data Protection Bill, 2021 & The Joint Parliamentary Committee Report: Key Changes, Mondaq, 25th January, 2022, available at https://www.mondaq.com/india/privacy-protection/1153588/data-protection-Bill-2021-the-joint-parliamentary-committee-report-key-changes .

32   Supra note 3.

33   Personal details of 6.1 million Facebook users in India leaked online, The Economic times, 06th April, 2021,available at https://economictimes.indiatimes.com/tech/technology/personal-details-of-6-1-million-facebook-users-in-india-leaked-online/articleshow/81916959.cms

34   David pegg and Sam cutler," What is Pegasus spyware and how does it hack phones?", The Guardian, July 18[th]2021, available at https://www.theguardian.com/news/2021/jul/18/what-is-pegasus-spyware-and-how-does- it-hack-phones.

35   Priyam Shukla," The state of data privacy in India amid hanging data protection Bill" The Outlook, 28th October, 2021 available at https://www.outlookindia.com/website/story/india-news-the-state-of-data-privacy-in- india/390849

36   State v. Helmbright (Judgement) (2003), United States Court of Appeals (Ohio).

37   Jonathan Kim, Fourth Amendment, Legal Information Institute (2022).

38   City of Ontario v. Quon(Judgement) (2010), Supreme court of the United States.

39   European parliament and of the council, General data protection regulation (GDPR), Regulation 2016/679 (April 27, 2016).

Chapter 3

A Comparative Analysis of Right to Data Portability Under Personal Data Protection Bill, 2019 with GDPR: Balancing Privacy and Competition Law Considerations in Digital Sector

Priyanshi
Academic Fellow
National Law University
Delhi, India

Abstract

With the unprecedented growth of the digital economy, users' data have emerged as an intangible asset for enterprises triggering the debate about data privacy and users' control over their data. 'Right to Data Portability' constitutes an essential element allowing users to receive and transfer their personal data and exercise increased control over such data. Experts have also suggested that portability can make the digital economy seamless, empower users, foster competition by allowing the flow of valuable data, ultimately contributing to consumer welfare. While users' control over their personal data is essential, the collection and processing of data are equally important for enterprises to derive value and gain a competitive edge. The conflict thus becomes apparent between the interest of the users and that of enterprises. The General Data Protection Regulations and established legal positions in the European Union can help draw Conclusions for an effective data portability provision. Essentially, categorizing the datasets that are subject to mandatory portability is important as its application to inferred, processed, or derived data, etc. alike, can depreciate the commercial value each possesses. Moreover, the position on trade secrets must be ascertained given the complexity involved in demarcating portable datasets from trade secrets due to continuous processing/data anonymization. This study attempts to critically analyse the right to data portability as a competition law remedy in the digital sector and deliberate upon effective portability provision, balancing the conflicting considerations. It aims to study legislations, policies, parliamentary discussions, scholarly papers, and articles to extract plausible solutions and propose a robust data portability provision.

Keywords: Right to Data Portability, GDPR, Privacy, Competition, analysis.

I. Introduction

The right to privacy under the Constitution of India has been recognized as one of the fundamental rights under Article 21. The Apex court's ruling in *K.S. Puttaswamy v. Union of India*[1] is a hallmark that established an individual's autonomy and recognized that privacy is intrinsic to the right to life and personal liberty. The court in its landmark opinion took note of the threat to one's privacy caused with the use of technology which comes at the cost of giving away personal information related to one's life and existence[2].

In recent years, with the expansion of the digital economy, the right to privacy has undergone a transformative interpretation. Technology, where on one hand has eased the lives of individuals, on the other hand, it has pervaded individuals' personal space, giving way to multiple challenges.

As we continue to progress as a technologically sound and advanced society, one obvious negative is being subject to continuous digital surveillance and monitoring by state and companies, especially the big technology giants[3]. Moreover, in the context of the digital economy, concerns pertain to the collection and use of data without users' consent/ without users' being aware of such tracking. Another significant challenge is the users losing control over their personal data.

For businesses today, access to relevant data has evolved as one of the key assets, required for any enterprise to operate and survive in the digital space. In simple terms, the collection of users' data online and data processing are inherent to the functioning of technology companies assisting them in modulating the results based on past interactions.

One of the objectives of the Personal Data Protection Bill, 2019 (erstwhile Bill) was to grant users, increased control over their personal data for which the right to data portability[4] was incorporated in the Bill. The right guaranteed users to obtain their personal data, provided by them to any data controller, reuse the data, or get it transferred to another controller[5]. In August 2022, the Bill was withdrawn by the government citing the need to revamp the legislation. In the backdrop of this, currently, a new data protection Bill is being contemplated by the government.

This Chapter aims to critically analyse the provision on right to data portability as included in the erstwhile Bill and further articulate the issues prevalent from the lens of competition law while drawing a comparative analysis of the Bill with European Union's General Data Protection Regulation (GDPR). This would assist in highlighting the issues surrounding the application of right to data portability as the government continues to deliberate on a new data protection law. It also aims to propose suggestions for a robust and effective right to data portability provision intended to be included with the data protection law of the country.

Data Portability is perceived as a pro-competitive remedy[6], allowing users to control their personal data, however, on the other hand, this collection and processing of relevant data are equally significant for the businesses operating in the digital economy. Enterprises are increasingly dependent on the users' data for their business operations. Moreover, users' data has been recognized as a huge competitive advantage for enterprise to carry out a quick analysis of the collected data through machine learning tools.[7] The digital markets are structured to collect, process, and make sense of a huge amount of data at a given point in time to enhance the products and services offered, hence its commercial value is immense.

II. The Conflict: Privacy vs. Competition

Having understood the competitive advantage data grants and the value attached to it as an asset, the conflict between privacy and competition is quite apparent. In the context of data portability, the right would have allowed users to exercise control over their personal data (including accessing, obtaining the data in a machine-readable format, reusing or transferring it to a different service provider), however, the retraction of datasets from the technology used by the enterprises, can also lead to adverse impact on the operations and the quality of output produced.

Both privacy and competition are ultimately functioning to enhance the welfare of the users, however, before achieving the ultimate objective, the portability of data can lead to conflicting interests between the users and the enterprises. Hence, it is pertinent to strike a balance between the two considerations given their impact on the users and the complexities involved in the technology sector. In addition to the need for a balance between the two, it is of relevance here that the provision on Data Portability[8] was quite similar to its European counterpart, the GDPR[9]. A closer look at the GDPR's portability provision and its implementation so far would also assist in deriving crucial takeaways for India.

III. Research Questions

The present research paper aims to enquire into the following questions:

(i)   How does Data Portability functions as a pro-competitive remedy and the related challenges?

(ii)   What are the issues surrounding the demarcation of boundaries of personal data in the digital sector for the purpose of portability?

(iii)   What are the primary takeaways for India, from the GDPR?

(iv)   What are the challenges relating to the successful implementation of the portability provision?

(v)   What can be the plausible solutions to the identified issues?

IV. Data Portability: Tracing its role as a competition law remedy and related Challenges

Data and its use by technology companies have been the center of attention for antitrust agencies. With the rapidly changing business landscape, Big Techs have emerged as giant data repositories, resulting from the interplay of direct/indirect network effects[10] coupled with first mover advantage[11]. In addition, the self-enforcing feedback loops[12] further contribute to raising of switching costs for the consumers[13], trapping them in a closed over-personalized digital space, and negating the chances of users switching to competing service providers.

Accumulation of huge data with a few players operating in the market is viewed as a major concern when it comes to competition in the digital markets. In addition, the inability of smaller players and new entrants to have access to the relevant data further reduces competitive force in the market.[14] For any new entrant in the business, limited access to relevant data and lack of processing can be a huge entry barrier, given that digital businesses already involve huge sunk costs.

Corroborating these views, as opined by Shoshana Zuboff that users today are caught in a continuous cycle of surveillance capitalism[15], referring to the exercise of the collection and processing of user's data by the technology companies, without their knowledge and express consent.

Citing these problems, data portability is seen as a remedy serving dual purposes: allowing consumers to switch between different competing service providers, exercising their right to choose between the providers; and providing enterprises including small and new entrants, with access to the relevant data and assisting them in producing desirable results for the consumers.

Data Portability is believed to make markets more contestable and competitive by allowing users to carry their data with them. Theoretically, it may appear achievable, however, its practical application appears complex.

Even though existing in the GDPR since 2018, data portability's practical and technical application remains unclear to date both on the touchstone of privacy and competition. The most important issue warranting clarity is, that the term 'Personal Data' as it exists in the digital sector. There are numerous kinds of datasets existing in the digital economy, broadly comprising of - volunteered data, derived data, observed data, and inferred data, each carrying its own specific commercial value for the enterprises in the digital economy[16].

It must be understood that due to the inherent structure of the digital economy, the boundaries between the different datasets are blurred over time as the data undergoes processing through the application of machine learning tools and analytics. Datasets are rampantly analyzed for numerous business objectives. For business, data is frequently shared by the enterprises with the business partners, for enhancing the advertisement value chain, and enriching the value created by any enterprise on its other parallelly running ventures, triggering machine learning and altering subsequent interactions according to the previous enterprise-consumer interactions.

Having understood the relevance of portability to privacy and competition law considerations, let us now delve into the intricate challenges involved in the successful implementation of Data Portability.

(i) Defining the Contours of 'Personal Data'

Under the erstwhile Bill, the right to data portability is applicable to users' 'personal data'. Clause 2 (28) of the Bill defined personal data as:

> *"Data about or relating to a natural person who is directly or indirectly identifiable, having regard to any characteristic, trait, attribute or any other feature of the identity of such natural person, whether online or offline, or any combination of such features with any other information, and shall include any inference drawn from such data for the purpose of profiling"*[17]

Upon perusing the definition of personal data under its counterpart, the GDPR, defines personal data as:

> *"Any information relating to an identified or identifiable natural person ('data subject') ; an identifiable natural person is one who can*

be identified, directly or indirectly, in particular by reference to an identifier such as a name, an identification number, location data, an online identifier or to one or more factors specific to the physical, physiological, genetic, mental, economic, cultural or social identity of that natural person"[18]

For further clarity on the scope of personal data under the GDPR, it includes data that can be related to an identifiable natural person including name, identification number, location data, etc. among others. The regime has kept it broadly worded and is interpreted widely too. The standard for inclusion in the definition of personal data is 'identification of a person through such data', it has been held in cases that personal data can include even less explicit information.[19] Moreover, the data should pertain to a natural person and does not include corporations, institutions, and other legal entities.

Relevant Recital 68 further provides that portability only extends to the data provided through the consent of the users. It provides that where the data concerns multiple persons, portability should not prejudice the rights of other persons involved.

The recital clarifies that the right does not extend to anonymized data and in the case of pseudonymization, the right would only apply where it is possible to identify the person through the data by applying existing and available technological tools[20]. This is relevant especially in cases of digital economy wherein even though the data cannot be directly attributed to a person, the same only goes under pseudonymization and can still be related to persons with the use of identifiers such as IP Addresses, Cookies, other identification tags used for creating profiles.[21]

Upon comparison, it is evident that while there are clear guidelines and interpretations of the term personal data under GDPR, however, the definition under the erstwhile Bill was extremely broad extending even to the 'inferences drawn' which are extremely crucial to the business interests of enterprises. It has been identified that operations of technology enterprises that rely on data analytics, behavioral advertising, and social media particularly will be affected by the broad interpretation of personal data.

Further, 'Inferred data' is a category of data generated through continuous users' interaction with the interface and is generated over the course of time by running algorithms on the data expressly/impliedly provided by the user. Inferences can be drawn from history, patterns of online consumption, recorded impressions of activities on the interface, etc.[22] It is imperative to note that while the Bill included inferences drawn under the definition of personal data, it nowhere defined inferred data, leaving scope for wide and ambiguous interpretations.

The digital sector runs on data collection and its processing, and the datasets are intangible assets possessing immense value porting which can render the technologies useless and incapable of producing any value. Withdrawal of commercially relevant data can lead to loss of commercial value, and this holds true for other datasets as well which too have been included within the scope of portability.

The clause in the erstwhile Bill included the following category of datasets -

a. **'Provided Data'** – The data voluntarily provided by the users themselves while interacting with the interface. This may include data like name, address, emails, and numbers, to name a few.

b. **'Generated/Derived Data'**[23] – The data which is observed by the interface about any individual and is generated from other data, after which they become new data elements related to a particular individual. Derived data are said to be created in a fairly "mechanical" fashion using simple reasoning and basic mathematics to detect patterns within a data set and create classifications.

c. **'Data forming part of the profile on the Data principal'** – This category can be immensely relevant to the business operating digitally for applying tools to the data existing on the profile and accordingly producing results for the users.

d. **'Data Otherwise Obtained'** – The category even though incorporated within the clause was nowhere defined, leaving gaps as to its interpretation.

(ii)   Data Anonymization and its challenges to portability

Data Anonymization is a technique applied to the digitally accumulated data through which the information is converted into a form that can no longer be associated with any identifiable natural person. It is carried out by removing the personal identifiers associated with the data. Given the impossibility of establishing any identity attached to such anonymized data, under the GDPR, anonymized data has been excluded from the requirement of allowing data portability.[24]

However, there's another side to the aspect of anonymization of data. In prominent jurisdictions like the US and EU, it has been opined that it is not possible to achieve absolute anonymization of data[25] and it can in one way or the other through application of identifiers and trackers, be attributed to a natural person.

Applying in the context of digital enterprises, given the conflicting opinions on the point and lack of clarity, it is unclear how the different data categories be distinguished from anonymous data for the purpose of allowing porting of the datasets. It is pertinent to note that neither the Draft Committee of the Bill (Sri Krishna Committee) nor the Joint Parliamentary Committee, the discussion of which would follow later in the work, have discussed the case of data that is rendered anonymous upon processing which certainly can include personal data as well. Moreover, the concerns surrounding the issue are further amplified due to the unique characteristics of the market we are dealing with.

In context of digital markets, wherein data collection and process are a never-unending continuous cycle, the personal data collected by the enterprises can be anonymized within seconds and while it can still be part of the personal data. For instance, any digital interface which has gathered basic information from its users and based on past user experience has also drawn inferences from the data so collected, in such a case distinguishing the personal data from the anonymized data could be extremely difficult with widespread implications for the privacy of an individual. In addition, the extraction of information, without careful identification of the categories of data, can affect the efficiency of the technologies employed and ultimately would be harmful to the business.

Hence, if the definition is kept too broad without any clear guidance, it may give rise to personal data being easily manipulated to be anonymized for the purpose of protecting it from mandatory portability requirements. Again, warranting the conflicting interest between privacy and competition to be balanced.

As per the erstwhile Bill, personal data also included information related to characteristics, traits, attributes, or any other feature of the identity of a such natural person. It must be noted that aspects like traits, attributes, and other identifiable features are inherent to the functioning of Artificial Intelligence and plays a major role in assessing the behavior of the users and calibrating the results for the users as per the assessment carried out by the technologies and for carrying out profiling of any user.

There are two aspects that require attention. Firstly, if users are allowed to retract only the personal data, identifying such data amidst continuous anonymization would place an excessive burden on the enterprises. This may also lead to subjective interpretation of portable datasets given the fact that enterprise would not be in favor of allowing porting the data carrying immense economic value and which can also include personal data as well. Secondly, the segregation of the datasets would require immense efforts to ensure that users are able to access their right to privacy of their personal data while also considering the interests of the enterprises in the data collected and processed, whose retraction can yield to negative welfare effect in the markets. Thus, identifying the datasets is one of the fundamental issues surrounding the practical application of data portability in digital markets.

(iii) Compliance costs burden and consumer welfare

The concerns related to compliance costs over the industry were raised by Sri Krishna Committee Report[26] citing the need for certain standards to be developed regarding data portability. Compliance with Data Portability is a costly affair. Under the GDPR, the technology/software sector is experiencing the second-highest growth in regulatory compliance costs (99%) between 2011 and 2017 after the healthcare sector.[27] In the US, compliance cost roughly is estimated to be around $510 million for companies annually for data portability alone, costing consumers and organizations cumulatively around $18 Billion annually.

The costs can vary depending on the kind of organization. Evidence shows that 'Born digital' organizations will be able to build the necessary infrastructure for compliance from scratch. Businesses relying on legacy systems, however, will be faced with suboptimal systems. This makes compliance costs higher for legacy systems.[28]

Hence, this implies that applying data portability to all enterprises alike without any differentiating criterion, can affect the functioning of small and medium sector enterprises, not having access to the required technological tools allowing portability. The costs associated with portability include development, authentication, processing and maintenance and comprise of allowing access, transfer, deletion requests from the users, to be borne by the enterprises.

The exemplary compliance costs are expected to be ultimately placed on the consumers in the long run affecting their welfare adversely. Before data portability becomes a mandate, it must be assessed as to how it can economically burden the industry with compliance. Increasing the costs of compliance will affect the quality of the services provided to the users. Additionally, it must also be considered that even though compliance is costly, the alternative (non-compliance) is even costlier for enterprises in terms of heavy fines across jurisdictions.

V. Clause 19 of Erstwhile Bill: A Critical Analysis from a Competition Law Perspective

The clause made data portability applicable to the personal data of the data principal, processed through automated means, i.e., any equipment capable of operating automatically in response to instructions given for the purpose of processing data[29]. In the context of digital markets, data irrespective of its different categories are all processed through automated means, making portability applicable to all data alike. The Clause allowed users to receive the following data in machine-readable format[30];

"(i) the personal data provided to the data fiduciary.

(ii) the data which has been generated in the course of the provision of services or use of goods by the data fiduciary; or

> *(iii) the data which forms part of any profile on the data principal, or which the data fiduciary has otherwise obtained;"*

The gap in the clause existed in the fact that the Bill nowhere defined the scope of 'generated data' and 'data otherwise obtained', leaving wide scope for interpretation.

In digital markets, for instance, information derived from the profile and activities of any user can also fall within the scope of generated data when it is put to processing for the purpose of generating information, to be used for service improvements for subsequent interactions. This holds true for other categories of datasets as well, given the lack of clarity on how the information flowing from the user is used by the enterprises. In such a situation, identifying the sets of portable datasets that can balance the right to privacy along with the interest of the enterprises, appears quintessential.

If one studies the applicability of Article 20 under the GDPR, the provision only includes the data 'provided by the user' to the controller, and the 2017 guidelines on data portability categorically specify that portability should only pertain to the data provided by the user knowingly or actively and the data generated from his/her activity, defining the scope of data subject to portability.

The guidelines have clarified that data "provided by" includes personal data that relate to the data subject activity or results from the observation of an individual's behavior but does not include data resulting from subsequent analysis of that behavior.

It does not include 'derived' or 'inferred data', the same being outside the scope of 'provided by the user'. The guidelines clarify that results produced through the running of algorithms fall outside the scope of portability. In contrast, the erstwhile Bill has enlisted the data categories including the data provided, generated, forming part of the profile, and otherwise obtained.

Hence, the categories of portable datasets and applicability of the provisions must be studied in the light of striking a balance between consumer – enterprises interests, clearly defining the different kinds of datasets and understanding its implication on both the consumers and enterprises.

The clause provided for certain exceptions for data portability -

For our research, the following exceptions seem of relevance. The provision does not apply to "where it would relate to a trade secret or providing the information would not be technically feasible."

The first exception pertained to the portability of information/ data involving trade secrets in which case portability requests can be denied by the data fiduciaries. The data collected through algorithms, if provided to a user to be transferred to another competing service provider can reveal secrets that would not be in the interest of the enterprises although it provides users with the option to switch between service providers.[31] Granting right to data portability at the cost of adversely affecting the current service provider's operation doesn't seem to be a preferred solution.

The GDPR tries to maintain a balance between the right of the user and the intellectual property/trade secret of the enterprises. Under the GDPR, portability requests can be denied if it relates to trade secrets of the enterprise however, the result should not be the refusal of the entire request, and efforts must be made to provide information without revealing trade secrets/IPRs, wherever possible.

Under GDPR, algorithms can be protected as trade secrets given the following are fulfilled[32]:

"i) the information is not known either by the public at large or by the experts of the sector in question.

ii) the information has commercial value

iii) you have taken steps to keep the information secret: for example, you keep it in safe storage, and you have signed non-disclosure agreements with anyone that has access to it or with whom you have shared the information."

In the context of the digital sector, for instance, algorithms applied to the collected/observed/derived data can be covered under trade secrets, being commercially relevant and sensitive to the business. However, when it comes to the Indian regime, there is no statute that deals with the protection to trade secrets.

On the other hand, in interpretations given by Indian Courts, trade secrets are considered confidential information and are dealt with under contract law, confidentiality clauses, and commercial agreements. Moreover, protection to trade secrets has not been recognised under any law in India and mostly remains incorporated in the form of confidential agreements between the parties.

Hence, it is essential to first ascertain how the protection of algorithms can be enforced in the Indian regime. The ambiguity surrounding trade secrets must be solved before their references are included in the proposed data protection laws as an exception.

(i) Algorithms as Intellectual Property or Trade Secrets

When one tries to study the functioning of digital businesses, the primary force behind their technologically incentivized business is the 'Algorithms'.

In simpler terms, *"algorithms are a set of systematic operations to be followed in calculations and which offers a solution to a problem."*[33]

Digitally, algorithms are operations that assist technology in analysing the gathered database on parameters like habits, behaviour, likes, experiences of the users. This is also known as audience segmentation.[34]

Algorithms are intangible assets that create value for businesses and primarily assist in processing, interpreting and analysing the data and generating economic value.[35] Primarily employed to target the users, price differentiate, optimize the Ad Placement strategies, etc. Given that algorithms are extremely crucial for a business to gain competitive advantage; they have been accorded IP protection under the GDPR.

However, India's stance on whether IP protection extends to algorithms or not, is still not clear given the ambiguity surrounding the applicability of patents and trade secrets to algorithms.

Data Portability and grant of the trade secret to algorithms have a rather direct connection. To put it in simpler terms, the raw data collected from the users are put to processing and segmentation with the use of algorithms. The categories of data for instance, observed, derived and inferred data are all results of the application of algorithms.

Hence, it seems true that if an exemption is created for trade secrets, enterprises would be able to protect a massive amount of data, open to being manipulated by the enterprises, under the garb of trade secrets defeating the purpose of data portability. The same has been opined in the report of the Committees constituted to draft and study the Bill.

(ii)   Analysis of the Report of Justice B.N. Srikrishna Committee

Justice B.N. Srikrishna Committee report on Draft Data Protection Bill, 2018, while proposing the Bill, opined that portability is critical in making the digital economy seamless.[36] The report further argued that the users should be able to access and transfer the data that they have provided, in a machine-readable format.[37] In Committee's view, portability, in addition to greater control of the data, eases the transfer of data from one enterprise to another if users wish to retract and switch and enhances fair competition which ultimately is beneficial for consumers and yields their welfare.

The committee also took note of the possibility of revealing trade secrets since the right extends to wide categories like generated data, data from profiles of the users, and even otherwise obtained. It further stated that portability should be allowed only if it does not reveal any trade secrets and requests should be denied if it's impossible to provide without revealing a trade secret. This view is in synchronization with the guidelines under the GDPR. The committee advised establishing a code of conduct for the enterprises to be adhered to and minimizing the instances of denying the users their right to obtain and transfer their data.

One of the most important implications highlighted by the committee was the heavy compliance costs burden associated with the mandate of data portability. Allowing portability of the data collected, generated, otherwise obtained and inferred data can significantly increase the costs for the data fiduciary since processing portability requests and categorizing the datasets from the anonymized datasets, can burden the enterprises and jeopardise their functioning. The committee suggested that fiduciaries be allowed to charge a fee while catering to portability requests from the users.[38] The Committee opinion appeared far more sighted when it comes to roadblocks/challenges in the way of data portability.

(iii) Analysis of the Report of the Joint Parliamentary Committee (JPC)

Contrary to the opinion of the Srikrishna committee, the JPC suggested doing away with the exception of the trade secrets citing high chances of manipulation. In its opinion -

In JPC's opinion, since the definition of trade secrets cannot be certainly formulated, it gives a scope to the data fiduciaries to deny portability by resorting to the exception of trade secrets. It would be easy for fiduciaries to conceal their actions by using the exception.[39] Recommendation No. 40 of the report proposed to remove the exception of trade secrets from the clause.

As far as technical feasibility was concerned, while it was suggested to be retained under the law, it was suggested that the same must have been governed by separate regulations.

In its report, the JPC restructured the Clause 19(2)(b) as -

"The provisions of sub-section (1) shall not apply where –

a) ......

b) compliance with the request in sub-section (1) would not be technically feasible, as determined by the data fiduciary in such manner as may be specified by regulations"

The analysis of the recommendation reveals that the suggested phrase *"as determined by the data fiduciary in such manner as may be specified by regulations"* would give wide discretion in deciding the technical feasibility and can be manipulated by the data fiduciaries. Establishing precise guidelines and their robust implementation beforehand is the sine qua non for both privacy and competitive intensity to be balanced in the digital sector.

In addition, the report also enlisted the concerns of the stakeholders raised in their submission to the JPC[40], let us delve into the concerns and analyse them from the lens of privacy and competition -

a.   'Portability should not be limited to large entities and automated processes alone and should cover all the data.'

The suggestion was to make portability requirements equally applicable to all enterprises and not alone the large entities. While it is important in one way to apply the requirement over all the entities to allow users to exercise their choice and ensure the flow of data, it must be kept in mind that enterprises operating in the medium and small sectors (SMEs) including the other new entrants require collection and processing of the data in order to effectively survive and compete against the tech giants operating in the market.

Moreover, if imposed equally on all, it would place a significant burden on the small enterprises in the form of huge compliance costs while also losing out on the data required to function. Covering all the data and all the enterprises can reverse the welfare effect sought to be achieved through the suggested remedy of portability.

b.   'The meaning of Machine-readable to be clarified.'

The Bill nowhere defined the term 'Machine-Readable'. Similarly, the GDPR too nowhere defines the term. The existing literature on the same suggests that Data is structured and machine-readable if it can be automatically read and processed by a computer. This is undertaken by means of an application programming interface ("API") for facilitating data exchanges with data subjects[41]. The machine-readable format ensures that the data can be reused by the users over other service providers, promoting interoperability, minimizing the technicalities and difficulty involved in reproducing the data over other enterprises while switching.

c.   'The meaning of Trade Secrets to be clarified.'

As already discussed, there is no specific law protecting trade secrets in India. Although courts have dealt with cases involving disputes over trade secrets and have rendered observations. The Delhi High Court in Burlington Home Shopping Pvt Ltd v. Rajnish Chhibber (1995)[42] defined trade secrets as:

"Any information with commercial value, which is not available in the public domain and the disclosure of which would cause significant harm to the owner."

In addition, India is also a signatory to TRIPS Agreement (1995) which defines trade secrets as:

'Information which is kept secret has a commercial value and the owner of the information takes reasonable steps to keep it secret.[43]

However, whether information like algorithms, for instance, can be protected is still unclear and remains to be ascertained.

d.   'To include IPR along with trade secrets.'

While the committee opined that trade secrets be removed as an exception, the stakeholders on the other hand, expressed that in addition to the trade secrets, their intellectual property should also be covered under the exceptions.

If we look into the views of experts on the matter, it has been opined that where on one hand, while we are aware of what competition is and that it should be protected, in the realm of data, we are not sure what intellectual property is and whether it calls for protection.[44] Hence, the idea of granting IPR in the Internet of things mix is strongly opposed.[45]

Hence, the 'personal data' whose boundaries and extent are still big questions and unclear, if allowed to be protected through Intellectual property rights, would defeat the whole purpose of the right to data portability.

Granting intellectual property exceptions would promote the commercialization of data, restricting the welfare effects sought to be achieved through the free flow of data. It can also act as a huge impediment to the users' choice of retracting their personal data and switching between providers. Lastly, this would restrict the scope of portability by expanding the boundaries of protected data.

e. *'Portability should not be extended to inferred, processed, or derived data.'*

This argument caters to the observation made earlier that each dataset carries its own commercial value. Moreover, it is the processing of the 'volunteered' and 'observed data' which produces the inferred, processed, and derived data for the enterprises. The inclusion of each of aforesaid datasets would depend on the pre-condition of being attributed to a natural person and whether the person can be directly/ indirectly identified through such datasets.

The profile of a person, for instance, can be attributed to a natural person and the person can be identified, however, the same profile is also used for processing and drawing inferences from the data observed and gathered. It would not be wrong to suggest that they are of commercial interest and hence portability should not extend to these datasets in particular. It may too, enable the disclosure of enterprises' analysis and intellectual property.[46]

Research in this domain reveals that inferred data may or may not fall into the category of personal data.[47] If inferred data is ascribed to groups or categories, it may not be personal data. However, in microtargeting inferred data will often be ascribed to an identified or identifiable natural person, yielding personal data.[48] Inputs from one profile can be used for other processes, hence, adding to the complexity of identifying the datasets for the purpose of portability. Allowing portability of inferred, derived and processed data can affect the functioning as these datasets contribute to improving the completeness and correctness of the datasets[49].

If the contours of personal data are not defined with certainty, data portability can be further limited by the enterprises by deleting the original data and manipulating the datasets along the lines of different categories of datasets[50].

VI. Balancing the Considerations: Plausible Solutions

Data Portability is an extensively technical remedy. While in theory, it appears to offer a probable solution to both privacy and competition, in practice the apparent conflict between the two could be a hurdle as we proceed, warranting a balance to be struck before its enactment as a provision within data protection laws.

The provision as it existed was too broad in terms of the datasets it intended to cover, making successful implementation less likely. While envisaging a provision on data portability, considerations as discussed in the preceding section require thorough scrutiny.

Based on the analysis, the following takeaways appear crucial while we deliberate on the future version of the Bill:

(i)   Data portability should not be applied alike to all enterprises and the same must be assessed based on its position in the market and the number of databases processed by an enterprise. Further, the big techs, its competitors and new entrants should not be subject to the same portability requirements.

(ii) At the outset inclusion of the different categories of datasets can place unnecessary burden on businesses hence, a suggestion would be to apply the portability requirement only to the 'raw data provided by the users voluntarily'. Further, it is of utmost relevance that personal data be more precisely defined in context of the digital markets doing away with the broad definition as previously existed.

(iii) Moving forward, under the exceptions to portability, Intellectual property should not be included as an exception under the section. This can create greater possibilities of enterprises rejecting users' portability requests.

(iv) Exception to trade secrets appear appropriate given the commercial interest attached to the algorithms and AI and machine-led learning technologies, however, it is important first to clear the definition and position of trade secrets and establish it under the national laws. In this direction, specific guidelines for 'trade secrets' and 'technical feasibility' must be introduced for an effective interpretation and implementation.

Endnotes

1   (2017) 10 SCC 1

2   Ibid. Para 193

3   'Privacy and Information Technology', Stanford Encyclopedia of Philosophy *(Oct 30, 2019) available at* https://plato.stanford.edu/entries/it-privacy/#ConVsInfPrI *(last visited August 7, 2022)*

4   Clause 19, The Personal Data Protection Bill, 2019

5   Graef, I., Husovec, M. & Purtova, N. (2018). Data Portability and Data Control: Lessons for an Emerging Concept in EU Law. German Law Journal, 19(6), 1359-1398

6   OECD, 'Data Portability, Interoperability and Competition – Note by India', (2021) DAF/COMP/WD(2021)31, *available at* https://one.oecd.org/document/DAF/COMP/WD(2021)31/en/pdf (last accessed August 10, 2022)

7   Andrei Hagiu and Julian Wright, 'When Data Creates Competitive Advantage' *available at* https://hbr.org/2020/01/when-data-creates-competitive-advantage (last visited August 7, 2022)

8   Clause 19

9   Article 20, The General Data Protection Regulation

10 Giovannetti, E., Hamoudia, M. The interaction between direct and indirect network externalities in the early diffusion of mobile social networking. Eurasian Bus Rev (2022). https://doi.org/10.1007/s40821-022-00208-1

11 Tanya Sammut-Bonnici, Derek F. Channon, 'First Mover Advantage', (Wiley Encyclopedia of Management, 2015) 10.1002/9781118785317. weom060085, p.1

12 OECD, 'Abuse of Dominance in Digital Markets', (DAF/COMP/GF/ WD(2020)32) December 8, 2020, *available at* https://one.oecd.org/ document/DAF/COMP/GF/WD(2020)32/en/pdf (last accessed August 10, 2022)

13 Dan Prud'homme, 'How Digital Businesses can leverage the high cost for consumers to switch platform' London School of Economics, September 24, 2019) *available at* https://blogs.lse.ac.uk/businessreview/2019/09/24/ how-digital-businesses-can-leverage-the-high-cost-for-consumers-to-switch-platforms/ (last accessed August 10, 2022)

14 Miljana Todorovic, 'Propensity of Data Accumulation to Raise 'Barriers to Entry' (Institute for the Internet and the Just Society, April 11, 2021) *available   at*   https://www.internetjustsociety.org/propensity-of-data-accumulation-to-raise-barriers-to-entry (last visited August 8, 2022)

15 Shoshana Zuboff, The Age of Surveillance Capitalism: The Fight for a Human Future at the New Frontier of Power. New York: Public Affairs, 2019

16 Guidelines on Right to Data Portability, Article 29 Data Protection Working Party (2017) *available at* file:///C:/Users/Admin/Downloads/wp242_ rev_01_en_D8A6FCF6-9039-846A-0C8040819826D818_44099%20 (3).pdf (last accessed August 8, 2022)

17 Clause 2(28) The Personal Data Protection Bill, 2019

18 Article 4(1), The General Data Protection Regulations

19 Opinion of the European Court of Justice

20 Recital 26, The General Data Protection Regulations

21 Recital 30, The General Data Protection Regulations

22 Bart Custers, Profiling as inferred data. Amplifier effects and positive feedback loops, (Amsterdam University Press, 2018), *available at* https://

scholarlypublications.universiteitleiden.nl/access/item%3A2976051/ view (last visited August 9, 2022)

23 Derived data are data generated from other data, after which they become new data elements related to a particular individual. Derived data are said to be created in a fairly "mechanical" fashion using simple reasoning and basic mathematics to detect patterns within a data set and create classifications.

24 Id. at 17

25 Article 2(a), EU Data Protection Directive of 1995

26 B.N. Srikrishna Committee Report on Personal Data Protection Bill, 2018 p.73

27 Alan McQuinn, Daniel Castro, 'The cost of an unnecessarily stringent federal data privacy law' (August 5, 2019) *available at* https://itif.org/publications/2019/08/05/costs-unnecessarily-stringent-federal-data-privacy-law/ (last visited August 9, 2022)

28 Personal Data Protection Commission, Discussion paper on Data Portability, (February 25, 2019) *available at* https://www.pdpc.gov.sg/-/media/Files/PDPC/PDF-Files/Resource-for-Organisation/Data-Portability/PDPC-CCCS-Data-Portability-Discussion-Paper---250219.pdf (last visited August 9, 2022)

29 Clause 3(6), The Personal Data Protection Bill, 2019

30 Clause 19 (1)(a), The Personal Data Protection Bill, 2019

31 Kashish Makkar, Art. 20, GDPR: A case for Extensive Interpretation, (13 JUL 2018) available at https://lawschoolpolicyreview.com/2018/07/13/art-20-gdpr-a-case-for-extensive-interpretation/ (last visited August 10, 2022)

32 The European Union, 'Trade Secrets', *available at* https://europa.eu/youreurope/business/running-business/intellectual-property/trade-secrets/index_en.htm (last visited August 10, 2022)

33 DAAS Suite, 'What does an algorithm do in Digital Marketing and how to reap the benefits' *available at* https://daassuite.com/blog/en/definition-algorithms-how-to-make-the-most-of-them-in-digital-marketing/ (last visited August 10, 2022)

34 Ibid.

35 OECD, 'Digitalisation, Business Models And Value Creation' available at https://www.oecd-ilibrary.org/docserver/9789264293083-4-en.

pdf?expires=1660126129&id=id&accname=guest&checksum=5DCD-4F686B6D1F243F1924988FDC35BA (last visited August 10, 2022)

36 Ibid at 23, p. 75

37 White Paper of the Committee of Experts on a Data Protection Framework for India at p. 135

38 Ibid at 23, p. 75

39 The Report of The Joint Parliamentary Committee on Personal Data Protection Bill, 2019, *available at* http://164.100.47.193/lsscommittee/Joint%20Committee%20on%20the%20Personal%20Data%20Protection%20Bill,%202019/17_Joint_Committee_on_the_Personal_Data_Protection_Bill_2019_1.pdf (last visited August 10, 2022)

40 Ibid at p. 76-77

41 Golden Data Law, 'The right to data portability under EU Data Protection Law' *available at* https://medium.com/golden-data/what-is-the-right-to-data-portability-under-eu-data-protection-law-8efa509fc788 (last visited August 10, 2022)

42 *1995* PTC (15) 278

43 Article 39(2)(c), TRIPS Agreement, *available at* https://www.wto.org/english/docs_e/legal_e/27-trips.pdf (last visited August 10, 2022)

44 Herbert Hovenkamp, Signposts of Anti-competitive Exclusion: Restraints on Innovation and Economies of Scale, in Barry E. Hawk (ed), International Antitrust Law & Policy 2006 (Juris Publ., Fordham Competition Law Institute, 2007) p. 413

45 Drexl, Hilty, Desaunettes, Greiner, Kim, Richter, Surblyté and Wiedemann, Data ownership and Access to data, Position Statement of the Max Planck Institute for Innovation and Competition, Max Planck Institute for Innovation and Competition Research Paper, No. 16-10, (2016).

46 Emre Bayamlıoğlu, Irina Baraluic, Liisa Janssens and Mireille Hildebrandt (eds), 'Being Profiled: Cogitas Ergo Sum. 10 Years of Profiling the European Citizen', Amsterdam University Press, 2018

47 Ibid. at 13

48 Ibid. at 19

49 Ibid.

50 Robert Madge "Five loopholes in the GDPR" (My Data Journal, August 27, 2017), *available at* https://medium.com/mydata/fiveloopholes-in-the-gdpr-367443c4248b (last visited August 10, 2022)

Chapter 4

Data Privacy Trends in Medical Legislation: A Contemporary Perspective

Vasvi Talwar

Assistant Professor

Asian Law College, Noida, U.P., India

Nipun Goswami

Legal Consultant

Integron Company, Noida, U.P., India

Abstract

Human health is one of the crucial factors that concerns everyone in today's world. There is an emergence of technology as result of development in society. Thus, this technology will some or the other impact over the medical aspect of the society especially if we talk about privacy which is one of the essential components of the right to llife enshrined in the Constitution of India. Moreover, privacy plays a significant role at societal level for taking care of public health activities at a large level. Therefore, the need for data protection in health care comes into the picture as it plays a significant role with concern to privacy because we will observe in the current scenario that due to digitalization lot of data breaches and there were lot of cyber-attacks which have become common these days and patient information can be shared to others, this gives a rise for taking care of data privacy at a pampered scale. In this Chapter, we will be posting the best picture of the role of technology and medical laws to discuss the importance of data privacy in our day-to-day life as it plays a key role in maintain the trust and transparency between the patients and medical institutions. In our research paper, we will discuss about role and importance of technology with concern to data protection, how and why these practices are important in our day-to-day life, furthermore, a brief review of the rules and regulations which present in the current scenario to take care of privacy on any individual.

Keywords – Data protection, rules and regulations, health, digitalization, medical.

I. Introduction

Privacy is always considered as a crucial factor when we want to talk or wants to describe about the quality of life. If we observe, we will notice the fact that privacy is never enshrined explicitly in the constitution of India still it is considered as a crucial factor when we consider the dignity or personnel liberty. Privacy is nothing but an implied principle i.e., defined under the Article 21 of the Constitution of India.

According to Black Law Dictionary privacy means "the right to be let alone; the right of a person to be free from any unwarranted publicity; the right to live without any unwarranted interference by the public in matters with which the public is not necessarily concerned." Thus, if we try to interpret the said concept in wider sense it says that every individual has the right to reside alone in the society or we can say that every individual used to perform two kinds of activities. Firstly, the acts or the conduct he or she is performing in the public where he wants that his or her conduct needs to recognized by the society whereas if we talk about another aspect, we will observe that the very same individual also perform some acts which he never wants to disclose in public or wants that no opinion is framed upon him on the grounds of the conduct or performance of act.

II. History and Development of Right to Privacy

Now when we understood the definition of right to privacy, we need to answer the question that how the right to privacy comes into the picture and evolved itself with concern to the modern times whether it is in context to Indian Society or at the international level platform. Right to privacy is one of the rights which was got firstly recognized from the ancient times through Dharamshastras, Vedic culture, Upanishads, Manu Smritis and through various other ancient stories such as Ramayana, Mahabharata etc.

But it is pertinent to note the said concept got developed and gained its importance in the Era of British Rule. It was so because it had kept the foundation of privacy under the ambit of right to life through its interpretations. If we look at the making of the Constitution, Dr. B.R. Ambedkar particularly focused towards right to privacy because he understood the fact if the right to privacy was not given then despite the India had obtained freedom from bruisers, but people will not enjoy the freedom in true sense as there will be a lot of interference that will take from the side of authorities which somehow hampers the freedom of citizens. Thus, his prior intention was to create the constitution which could create the environment of democracy in its true sense.

Further, if take the look over the status of the right to privacy through many reports we will discover the fact how important role right to privacy had played in order to protect the rights of the citizen and in one of the report named as state and minorities it was stated that, *"the right of the people to be secure in their persons, house, papers and effects against unreasonable searches and seizures shall not be violated and no warrants shall issue but upon probable cause, supported by oath of affirmation and particularly describing the place to be searched and the persons or things to be seized."*

Further, it is the judiciary who had also played a key role to interpret the policy of right to privacy as we mentioned above that right to privacy is nowhere explicitly prescribed but present in the constitution through right to life enshrined in Article 21 of the Constitution of India.

In **MP Sharma v. Satish Chandra**, it was observed by the Supreme Court that right to privacy which provided to us through the Article 21 of the constitution of India is not an absolute right but comes with certain restrictions especially when it comes to the matter like search and seizure if done by the authorities. The Supreme Court of India further incorporated the similar approach in matter of **Kharak Singh v. The State of U.P.** where the question of surveillance is violation of Article 21 of constitution on the ground of right to privacy was raised. The Supreme Court in the above matter again stated the Fact that right to privacy is not a full fledge right enshrined in the Constitution but it the right which comes along with certain exceptions which was attached to it.

In **Govind v. State of M.P.**, it was observed by the Supreme Court that there is an absence of enactment on the part of legislature. Thus, the development of the right to privacy can only be achieved through precedents therefore it will be inadequate to consider any one single precedent with concern to the right to privacy.

In one of the landmark judgements **ADM Jabalpur v. Shivakanta Shukla**, where the Supreme court of India was adjudicating the question of law concern to Right to life and restrictions with concern to personnel liberty Justice Khanna had observed, *"Article 21 is not the sole repository of the right to personal liberty…no one shall be deprived of his life and personal liberty without the authority of laws follows not merely from common law, it flows equally from statutory law like the penal law in force in India."*

Further in various landmark judgements named as **Maneka Gandhi, R. Rajagopal v. Tamil Naidu, People's Union for Civil Liberties v. Union of India, Salvi and others v. State of Karnataka** and other judgements, the Supreme Court of India had tried it's best to explain the concept of right to privacy whether it is through explaining the evidentiary value or through creating a difference between the mental and physical right to privacy.

III. Right to Privacy in International Law

Till now, we had talked about Right to Privacy with context to Indian Society but now we need to understand the scope of Right to privacy with context to international law. So, when we talk about the international law the first thing that comes in our mind is that the international law is a law, or we can say a code of conduct that regulate the behavior of different nations at the common platform.

Thereinafter, if we talk about the concept of privacy it was recognized as one of essential concept when we talk about international and it is so because the concept of right to privacy is enshrined under many of the international documents such as international covenant on Civil and Political Rights, Universal Declaration of Human Rights, etc. If we look from time period where league of nations got failed and the concept United Nations comes into the picture, we will observe that there was no country who had adopted the right to privacy as the part of the constitution. After the incorporation of the concept of right to privacy, the sovereigns were providing the guarantee to protect the right to privacy.

While the international law aiming at the right to privacy to be considered as the human rights still the biggest challenge before the United Nations. Moreover, it will be the situation if the any of nations denies accepting it and lead to the situation like World War II. While considering the concept of right to privacy as the international human rights, we countered some questions like right to privacy has two aspects connected to it privacy with context to personnel life and right to privacy with context to society. Thus, while creating it as the human rights are we not binding the sovereign and compromising with the security issues with reference to external threats like terrorism etc. and we all knew that **security and peace** are only the two principles on which was the International Organization like United Nations is established.

So, when we talk about right to privacy with context to the international documents then we will be observing the fact United Nations came up with the Bills of rights which included Universal Declaration of Human Rights, International Covenant on Civil and Political Rights and International Covenant of Economic Social and Cultural Rights was the first document to be recognized right to privacy as the human rights. The reason behind these three documents was to declare and made them mandatory for each sovereign to comply with the said documents.

(i) Universal Declaration of Human Rights

It was the first document which states the importance of right to privacy to be considered as the human rights. According to above mentioned document it stated that, *"No one shall be subjected to arbitrary interference with his privacy, family, home, or correspondence, nor to attacks upon his honor and reputation. Everyone has the right to the protection of the law against such interference or attacks".* So, the intention of the drafting committee of the Universal Declaration of Human Rights can be explicitly visible that the framers want to protect the right to privacy with context to the honor of the individual, honor of the family and further it had barred by the sovereign from arbitrary interference in the matters of any individual or their families.

(ii) International Covenant on Civil and Political Rights

This document is second most important international document, and it is so because it denotes the protocols with concern to civil and political rights of the human rights which are declared by the Universal Declaration of the Human Rights. Thus, this document creates the mandatory for every sovereign to provide their assistance in order to provide the rights to the individual. This document again denotes the same concept that is stated in Universal Declaration of Human Rights as the framers want to protect the right to privacy with context to the honor of the individual, honor of the family and further it had barred by the sovereign from arbitrary interference in the matters of any individual or their families.

(iii) Convention on the Rights of The Child

They said Convention was made in order to protect the rights of the Child. Thus, the abovementioned document states that *"No child shall be subjected to arbitrary or unlawful interference with his or his privacy, family, home, or correspondence, nor to unlawful attacks on his or her honor and reputation. The child has the right to the protection of law against such interference or attacks."* Therefore, we can say that the said convention is an extension of above mention conventions the only difference here the provisions this convention counts children subject matter. Therefore, this convention is only extended to children only.

(iv) International Convention on The Protection of All Migrant Workers and Members of Their Family, 1990

This Convention deals with protection of all kinds of migrant workers and the rights with context to family who is depended upon the said worker, and it is considered to be important because migrant workers are generally of poor background and if they are also exploited then it will not only create the impact over the one person, but it will impact the whole family. This convention speaks that *"No migrant worker or member of his or her family shall be subjected to arbitrary or unlawful interference with his or her, privacy family, correspondence, or other communications, or to unlawful attacks on his or her honor and reputation. Each migrant worker and member of his or her family shall have the right to the protection of the law against such interference or attacks."*

(v) European Convention for The Protection of Human Rights and Fundamental Freedom

The European convention is made to maintain harmony between European states but with respect to the right to privacy the said convention states, *"everyone has the right to respect for his private and family life, his home, and his correspondence. There shall be no interference by a public authority with the exercise of this right except such as in accordance with the law and is necessary in a democratic society in the interests of national security, public safety, or the economic well-being of the country, for the protection of health or morals, or for the protection of the rights and freedoms of others."* The above said convention explicitly bounds every individual including the public authority to respect the privacy of the other person with certain restrictions applied to it with context to security or public safety issues.

(vi)   American Convention on Human Rights

The said convention is made in order to create the legal system where the goal of social justice could be achieved. This convention was more oriented towards the concept of social justice with context western ideologies states. Thus, the abovementioned convention states that *"Everyone has the right to have his honor respected and his dignity recognized. No one may be the object of arbitrary or abusive interference with his private life, his family, his home, or his correspondence, or of unlawful attacks on his honor or reputation. Everyone has the right to the protection of the law against such interference or attacks."*

(vii)   Cairo Declaration on Human Rights in Islam

It is considered to be one of the important declarations with context to the human rights that were accepted at the international level with regard to Islamic community the abovementioned convention states that, *"(a) Everyone shall have the right to live in security for himself, his religion, his dependents, his honor, and his property. (b) Everyone shall have the right to privacy in the conduct of his private affairs, in his home, among his family, about his property and his relationships. It is not permitted to spy on him, to place him under surveillance or to besmirch his good name. The State shall protect him from arbitrary interference. (c) A private residence is inviolable in all cases. It will not be entered without permission from its inhabitants or in any unlawful manner, nor shall it be demolished or confiscated, and its dwellers evicted".*

(viii) Arab Charter on Human Rights

The above-mentioned convention was made to accept and regulate the human rights with context to states who had direct relation with context to the human rights. The said convention is not only oriented to right to privacy concept which was given but its disregards the surveillance issues as well. The said convention states, *"privacy shall be inviolable, and any infringement thereof shall constitute an offence. This privacy includes private family affairs, the inviolability of the home and the confidentiality of correspondence and other private means of communication.".*

Thus, if we try to analyze the above-mentioned Conventions and their provisions which context to concept of Right to privacy, we will observe the fact that there were different conventions that were made at the international level with different aims and objectives but the intention with reference to privacy remains the same.

Now, we need to understand the fact that international law is one of the laws which is not stable as it is growing and developing itself day by day due to development in society, globalization and etc. thus, the ambit of right to privacy also had changed with context with development of technology and their usage whether it is IT field, medical field, or any other field.

Now, if we consider only the technological development with context to right to privacy then we should the example of the emergence of phone. Now, what exactly happens while using the smart phone when we install any application which authentic application or third-party application. The said application while getting installed seeks the permission to access to the resources which are confidential and by providing them the access the owner of the application can have a look over the abovementioned information which he might sell to private companies to earn revenue. Thus, the question here is - are there any stringent laws domestically or internationally which can regulate the usage of these technologies.

The answer is partially yes and partially no. This is so because when we talk about the usage of technology, we have different Acts with context to different technology for example we have Information Technology Act which regulates all kinds of electronic technologies which we are developing these days like laptops, phones etc. but if we look over the laws with concern to the concept of Rights to Privacy, we will discover that there is no present law with reference to it. Even the legislature had drafted one of the Bills which was remained pending before the parliament since the long time, but it is pertinent to note that the said Bill had also been withdrawn by the legislature.

Now, if we try to understand we will be observing the fact that technology is not only develop in the electronic gadget but had also developed in terms of medical sciences. Thus, the ambit of the Right to Privacy would have also increased but the question now what role do technology had played in development of medical science with context to Privacy.

While answering the above question we will be observing that one of the landmark developments that took place is emergence of the concept named DNA. DNA stands for deoxyribonucleic acid. DNA is a basis unit that stores all the vital information of our body thus even the person are twins then also there might be minute difference with concern to their DNA. Therefore, we can say that DNA can act like biometric identity or identification card through which we can recognized the individual.

The same concept was used by the Indian Government through their Aadhaar Card policy which was later challenge on the grounds of violation of Right to Privacy enshrined Article 21 of the Constitution. The question of law before the judiciary was whether public authority has the power collect and store the vital information of the individuals or it is the arbitrariness of the public authority to store and collect the information of the people. Therefore, in the ***Justice K.S. Puttswamy (Retd) vs. Union of India*** it was declared that right to privacy is a fundamental right and state has the duty to protect the rights of the individuals.

Through the above-mentioned circumstances, we will be observing the fact that it is the Legislature role which is important to regulate the modern technology in medical science with concern to right to privacy enshrined in Article 21 of the Constitution of India. So, we will be discussing some of the important Acts that are formulate by the legislature to regulate the technology.

IV. Indian Laws and Regulations

In India, privacy of medical records spread over a number of laws:

(i) Criminal Procedure (Identification) Act, 2022

The said Act was introduced by the legislature in order to create a biometric database of the accused, victim, convict, etc. for the purpose to provide legal aid to the investigation in order to resolve or settle the matter in an effective and efficient manner but the question here arises that whether the above-mentioned Act is intervening the ambit of privacy of an individual or simply we could say that what is the constitutional validity of the above mentioned Act when we overview the whole Act with Right to Privacy. The said question here also arises because we are taking the consent of the accused while collecting his DNA but we are creating a compulsion that if we even try to refuse the said provision makes the said refusal as the offence of Section 186 of Indian Penal Code. Moreover, in the said matter the police-in-charge as well as the magistrate has the power to direct to collect data from any person. The said data which will be collected will be stored for 75 years after which the data will be deleted further another method through the data can be deleted is through the order of the magistrate. Whereas the concern data will be secured by the National Crime Report Bureau i.e., NCRB which is again a public authority. Thus, we can say that the Public Authority is somewhere interfering in the private life of the individual which was one of the international obligations that state needs to perform.

The most critical issue with concern to the said Act is that there is no back up that will be available to us as we know that the India is lacking in the cyber security then there will be a lot of issues with concern to tampering of data, leaking of data, deletion of data, etc. The said Act is maintaining the database of the accused who had committed the heinous offences against the Humanity or having punishment of seven or more than seven years rather including petty offences and moreover the said Act is also violating the ambit of self- incriminatory enshrined in Article 20(3) of the Constitution of India if the said data is used as the evidence in the pending matter.

(ii)   DNA Technology (Use and Application) Regulation Bill, 2019

The said Bill is nothing but a supplementary provision to the Criminal Procedure (Identification) Act, 2022 and it is so because if we try to understand the ambit of the said Bill, we here is dealing with the collection of DNAs, storing of DNA and further deals with the how the DNA profiling of the accused needs to be done or deleted. So, they said Bill is dealing with the regulation with reference to the use and application of DNA Technology.

The said Bill is providing us schedule through which we can determine that whether we can DNA Technology can be used or not. The said Bill further states regional and sub- regional DNA Data bank where the collected samples will be stored. The said Bill also establishes the DNA regulatory Board who has the responsibility to regulate the labs and other facilities to collect and reserve the data which is collected by the investigating authorities. The important factor which is incorporated by the said Bill that if any person found to be guilty of the offence of disclosure of identity or use the said data without proper authorization then that person will be liable for the punishment up to 3 years and fine up to the amount of One Lakh Rupees.

(iii)   The Indian Medical Council, 1956

The said Act was introduced by the legislature to define the conduct of the doctors with concern to the patient they are dealing with. Thus, when we talk about the right to privacy with concern to the medical ethics, we will be observing the doctor while following up his medical ethics he needs to discuss everything with his patient but in some cases the discussion needs to be done with the family members. Thus, the discussion of the respective information of the patient is subject to the Privacy.

Furthermore, the said act deals with the medical records which is made by the medical authorities thus it is important when we are discussing the concept of right to privacy because it contains all the vital information with concern to the patient which he or she might or might not share with the others. Therefore, the said Act deals with the fact how the records need to be maintained and if it is destroyed then how it will be destroyed.

In the **Raghunath G. Raheja v. Maharashtra Medical Council** it was stated that "we do not think that the hospitals or the doctors can claim any secrecy or any confidentiality in the matter of copies of the case papers relating to the patient. These must be made available to him on demand, subject to payment of usual charges. If necessary, the Medical Council may issue a press note in this behalf giving it wide publicity in all the media."

In the matter of **Kanayama Ramanlal Trivedi v. Dr. Satyanarayana Vishwakarma**, it was held that the hospital authorities as well as the doctors were held the liable deficiency of service because the medical records were unable to be produced.

In various other judgements like **S.A. Qureshi** and **Dr. Shyam Kumar,** it was held that it is the responsibility of the Hospital Authority to maintain and disposing of the medical records.

V. Right to Information and Right to Privacy with Context to Medical Aspect of Indian Society

Right to Information Act, 2005 was introduced by the legislature for the development of the society. The said act was formulated to so that the transparency as well as the accountability could be maintained when it comes to development. The legislature understood the importance transparency and accountability from framing of the constitution but with respect to time development had taken place in all the senses that whether it is societal or corruption. But if we try to understand the mindset of the legislature, we will observe the fact that legislature had bought the said legislation so that the corruption could be reduced. Moreover, the public could gain the more information with concern to work or the status of concern work. So, this Act was enacted in a manner that the people could be aware about the policies or other information that government of India is providing them for their economic and social development.

If we just interpret the meaning of the name of Act, we will observe that the name of Act is completely contrary with the concept of Right to Privacy. If we investigate the provision of the Act, we will be observing that the said Act had never creating the contrary provision, but it is creating the parameter for both the Rights so that when information can be accessible it will be provided but if the information is personnel or issues like national security, it will not be disclosed.

Now, when we talk about right to information with concern to the respective above-mentioned concepts, we will be observing that there is no strict law that is provided through which it can be clear cut stated that the information can disclose or not when it comes to information with concern to medical patient or other medical facility could be disclosed or not.

So, question is when there is no law with concern to right to privacy how we will determine that the said information is falling under ambit of Right to Privacy or Right to Information. The answer the above said question is role of Judiciary comes into picture where we already had observed that how right to privacy was evolved through precedents, but judiciary here will interpret the fact whether the information which we are seeking for the public welfare or simply.

We can say that it will benefit the society at large whereas if it is used for the mental harassment or interfering the personnel sphere of the individual. For example, in the year of 2007 there were many judgements that were passed with concern to Central Information Centre and State Information Centre where the Hon'ble Court had directed not to disclose the confidential information of the person till the time the matter is of public safety.

VI. Conclusions

The ethical health research and privacy protection will benefit to society a lot, the health research was vital in improving our health care sector, and will protect the individuals, which are involved in research from any harm and will preserve there right, which was really essential to conduct the ethical research, the main function of keeping privacy was to protect the interest of the individuals, but it was very important to keep privacy at social level, in order to keep individuals dignity. A major goal of privacy was to ensure that all individual health problems are properly maintained and the major aim of privacy rule was to ensure that individuals health information are properly protected in order to promote high quality health care.

Then the general population, the stress which was related to fertility problems which can be there due to increase martial conflict. Over the past, there were remarkable progress which have been made in modern reproductive technology to solve all the problem. However, the families which are created through assisted reproductive technology have raised a number of concerns about various potentially consequences, for parenting and child development. An ART is a definitely a dynamic field which talks about the medical improvement, new treatment modalities, ethical issues and cost benefit analyses for allocation of resources.

References

1   Black Law Dictionary.

2   Rama Narayan Dutta Shastri Ram (translated by) Valmiki Ramayana, Yudai Kind, Shripad Damodar Stealer (ed.) Mahabharata, Adi Parva, p.one thousand; Har Govind Sastry (ed.) Manu smriti..

3   Ambedkar B. R., State and Minorities. Article II Section I. Fundamental Rights of Citizen.

4   Sangam Thapa, "The Evolution of Right to Privacy in India", volume 10 Issue 2 Ser. I in International Journal of Humanities and Social Science Invention (IJHSSI), (PP 53-58).

5   M. P. Sharma and Ors. v. Satish Chandra District. Magistrate, Delhi and Ors., AIR 1954 SC 300.

6   Kharak Singh v. State of Uttar Pradesh, AIR 1963 SC 1295.

7   ADM Jabalpur vs Shivkant Shukla, *AIR 1976 SC 1207.*

8   Govind v State of Madhya Pradesh, AIR 1975 SC 1378.

9   Maneka Gandhi vs Union of India, AIR 1978 AIR 597, 1978 SCR (2) 621.

10 R. Rajagopal and Ors. vs. State of Tamil Nadu and Ors., AIR 1995 SC 264.

11 People's Union *for* Civil Liberties's case, AIR 1997 SC 568.

12 Smt. Selvi and Ors v. State of Karnataka, AIR 2010 SC 1974.

13 ECOSOC, Commission on Human Rights and Sub-Commission on the Status of Women, 22 February 1946, E/27, section A at para 2(a); and ECOSOC Res 9(II), 21 June 1946, E/RES/9(II).

14 Drafting Committee on an International Bill of Human Rights, Documented Outline, 11 June 1947, E/CN.4/AC.1/3/Add.1 ('Drafting Commission Documented Outline').

15 Oliver Dieleman and Maria Nicole Clees, "How the Right to Privacy Became a Human Right", Human Rights Law Review, 2014, 14, (441–458).

16 Commission on Human Rights, 2nd Session, Summary Record of the 28th Meeting, 4 December 1947, E/CN.4/SR/28 ('Commission Summary Record 28').

17 Universal Declaration of Human Rights.

18 UN Secretary-General, Annotations on the Text of the Draft International Covenants on Human Rights, 1 July 1955, A/2929.

19 International Covenant on Civil and Political Rights.

20 Convention on Rights of the Child.

21 International Convention on The Protection of All Migrant Workers and Members of Their Family.

22 European Convention for The Protection of Human Rights and Fundamental Freedom.

23 American Convention on Human Rights.

24 Cairo Declaration on Human Rights in Islam.

25 Arab Charter on Human Rights.

26 Xingu, M., Policy Recommendations for Surveillance Law in India and an Analysis of Legal Provisions on Surveillance in India and the Necessary and Proportionate Principles, CIS. Available at: http://cis-india. org/internet-governance/blog/policy-recommendations-for-surveillance-law-in-india-and-analysis-of-legalprovisions-on-surveillance-in-india-and-the-necessary-and-proportionate-principles.pdf.

27 Justice K.S.Puttaswamy(Retd) vs Union Of India, (2017) 10 SCC 1.

28 Criminal Procedure (Identification) Act,2022, India, *available at:* https://www.clearias.com/criminal- procedure-identification-act-2022/ (last visited on November 03,2022).

29 Thompson, William C. "The potential for error in forensic DNA testing (and how that complicates the use of DNA databases for criminal

identification).” In council for responsible genetics (CRG) National Conference: forensic DNA databases and race: Issues, Abuses and Actions.

30   The Law Commission of India, Human DNA Profiling, Report No. 271.

31   Himanshu Pandey & Anita Tiwari, Evidential Value of DNA: A Judicial Approach, Bharti Law Review (January- June 2017).

32   Khaleda Parveen, Forensic Use of DNA Information vs. Human Rights and Privacy Challenges, 17 U.W. Sydney L. Rev. 41 (2013).

33   Samat. H. Joshi, “The Indian Medical Council Act, 1956” available at: https://lawyerslaw.org/the-indian- medical-council-act-1956/ (last visited on November 03,2022).

34   Raghunath G. Raheja vs The Maharashtra Medical Council, AIR 1996 Bom 198.

35   Kanaiyalal Ramanlal Trivedi v Dr. Satyanarayan Vishwakarma , 1997 CPJ 332 (Gujarat).

36   SA Quereshi v. Padode Memorial Hospital and Research Centre, II 2000 CPJ 463.

37   Dr. Shyam Kumar v Rameshbhai, Harmanbhai Kachiya, 2002 1 CPR 320.

38   Swayam Shree Mishra, “RTI and social accountability” available at: https://www.governancenow.com/views/columns/rti-and-social-accountability (last visited on November 03,2022).

39   Constitution of India.

40   Right to Information Act, 2005, (Act 22 of 2022).

41   Krishna Lasya, “Right to Privacy vis-a-vis Right to Information. A Quintessential Circumspection” available at: https://www.legalserviceindia.com/legal/article-4557-right-to-privacy-vis-a-vis-right-to-information-a- quintessential-circumspection.html (last visited on November 03,2022).

Chapter 5

Right to Privacy and Data Protection Under Indian Legal Regime

Dr. Kush Kalra

Assistant Professor

NMIMS

Mumbai, Maharashtra, India

Riju Raj Jamwal

Ph.D. Research Scholar

University of Petroleum and Energy Studies

Dehradun, Uttarakhand, India

Abstract

The online landscape is dynamic and constantly changing across the globe. New platforms are rapidly emerging, existing platforms are simultaneously evolving. This Chapter views data protection and privacy problems in India through the lens of existing policies that facilitate the current online ecosystem. A substantial portion of the Indian population lacks adequate digital literacy, making them vulnerable to online privacy risks. Many individuals may not fully comprehend how their data is being collected, used, and shared, making informed consent perplexing. The study explores the stakes, challenges, and fragmented policies in the data protection while examining the charismatic political influence of the political leadership and role of government agencies and private corporations in the world of unauthorized access and misuse of data. Additionally, the study reviews the draft Digital Personal Data Protection Bill (DPDP) Bill, 2022 and argues that data ownership and consent for protecting the civil rights is paramount in the face of technological changes in the global digital environment. As technology continues to advance at an unprecedented pace, it has the potential to greatly impact our privacy in various ways.

Keywords: Protection, Digital, Personal ,Data, Bill, Right to Privacy.

I. Introduction

In a data-driven world, the significance of data protection and privacy is becoming more widely understood with the increase in social and economic activities online. India is ranked as the second largest social network user (Ruby, 2023) and holds eight in e-commerce markets (ecommerce DB, 2022) and ranked third amongst nations that experienced cyberattacks in Asia according to Global Cyber security Index (GCI) 2020. The increasing collection, use, transfer, and disclosure of personal data to third parties without the data owner's knowledge or consent rekindle the discourse on the importance of sound public policy. This paper highlights existing global policies and technologies that could facilitate the online ecosystem to address data protection and privacy concerns in India.

The review of the proposed Digital Personal Data Protection Bill (DPDP) Bill, 2022, highlights that it lacks clarity in privacy protection. In the context of the unclear legal framework, there is fear of policy capture by the Central Government and private companies which could threaten democratic principles, civil rights, economic progress, and investors' trust in the government. Technological advancements are leading to the production of an exponentially increasing collection of data and information known as "Big Data". Social media, crowdsourcing platforms, applications (apps), Internet of Things (IoT) devices are producing a lot of real-time intelligence from Billons around the world (Laurie Schintler, 2022). Data encompasses any representation of information, facts, concepts, opinions, or instructions that are appropriate for human or automated interpretation, communication, or processing. Data protection and privacy refer to the collection, storage, use, and sharing of personal data in a way that respects individual's privacy rights and protects their personal information from unauthorized access or misuse. The goal of data protection policies is to limit the infringement on an individuals' privacy when it comes to the personal information or data frequently processed by government agencies or any private business.

Currently, there is no explicit legislation protecting data protection or privacy in India. The fundamental right to privacy is not explicitly guaranteed by the Indian Constitution (Tripathi, 2017). The right to privacy has been included by the courts in the other fundamental rights that already exist, including the freedom of speech and expression under Article 19(1) (a) and the right to life and personal liberty under Article 21 of the Indian Constitution. However, as stated in Article 19(2) of the Constitution, the State may place reasonable limitations on these Fundamental Rights under the Indian Constitution (Supreme Court, 2017). Currently, the Indian Contract Act, of 1872 and the Information Technology Act, 2000 are the pertinent legislation in India that deals with data protection.

The state is part of every aspect of people's daily life. They keep records of marriages, births, and deaths. They manage the flow of immigration, collect taxes, deliver public services, and govern what people can buy, consume, and watch. Technology advancement is a revolution and makes it easier for addressing coordination failures due to its velocity, scope, and systems impact (Kavanagh, 2019). Nevertheless, the government has immense authority to set agendas through persuasion strategies. The policy fragmentation and lack of a comprehensive legal framework exposes India to a surveillance state and commercial exploitation.

II. Data Sovereignty

India faced uncertainty in terms of privacy concerns and data protection in the absence of a comprehensive privacy law, affecting the collection, purpose, and storage of data in India. The government introduced the revised draft Data Protection Bill, 2019 only to be withdrawn in August 2022. The fourth version, a new draft was released on 18 November 2022. The proposed legislation, known as the Digital Personal Data Protection Bill, 2022 (DPDP Bill, 2022), available for public comment and was anticipated to be submitted in Parliament during the 2023 budget session, which however was not listed for introduction. The Bill enshrines into Indian law some key proposals around the collection and uses limitations for businesses operating within India; an accurate accounting for user consent indicating why exactly a particular service is required from them along with limitations on how long that personal/sensitive info can be retained; obligations acquirement upon any change management in ownership following company acquisitions among others. It also goes beyond notifying firms but provides individuals access control mechanisms on how their private info is being used regarded as applicable occasions solely regulated by the customer itself. The proposed Bill outlines the duty of the Data Fiduciary to use the gathered data legitimately as well as the rights and responsibilities of the citizen (Digital Nagrik). As part of the compliance framework, it calls for the creation of a Data Protection Board of India to ascertain when the provisions of the draft Bill are not being followed, impose penalties when necessary, and carry out any other duties that the Central Government may delegate to it following the draft Bill's or other laws' provisions. A body corporate or any person collecting, receiving, possessing, storing, dealing, or handling information on behalf of the body corporate is currently required to observe the security practices and procedures outlined in the Information Technology (Reasonable Security Practices and Procedures and Sensitive Personal Data or Information) Rules, 2011, which were made by the Central Government in the exercise of its powers under the Information Technology Act 2000. The proposed DPDP Bill, 2022 will push regulators, legislators, the judiciary, civil society, and industry to understand why the state should decide on the level of degree an individual privacy right, with the wide exemptions of data fiduciary that it has proposed in the Bill. Did the central government just empower itself to have immense control and power to access its citizens' data?

The main challenges associated with this Bill include its potential impact on businesses' ability to collect user data, compliance costs related to meeting the new standards set by the law, and enforcement issues due to a lack of resources or expertise within government agencies responsible for enforcing it. Additionally, there are concerns about how effective this Bill will be at protecting users' privacy given existing gaps in Indian laws regarding online surveillance and cybercrime prevention.

To mitigate cyber security threats and for knowledge transfer through evaluating the impact of algorithmic decision-making on their citizen, India advances the idea of mandating the disclosure of source code (UNCTAD, 2021). This is opposed by the developing countries as it could negatively impact business interests. So, the debate is on the central issue of international trade: How much of its sovereign policy space should a country give up to profit from the system of free trade? (Basu, 2021).

The meaning and scope of sovereignty in the digital sphere both in terms of statutory requirements and claims of strategic autonomy is at the center of discussions about data governance (Christakis, 2020). India will likely play a significant role, and New Delhi needs to appreciate how important it is for the strategic interests of the nation for these debates to be shaped. With broad vision of data sovereignty, three pillars of Indian diplomacy have been the configuration of data for development, the assertion of sovereign writ on cross-border data flows, and the use of data as a strategic tool. India has made its opinions on several matters apparent, motivated by a strong strategic interest in forming norms on all three pillars (Basu, 2021). Moreover, it also helps protect individual privacy from unauthorized access or intrusion by external parties. For this reason, many organizations now have stringent compliance requirements for businesses operating in India that are linked to particular industries such as healthcare and financial services.

III. Censorship and Surveillance

Censorship and surveillance have become widespread in India, leading to issues such as the erosion of civil liberties, curbs on freedom of expression, and a lack of privacy. Citizens are being subjected to online monitoring as well as intrusive searches by government agents without reasonable cause or suspicion. This has led many citizens to criticise the Indian government's approach towards censorship and surveillance. In India, surveillance is governed by the 1885 Telegraph Act, along with the 2000 Information Technology Act (Noorani, 2019). Even though the Supreme Court has twice stated, in PUCL Case 1997 and in Puttaswamy Case 2017, that an order of surveillance can be passed only when strictly necessary and if there is no alternative, the lack of independent scrutiny and effective reporting mechanisms results in a lack of accountability (Vasani, Kuruppath, and Pednekar, 2019).

India witnesses a significant extension of policing and surveillance using personal data, which has created a pervasive sense of fear and a breakdown of social responsibility. The government is tightening up the law and using technologically advanced monitoring techniques to expand restrictions and increases policing and micro-management of society. Thus, the state infringes on privacy and introduces new restrictions on life, liberty, and livelihood in name of national security. The curbs on civil liberties of political oppositions, independent institutions, non-profits, journalists, academics, and other civil society activists have focused on data manipulation.

The central government is abusing its democratic legitimacy by seizing control of surveillance. According to the Centre for Internet and Society, the Indian government has proposed several surveillance-based intelligences gathering projects, especially after the 2008 Mumbai terrorist attacks, such as National Intelligence Grid (NATGRID), Crime and Criminal Tracking Network and Systems (CCTNS), Lawful Intercept and Monitoring (LIM) systems, Networking Traffic Analysis (NETRA), Central Monitoring System (CMS), and the National Cyber Crime Coordination Centre. The Central Monitoring System ("CMS"), official India's surveillance programme, poses a serious threat to personal privacy and democratic freedom of expression.

Through a system of widespread self-censorship among individuals whose speech ate restricted by CMS, it brings about a paradigm shift in the way speech is governed in India (Addison, 2015). The accountability, supervision, and transparency of India's surveillance system need to be significantly improved. Although there is a normative framework for the oversight of surveillance activities under the pertinent pieces of legislation, the "leakages" in the system are not being caught by the current mechanism, which seems to indicate that there is a need to harmonize and modify the current surveillance regime. However, it is impossible to get official information regarding precisely where the system is failing because of the opacity with which governmental surveillance activities are conducted.

India lacks systems to oversee and give checks on state surveillance in light of the introduction of some of the state's projected surveillance initiatives and accommodate new technologies while also considering those that threaten current privacy standards. Internet governance refers to the development and application by Governments, the private sector, and civil society, in their respective roles, of shared principles, norms, rules, decision-making procedures, and programs that shape the evolution and use of the Internet. (MEIT, 2022) These defined factors influence the development and use of the internet, including tasks related to the coordination and development of technical standards, the maintenance of vital infrastructure, and concerns with public policy.

The right to privacy was deemed a basic right under the umbrella of the right to life (Article 21) in the Constitution by the Supreme Court of India in a landmark decision issued in 2017. (Wire, 2017) This provides the right to privacy against the state and private parties. On grounds of social welfare and national security, the state continues to collect the data of the citizens.

Every second, someone is collecting data in the private sector (Domo, 2020). Individuals do not comprehend how this data is used by government agencies and companies, corporations. Additionally, there are several government data breaches: Aadhaar in March 2018, SBI in October 2019, COVID-19 Test Results in January 2021, Police Exam in February 2021, and Pegasus 2021. Government surveillance has been brought increasingly under public scrutiny, with proponents arguing that it increases security, and opponents decrying its invasion of privacy. (Michelle Cayford, Wolter Pieters, 2018). The government requires the trust of its citizens, and it does not have the constitutional authority to impose compulsion on its people.

IV. Need for Regulation of Data Protection

It is common knowledge that the free movement of information has aided in the globalisation and virtualization of society, raising issues with privacy, security, and the observance of fundamental rights. The usage of new technologies and the convergence of their application make it easier than ever for the public and commercial sectors to keep records on individuals for diverse purposes and to run the danger of violating their privacy. Many of the unidentified calls that people receive from companies today who offer their products over the phone and via email based on the data they obtain from sources that are not disclosed to customers are an example of such a violation of privacy. Therefore, an active policy and awareness by and on behalf of citizens is constantly a necessity.

Calls from telemarketing representatives on behalf of banks, financial institutions, mobile phone companies, etc. now disturb the privacy of a person's home or the ordinary conduct of business in an organization. These calls unquestionably violate people's privacy. Recent allegations of privacy violations by banks and mobile phone service providers were made in a public interest writ petition against the respondents by a recipient to the Supreme Court.

A core problem in this respect concerns what forms of regulation actually benefits citizens and how their interests can be determined. Further, as data protection is in the interest of the citizen this regulation must, as a starting point is acceptable. However, there are several conflicting interests that are active within this field, and it is a constant battle to ensure that these interests are balanced and that those of citizens are sufficiently protected. In view of this, it is further important to look at the efforts made for regulation and protection of data internationally.

V. Law enforcement and regulatory oversight

When it comes to government data, an independent supervisory or regulatory authority frequently has oversight over data protection and privacy in general, to ensure compliance with privacy and data protection law, including safeguarding individuals' rights. The supervising authority could be a single government representative, an ombudsman, or a group of people (Julia Clark, 2019). Real autonomy of such an authority is a vital aspect, and independence is measured by structural factors like the authority's composition, the process used to appoint members, the authority's authority and the timeframe for exercising oversight functions, the allocation of adequate resources, and the authority's capacity to make important decisions without interference from outside parties (World Bank, 2021). Law enforcement and regulatory oversight are essential for ensuring that data remains secure and protected from malicious actors.

Law enforcement plays an important role when it comes to protecting personal information online. The Indian government has established several laws such as the Information Technology Act of 2000 which provides legal remedies against cybercrimes like hacking or unauthorized access to private information stored on computers or networks. Additionally, there were various other laws related to privacy such as the Right to Privacy Act 2017 which provides citizens with certain rights regarding their data collected by companies operating within India's borders. These laws help protect individuals from having their sensitive information misused without consent or knowledge by providing criminal penalties for those who violate them.

Regulatory oversight also plays a key role in safeguarding user data online, especially since many organizations collect large amounts of customer information through their services or products offered over the internet. To prevent misuse of this data, regulations have been put into place requiring companies to take steps towards securing any personally identifiable information they possess about customers or users before sharing it with third parties outside of their organization structure. In India, there is no specific agency in charge of data protection. The IT Act proposes the employment of adjudication officers for adjudicating whether provisions of IT Act have been contravened. However, the reality of this mechanism's application in terms of data protection has been rather dismal. The Digital Personal Protection Bill (DPPP Bill, 2022) calls for the establishment of the Data Protection Board of India to carry out its provisions.

VI. Access to large amounts of personal data by private companies

The evolution of database management systems and technology since mid-1960s has been a crucial factor in the development of modern society (Foote, 2021). Database management systems have become increasingly important for businesses, governments, and individuals to store data securely and efficiently. Private companies play an essential role in India's persona; data protection landscape by providing secure storage solutions that meet industry standards with their efforts in protecting user privacy.

Personal data is becoming a more important source of competitive advantage for many business-to-consumer (B2C) organisations via targeted advertising, personalization, and customization. A growing number of traditional companies in the industrial, financial services, fast-moving consumer products, and other sectors are also embracing a data-centric approach to decision-making (Roopa Kudva, 2019).

Consumer Trends and industry trends have been influenced by technological advancements, with increasing data connectivity (4G/5G), AI, growth in Internet of Things (IoT) products, rising social media engagements and online services provided by e-commerce platforms.

State and central government agencies are developing platforms all throughout India to digitise documents and provide citizens with online services. The creation of platforms and their online database like Parivahan Sewa, DigiLocker, Aadhaar, IRCTC, and Bharat Interface for Money (BHIM) are some of the initiatives that pave the way for a more thorough digital footprint of people. Some of the companies in India that have access to large private data include Reliance Jio, Paytm, Flipkart, Amazon India, Ola Cabs, and Zomato, and foreign companies like Google, Microsoft, Amazon Web Services, IBM Watson, Oracle Corporation and SAP. However, the privacy concerns associated with collecting and using personal data by private companies, government agencies cannot be overlooked.

VII. Data Integrity

Data integrity refers to maintaining the accuracy, completeness, consistency, and reliability of data over time (Bigelow, 2022). Therefore, data integrity ensures that data remains accurate, reliable throughout its lifecycle, and that all information stored in a database or other system is correct and up to date. This includes verifying the source of the data; making sure there are no errors when entering new information into a system; regularly backing up files; using encryption techniques for sensitive data; and implementing access control measures to prevent unauthorized users from accessing confidential records. It is vital that proper data integrity is observed at all times and that government agencies and organizations must have clear policies in place regarding how their data systems should be managed. In comparison to the

European data protection policies, in Indian, data related policies need to reflect comprehensive guidelines on who can access what type of information within the organization's network as well as procedures for updating existing records with new ones.

Additionally, regular audits should be conducted by qualified personnel to check whether these policies are being adhered. Data integrity is primary enforced as law enforcement agencies and regulatory bodies play an integral part when it comes to protecting user's data online in India; however, each entity needs support from one another given the current fragmented policies of data protection, in order to achieve success in preventing malicious activities targeting vulnerable populations across different sectors including banking, finance, healthcare, and education. Meanwhile, collaboration among various ministries, technical and security agencies, policy makers and civil society can create sound policy, procedures, guidelines of international standards making sure everyone involved understands what responsibilities come along with handling someone else's sensitive info responsibly.

VIII. Conclusions

It's paramount to address the challenges that are evolving in the fragmented data protection and privacy eco system in India. A sound policy will create more secure environment where people feel comfortable using digital platforms knowing that proper measures are being taken to safeguard them against harm resulting from careless actions committed either knowingly or unknowingly by others around them. The Ministry of Electronics and Information Technology (MeitY) recently informed in the Lok Sabha that during the previous five calendar years, as many as 47 incidences of data leak and 142 incidents of data breach have been documented. Meanwhile, the possibility of cyber war is significant in India, and because these attacks have no territorial borders, a hacker might be located anywhere in the globe and use the stolen systems to acquire sensitive information that could be harmful to all parties involved. This is further manifested by a recent alert from the Indian Cybercrime Coordination Centre (I4C), the security agency for the Ministry of Home Affairs, reported that hackers are targeting 12000 Government websites.

In 2017, the Indian Supreme Court ruled that privacy is a fundamental human right. This ruling was seen as a major victory for civil liberties in India. The court's decision recognized that individuals have an inherent right to control their personal information and protect it from misuse or abuse by others. The court's recognition of privacy as a fundamental human right has far-reaching implications for citizens across India; people now have greater protection against government surveillance, data collection practices, and other forms of intrusion into their private lives. Thus, it overruled the verdicts of M.P. Sharma case in 1958 and the Kharak Singh case in 1961, where right to privacy was not protected under the Indian constitution.

Additionally, this ruling provides legal recourse should any individual feel like their rights are being violated due to a lack of privacy protections. By recognizing privacy as a fundamental human right, governments can ensure that they are protecting both public safety and individual freedoms at the same time. Overall, this historic ruling sets an important precedent for future cases involving digital security issues such as online harassment or cyber bullying. As technology continues to evolve rapidly in today's society so too must our laws surrounding these new technologies if we wish to maintain our freedom while also ensuring everyone 's safety.

India's data privacy and protection policies require data integrity, data ownership along with the consent of the public as core to sound public policy making to ensure. The proposed draft Digital Personal Data Protection Bill (DPDP) Bill, 2022 needs review to ensure that the

Central Government has limited powers and that safeguards are put in place to protecting the civil rights in the face of technological changes in the global digital environment.

References

1   Aadeetya, S. (2023, April 14). Retrieved from News18: https://www.news18.com/tech/12000-indian-govt-websites-targeted-by-hackers-fromindonesia-report-7543183.html

2   Abbruzzese, S. (2001). Social Aspects of Charisma. International Encyclopedia of the Social and Behavioral Sciences, 1653-1659.

3   Acquisti, A. (2010, December 1). OECD. Retrieved from https://www.oecd.org/: https://www.oecd.org/sti/ieconomy/46968784.pdf

4   Addison, L. (2015). The State of Survillence in India: The Central Monitoring Chillind Effect on Self Expression. Washington University Global Studies Law Review.

5   Ahluwalia, S. S. (2019, January 2). Identity Theft. Retrieved from UIDAI: https://uidai.gov.in/images/loksabha/LSPQ_3600_(Unstarred).pdf

6   Al-Zaman, M. S. (2021, February 28). Social Media Fake News in India. Retrieved from Asian Journal for Public Opinion Research: https://www.ajpor.org/article/19049-social-media-fake-news-in-india

7   Anand, M. (2022, December 2022). Politico. Retrieved from https://www.politico.eu/: https://www.politico.eu/article/opportunities-and-challenges-of-india-pm-narendramodi-g20-presidency/

8   Andi, S. (2021, June 23). How and why do consumers access news on social media? Retrieved from Reuters Institute for the Study of Journalism: https://reutersinstitute.politics.ox.ac.uk/digital-news-report/2021/how-and-why-doconsumers-access-news-social-media

9   Anirudh Burman, U. S. (2021). How Would Data Localization Benefit India? Retrieved from Carnegie India: https://carnegieendowment.org/files/202104- Burman_Sharma_DataLocalization_final.pdf

10 Anmol Dhindsa, S. K. (2020, May 19). The Constitutional Case against Aarogya Setu. Retrieved from SSRN: https://papers.ssrn.com/sol3/papers.cfm?abstract_id=3610569 11. Ayyangar, N. R. (1962, December

18). Kharak Singh vs The State of U. P. and Others. Retrieved from https://indiankanoon.org/: https://indiankanoon.org/doc/619152/

11 Balakrishnan, P. (2022, December 2). Retrieved from The Telegraph: https://www.telegraphindia.com/india/aiims-unprecented-data-breach-sparks-fearshackers-could-misuse-information/cid/1901550

12 Baptista, E. (2023, March 7). Reuters. Retrieved from Reuters.com: https://www.reuters.com/world/china/china-restructure-sci-tech-ministry-reach-selfreliance-faster-state-media-2023-03-07/

13 Basu, A. (2021). A 2030 Vision for India's Economic Diplomacy. Delhi: ORF.

14 Bhardwaj, K. (2020). Digital Surveillance Systems to Combat COVID-19 May Do More Harm Than Good. Economic and Political Weekly, Vol. 55, Issue No. 23.

15 Bigelow, S. J. (2022). Retrieved from www.techtarget.com: https://www.techtarget.com/searchdatacenter/definition/integrity

16 Bishnu, I., and Aakulu, S. (2019, May 13). Data Protection in The Indian Insurance Sector – Regulatory Framework Part I. Retrieved from https://corporate.cyrilamarchandblogs.com/: https://corporate.cyrilamarchandblogs.com/2019/05/data-protection-indian-insurancesector-regulatory-framework-part-1/

17 www.manupatra.com/donloads/2005-data/TelePIL.pdf

Chapter 6

Studying the Intricacies of Privacy and Data Protection in India – A Critical Analysis of the Indian Laws

Satyam Singh

PhD Research Scholar

Department of Law

DDU Gorakhpur University

Gorakhpur, UP, India

Dr. Shailesh Kumar Singh

Assistant Professor

Department of Law

DDU Gorakhpur University

Gorakhpur, UP, India

Abstract

Public authorities these days are challenging privacy in the name of "Procedure Established by Law" or "Public Duty," despite the fact that it is arguably the most crucial component of humanity's existence on this planet. For a moment, consider what life would be like for someone who had no privacy rights, which includes all private rights related to family, employment, relationships, and so on. In a fundamental way, privacy is as essential to a human body as oxygen is to a human body;

it is the means by which a person may actually live a peaceful life with dignity and freedom, which is the core of Article 21 of our Indian Constitution. Gradually, as our nation moves towards digitalization, it is not incorrect to refer to it as a "Cyber Era," with the increased use of social media and the Internet in various spheres, data security and data protection have become crucial components in terms of privacy, as your digital footprint is both a national issue and a national obligation to protect. Data security and privacy are inextricably linked, and they now occupy a sensitive and important area of law. Due to the spread of the Covid-19 epidemic and its related constraints, the research paper is based on the analogical method of research since secondary sources are used to gather information and then translate it into a precise piece of knowledge.

Keywords- Privacy, Rights, Data, Digital Footprint, Cyber Era.

I. Introduction

In India, the judiciary has referred to the right to privacy as an "***Instinct Right***," as it is important to note that developed countries such as the United States, the United Kingdom, and Russia, among others, have recognized the right to privacy as a fundamental right for several decades, as they were aware of the critical role that privacy plays in the nation's long term development and in all aspects of a peaceful life, which directly leads to the country's prosperity. There are various laws in India, including the massive Constitution, that does not define or define the specific meaning of privacy, but their scope is broad enough to guarantee and protect any Indian citizen's «*Individual Privacy.* "There are a few well-known Landmark Judgements in the Indian Judiciary, such as the ***Auto Shankar Case*** and the ***Puttaswamy Case***, in which the Hon'ble Supreme Court of India declared the Right to Privacy to be a basic right.

The concept of data protection is now the most important factor that is co-extensively tied to the Right to Privacy. Because a person's physical presence and existence are now more dependent on the Internet, social media, and E-space than ever before, data protection is equally critical and crucial. According to a study, the advanced time of law will undoubtedly be founded solely on Artificial Intelligence (AI), posing new obstacles and impediments in the way of the Right to Privacy and Data Protection in India and around the world. We can see how technology can infringe on your privacy and cause mistakes in your life. For the purposes of the law, investigating agencies can resurrect all deleted chats, messages, and recordings from stored backup, which is known as the "***Digital Footprint***" of any time, and it is regarded as an exact replica of the individual on the servers. Various Indian celebrities have recently been investigated by the Hon'ble Narcotics Control Bureau (NCB) on drug charges after the NCB recovered their old deleted chats from three years ago; however, because it was done by the Investigating Authorities, it was legal; however, in all other circumstances, except the consensual agreement, such action falls under the bare violation of the Right to Privacy, which has been recognized as a fundamental right under Article 21 in India since 2017.

II. International Scenario for Data Protection and the Right to Privacy

Apart from India, the right to privacy and data protection is a fundamental and legally recognized right in numerous countries of the world, including:

(i) **The USA-** The Right to Privacy is a judicially recognized right in both tort law and legislation, enacted by the Parliament in the form of Constitutional Amendment Acts of the United States of America (USA), which is the world's most powerful, resourceful, and developed nation. The person who has been wronged can file a lawsuit for "*Invasion of Privacy*" against the person who has infringed on his right to privacy or has released private information without his consent or otherwise. The Fourth Amendment to the United States Constitution protects citizens from unwarranted and arbitrary searches and seizures by government officials, for which one can sue in court for appropriate damages, and the *Fourteenth Amendment to the United States Constitution* of due process of law recognizes family, marriage, motherhood, procreation, and child-rearing as part of the general right to privacy.

(ii) **The UK-** The Right to Privacy did not exist in the United Kingdom (Britain) before the year 2000, but it was recognized as a legal right in the UK after the enactment of the Human Rights Act, 2000, which recognizes various human rights that are essential for mankind, including the Right to Privacy as a legal right in the UK, with several remedies on the abrogation of the same. Article 8 of the European Convention on Human Rights, of which the UK is a member, is widely incorporated into the Act, which encompasses the Right to Family, Bodily Liberty, and Privacy as a general right.

(iii) **Russia-** As a fundamental right, Article 23 of the Russian Constitution provides the right to privacy in one's personal life and family. Interception of communication, messaging, and surveillance is only permitted with the explicit consent of the individual involved unless it is carried out by the State or Investigating Authorities for a valid and lawful purpose. Article 24 of the Russian Constitution makes it illegal to collect, store, or publish any information about a person's private life or family without that person's explicit consent; otherwise, the data operator may be held liable under Article 13.11 of the Russian Federation's Code of Administrative Offenses and Criminal

Code, which carries a criminal penalty of up to two years in prison and a fine of up to 200,000 rubles.

(iv) **South Africa-** The Right to Privacy is a constitutionally protected right in South Africa. According to Section 14 of the Constitution of the Republic of South Africa, 1996, *"Every person has right to privacy and immunity from having his home or property searched and from having the privacy of their communication infringed."* As a result, if a person's fundamental right to privacy is violated, he or she can directly approach the Court to seek redress. Apart from that, there are a number of other laws, such as the Protection of Personal Information Act 4 of 2013 (POPI), which includes Data Protection as a key component, the Regulation of Interception of Communications Act, and the Electronic Communication and Transactions Act 25 of 2002 (ECTA), which governs data encryption specifically. The Protection of State Information Bill (POSIB), as well as the Cybercrimes and Cybersecurity Bills, are two legislations dealing with privacy and data protection that is now pending in the South African Parliament and will improve the country's fundamental right to privacy and data protection.

(v) **Japan-** In order to construct a welfare state with happiness, life, and liberty as its essence, Chapter III of the Constitution and the Civil Code provide specific rights and duties to the people in Japan. When interpreting the civil code in conflicts involving private individuals, Japanese courts are also overly sensitive to criteria of the constitutional right to privacy. Article 13 of the Constitution states that citizens' liberty and privacy are safeguarded against the arbitrary exercise of public power and that each individual has the right to protect his or her own personal information from being divulged to a third party without reason or agreement. Even if someone takes your photograph in Japan against your will, you can rightfully ask him to delete it or seek appropriate compensation from the concerned individual under the Tort Law.

As a result, it is clear that the Right to Privacy and Data Protection is not only an Indian concern, but a global concern, as many developed and developing nations are eager to focus on the Right to Privacy and Data Protection, as data, whether in any sphere of life, is the greatest strength as well as the greatest weakness of any person, corporation, or firm in today's era.

III. India's Legislation Regarding Privacy and Data Protection

My able readers should be aware that the terms *"privacy"* and *"data protection"* are not specified in any statute, law, order, or notification, but the extent and ambit of our Indian Constitution and other statutes are broad enough to include privacy and data protection. The following are the laws that regulate the situation:

(i) Constitution of India

With about 395 articles, twelve schedules, and 22 parts, the Indian Constitution is the biggest constitution in the world. It also includes a preamble, which is a crucial piece that guarantees certain Fundamental Rights to Indian citizens in PART III. Even India's Preamble gives all of its residents the freedom of opinion, speech, belief, and worship, as well as the individual's right to privacy within its purview to rule and govern and administer justice in cases of infringement.[9] A threat to or unauthorized access to any person's data without that person's express consent directly and gravely amounts to a violation of their right to privacy, for which they can directly approach the Hon'ble Supreme Court of India under Article 32 of the Constitution. In 2017, the Hon'ble Supreme Court of India declared the Right to Privacy to be a Fundamental Right, which broadly includes the idea of Data Protection.

(ii) **The Indian Penal Code**, enacted in 1860, is primarily concerned with Indian criminal law, enumerating the penalties and punishments for numerous crimes committed inside India's borders. Some of the laws relating to the Right to Privacy and Data Protection are as follows:

a. **Section 354-C**: Voyeurism is a crime that is defined as a man watching a woman engage in a private act where privacy is expected, capturing her images, and publishing them in any way. It is punishable by imprisonment for a minimum of three years and a maximum of seven years after a second conviction, with a minimum sentence of three years.

b. **Section 354-D**: Stalking is defined as when a man follows a woman without her permission, follows her on social media, or sends her offensive messages despite having no interest in her. For the first conviction, he faces three years in prison and a fine; for the second, he faces five years and a fine. This is also against the law according to the Information and Technology Act of 2000.

c. **Section 379**: If someone unlawfully obtains or copies any private information belonging to a person dishonestly from his possession without that person's consent, it is theft. It carries a maximum three-year jail sentence, a fine, or both.

d. **Section 383**: If someone threatens to defame them if they surrender any important documents or data to him, it is extortion. The violation is punished by up to three years in jail, a fine, or both if the data or documents are crucially confidential.

e. **Section 471**: If it violates a person's right to privacy or data protection, it is a felony to use a forged document or electronic record as genuine and is punishable by up to two years in prison, a fine, or both.

(iii) **The Information and Technology Act of 2000 (IT Act):** The Act specifically addresses Cyber Crimes, Frauds, and E-Commerce Webs; its provisions are broadly relevant to preventing crimes in cyberspace, and are closely tied to Data Protection and Privacy; they are:

a. **Section 66**: According to its broad definition, "hacking" refers to unauthorized access to computer resources or data without the owner's consent that harms that people or a body corporate's reputation, or goodwill as defined by Section 43A and is punishable by up to three years in prison, a fine of up to five lakh rupees, or both.

b. **Section 66C**: Today's computer resources contain private data that is not expected to be observed or recorded by an anonymous individual without permission, making it illegal to fraudulently use someone else's password. If he does so, he faces up to three years in prison or a fine of up to Rupees one lakh.

c. **Section 66 E**: It defines a violation of privacy as anyone who willfully takes, transmits, or publishes the image of another person without their knowledge, and is punishable by up to three years in prison or a fine.

d. **Section 67:** According to the law, publishing or transmitting objectionable material via any electronic form that induces someone to commit a crime or neglect to perform a legal obligation is a crime punishable by up to three years in prison on the first conviction and up to five years on the second, as well as a fine of up to ten lakh rupees.

e. **Section 67A-** It outlines the offence of publishing or transmitting a sexually explicit piece of content electronically, which carries a maximum five-year prison sentence and a maximum rupee ten-lack fine, as well as a maximum seven-year sentence and rupee ten-lack fine for a second conviction. The Copyright Act of 1857 was passed before to our country's independence, but it was incorporated into our legal system and currently safeguards the creator's intellectual property rights. It primarily safeguards the author's written, performed, musical, and visual works while the author is alive and for 60 years after death. The Act offers both civil and criminal consequences for anybody who takes another person's work and reproduces it or uses it without the author's consent for financial benefit or any other reason, including publishing, distributing, or transmitting. In the former, the victim may ask the tortfeasor for injunctions and damages, while in the latter, the tortfeasor may be sentenced to up to five years in jail if proven guilty.

(iv) **The Indian Contract Act, 1872**, guarantees the right to privacy and data protection in part but not entirely. The Conduct's provisions provide the legislation governing contracts, which are agreements between two or more people to perform a certain act in exchange for a monetary reward. Depending on the nature of the activity, the parties to a contract can include or exclude elements relating to privacy or data protection, which are entirely protected and governed by this Act, thereby partially protecting, and safeguarding the Right to Privacy and Data Protection in India.

IV. Judicial Interpretations of Privacy and Data Protection

The judiciary is always keen and sensitive in deciding cases involving infringement of an individual's Fundamental Right, as enshrined and protected by Part III of our Indian Constitution; however, from 1954 to 2018, the judiciary and Hon'ble Supreme Court of India have presented several opinions, fundamentally differing on the question of whether Right to Privacy is a Fundamental Right or Not. It is critical that we recognize that privacy is an inherent and inalienable right of humans on this planet, encompassing any activities or acts in which appropriate privacy is required or expected by any human, such as property, relationships, and sexual orientations, travel, and so on. However, while the declaration of the Right to Privacy as a Fundamental Right is beneficial to citizens, it also increases the State's obligations to protect the privacy of each and every individual residing in India's territory, as the Government may be held liable for any infringement of the Right to Privacy by any private individual. So, in order to fully comprehend the relevance of the Right to Privacy and Data Protection, we must first comprehend the Judiciary's position on the subject at various points throughout certain key and major Judgements, which are as follows:

The subject of whether or not privacy is a basic right was presented for the first time before the Hon'ble Supreme Court of India in this case. In the District's case, following an FIR, a Magistrate issued warrants and searches for the disputed documents of the Dalmia Group of Delhi, in accordance with the Code of Criminal Procedure, 1973. The group filed a Writ Petition in the Supreme Court challenging the legality of the search and seizure, claiming that the action was arbitrary and infringed on their right to privacy and fundamental rights as entrenched in **Article 19 (1) (f)** and **Article 20 (3),** i.e., self-incrimination. The case was referred to the Court's Eight-Judge Constitutional Bench, and after a thorough and thorough adjudication, it was determined that the Constitution does not intend to bring search and seizure within the ambit of privacy and that it is not invalid or unconstitutional if it is carried out in accordance with the procedure established by law, and also by following the adopted procedure of law, i.e. FIR and subsequent orders of Hon'ble District Magistrate. The Court further stated that, based on the Constitution's text, which does not refer to the Fourth Amendment of the United States

Constitution, it does not intend to recognize the Right to Privacy as a Fundamental Right, and hence it is not a fundamental right in India.

The case is well known as the "*Telephone Tapping Case*" since it involves the tapping of telephones, which is directly linked to the Right to Privacy and Data Protection. A Public Interest Litigation (PIL) was brought by a non-profit group in response to the Central Bureau of Investigation's (CBI) telephone tapping of politicians in the guise of public security and public safety. Telephone tapping, according to the Petitioners, is a major invasion of one's right to privacy and a flagrant violation of Article 21, which guarantees one's right to life and personal liberty. The provisions of **Section 5 of the Indian Telegraph Act**, 1885, which gives the Central and State Governments the ability to record phone calls if certain circumstances are met, were challenged. The Hon'ble Supreme Court of India held that phone tapping is a serious violation of Article 21 and the Right to Privacy and that it should only be used when there is a question of public safety, which means a situation of grave danger to the public at large or a public emergency; in all other cases, the Central Government cannot use phone tapping, even if the authorities are satisfied that India's sovereignty or integrity is threatened. Finally, the Court established specific standards to be followed when using Section 5 (2) of the Indian Telegraph Act, 1885, such as appointing a review committee and evaluating the Order of Phone Tapping every two months, among other things.

In this Case, the Central Government's *«Aadhar Card Scheme*," which was utilized in a variety of situations, including cylinder booking, banking, and electoral voting questioned. The Petitioners claimed that the Aadhaar Cards are sought from residents on numerous occasions, containing their sensitive and private data, including biometric fingerprints, which has no security and is blatantly in violation of Article 21's Right to Privacy. The Supreme Court of India's nine-judge Constitutional Bench overturned the 1954 decisions in the MP Sharma case, holding that the right to privacy is a fundamental right protected by Article 21 of our Constitution, which cannot be abridged or infringed except through the due process of law, and that the Aadhaar Card Scheme is unconstitutional and infringes on the right to privacy in India. The Central Government has created a committee to work on the preservation of sensitive data and citizen privacy, as breached by the Aadhaar Card Scheme, and a report would be submitted in this regard, as directed by the Hon'ble Supreme Court.

As a result, the Right to Privacy is now a fundamental right in India, protected under Article 21 PART III of our Constitution, as of 2017. As a result, if any citizen residing in India is aggrieved or his Right to Privacy is infringed by the State, he or she can either file a Writ Petition under Article 32 before the Hon'ble Supreme Court of India, which is only in the case of infringement of a Fundamental Right; otherwise, one must follow the hierarchy of Courts before approaching the Apex Court of the Nation, or he or she can approach the Hon'ble High Court of the concerned State.

V. Exceptions to the Right to Privacy and the Protection of Personal Information

According to Article 21 of our Constitution, the right to privacy is a basic right, but it's crucial to remember that a fundamental right is not an absolute one; as a result, it may be limited or suspended under special conditions. The same is true of the right to privacy, but Article 21 of our Constitution cannot be suspended, even in an emergency, according to the Hon. Supreme Court of India. There are, however, a few privacy exceptions, which are as :

(i) **The procedure established by law**- Article 21 states that no one shall be deprived of their right to life except in accordance with the procedure established by law, which indicates that the right to life is fundamental but not absolute and that there are certain exceptions to it that are unquestionably legal. If it is declared an absolute right, there is a high risk of abuse, as observed in a number of foreign countries, leading to a miscarriage of justice and destroying the whole objective of the legislation.

(ii) **Public Security-** According to several Judicial Pronouncements, public security refers to a situation of impending danger where the general public is at risk and it is necessary for the investigating authorities and the State to act immediately; therefore, if the Indian Telegraph Act, 1875 infringes on any individual's right to privacy, it will not be viewed as such.

(iii) **Public Duty-** When a public servant is acting in an official capacity, performing sovereign functions, or acting on the orders of his higher officials, he is always legally immune and thus no charge of invasion of privacy can be brought against him. However, if it is proven that the action was arbitrary in nature, biased, or motivated by personal animosity, and thus violates the rule of law, a separate action can be brought against him.

(iv) **Public Safety-** Section 5 (2) of the Indian Telegraph Act, 1875, establishes the grounds for intercepting communications and messages, acting as an exception to the right to privacy. According to the courts, public safety means that when the safety of the general public is at risk or when India's sovereignty is at risk, the action of intercepting communications will not be regarded as an infringement of the right to privacy.

VI. Conclusions

Finally, in the aforementioned Research Paper, we examined the importance of privacy for humans and humanity in order to live in peace, and the same is true for protection. If we think back to our ancestors and forefathers, we can see that they focused on and taught us about the importance of knowledge, as knowledge is regarded as the key element for success and becoming something great in life. It is therefore not surprising that in ancient times, the person with the most knowledge was regarded as the King of the place, whether it was a city, country, or profession. Now, the same thing is replaced with 'Data.' The most important thing that any person has with him today is not his ornaments, valuables, or money because they can be re-earned if they are lost, but it is the data that directly affects your existence and life if it falls into the wrong hands, misuse of which is unimaginable and inexplicable. The importance of data and privacy can be found in every aspect and sphere of human life, whether it is in business administration, where the SWOT (Strengths, Weaknesses, Opportunities, Threats) Analysis is based primarily on data compiled by your competitors, or in the health or legal sectors, where data plays an extensive and extremely important role in determining whether the results are in favor or against the person or organization at large. As a result, it is not incorrect to call privacy and data *"A Determining Element"* of the success or failure of any person, firm, or corporation in today's world, and the same is true for a country. If a country wishes to flourish, the right to privacy and data protection must be a top priority.

In order to provide speedy relief to the aggrieved and act in furtherance of justice, some fast-track courts should be established in the country specifically dealing with cases of privacy infringement, child abuse, and sexual abuse of children in which social media is used as a platform by the perpetrator of the crime, and the earliest disposition of the case with providing stringent punishment to the wrongdoer must be ensured.

References

1   G. Greenleaf, "India's national ID system: Danger grows in a privacy vacuum," *Computer Law and Security Review*, vol. 26, no. 5, pp. 479–491, Sep. 2010, doi: 10.1016/j.clsr.2010.07.009.

2   R. K. G. Goyal, The Right to Privacy in India: Concept and Evolution. Lightning Source, 2016.

3   Columbia University, *R. Rajagopal v. State of Tamil Nadu.* 1994. Accessed: Jun. 06, 2022. [Online]. Available: https://globalfreedomofexpression. columbia.edu/cases/r-rajagopal-v-state-of-t-n/

4   Columbia University, *Puttaswamy v. Union of India (I).* 2017. Accessed: Jun. 06, 2022. [Online]. Available: https://globalfreedomofexpression. columbia.edu/cases/puttaswamy-v-india/

5   J. A. Dar, *Privacy and Data Protection Laws in India, USA and European Union.* Walnut Publication, 2019.

6   L. Orbis, Life, Liberty and Privacy: The Indian Jurisprudence on Individual's Data Privacy, 1st edition. Notion Press, 2020.

7   J. Yi, Y. Du, F. Liang, W. Tu, W. Qi, and Y. Ge, "Mapping human's digital footprints on the Tibetan Plateau from multi-source geospatial big data," *Science of The Total Environment*, vol. 711, p. 134540, Apr. 2020, doi: 10.1016/j.scitotenv.2019.134540.

8   R. M. Duffy and B. D. Kelly, "Privacy, confidentiality and carers: India's harmonisation of national guidelines and international mental health law," *Ethics, Medicine and Public Health*, vol. 3, no. 1, pp. 98–106, Jan. 2017, doi: 10.1016/j.jemep.2017.02.018.

9   P. K. AGRAWAL, *CONSTITUTION OF INDIA : The Constitution of India is the supreme law of India*, First edition. S.l.: Prabhat Prakashan, 2017.

10 J. K, B. H, T. H, and M. A. P, "A review of preserving privacy in data collected from buildings with differential privacy," *Journal of Building Engineering*, vol. 56, p. 104724, Sep. 2022, doi: 10.1016/j. jobe.2022.104724.

11 K.-W. Wu, S. Y. Huang, D. C. Yen, and I. Popova, "The effect of online privacy policy on consumer privacy concern and trust," *Computers in Human Behavior*, vol. 28, no. 3, pp. 889–897, May 2012, doi: 10.1016/j. chb.2011.12.008.

12 G. Greenleaf and B. Cottier, "International and regional commitments in African data privacy laws: A comparative analysis," *Computer Law and Security Review*, vol. 44, p. 105638, Apr. 2022, doi: 10.1016/j.clsr.2021.105638.

13 A. Karale, "The Challenges of IoT Addressing Security, Ethics, Privacy, and Laws," *Internet of Things*, vol. 15, p. 100420, Sep. 2021, doi: 10.1016/j.iot.2021.100420.

14 G. Liyanaarachchi, "Managing privacy paradox through national culture: Reshaping online retailing strategy," *Journal of Retailing and Consumer Services*, vol. 60, p. 102500, May 2021, doi: 10.1016/j.jretconser.2021.102500.

15 NLU Delhi, *M.P. Sharma and Ors. vs. Satish Chandra and Ors.* 1954. Accessed: Jun. 06, 2022. [Online]. Available: https://privacylibrary.ccgnlud.org/case/saroj-rani-vs-sudarshan-kumar-chadha

16 Columbia University, *People's Union of Civil Liberties (PUCL) v. Union of India.* 2003. Accessed: Jun. 06, 2022. [Online]. Available: https://globalfreedomofexpression.columbia.edu/cases/peoples-union-of-civil-liberties-pucl-v-union-of-india/

17 Supreme Court Observer, "Fundamental Right to Privacy- Justice K.S. Puttaswamy v Union of India," *Supreme Court Observer*, Oct. 25, 2021. https://www.scobserver.in/cases/puttaswamy-v-union-of-india-fundamental-right-to-privacy-case-background/ (accessed Jun. 06, 2022).

Chapter 7

Issues and Challenges in Smart Cards: How to Address Privacy Concerns

Poonam
Assistant Professor
School of Law
Shoolini University of Biotechnology & Management Sciences
Solan, Himachal Pradesh, India

Dr. Ritika Shukla
Assistant Professor
SSHSS, Sharda University
Greater Noida, UP, India

Abstract

In our increasingly digitalized world, smart cards have become indispensable tools, seamlessly blending convenience, and security. These versatile, compact devices store and process information, enabling secure access to a wide range of services, from financial transactions to building entry. Smart cards are now ubiquitous, influencing the way we engage with technology and systems. However, like any technological innovation, they come with their own set of problems and complexities that warrant careful examination. This research paper explores the multifaceted realm of smart cards, illuminating their intricacies, vulnerabilities, and potential opportunities. From their modest origins as magnetic stripe cards to their advanced microprocessor-based counterparts, smart cards have undergone significant evolution. Today, they play a crucial role in ensuring secure authentication, data storage, and transaction processing across various applications. While smart cards offer numerous advantages, including heightened security and convenience, they are not impervious to a range of challenges. These challenges encompass security breaches, concerns about data privacy, interoperability issues, and the ever-present threat of technological obsolescence. As we increasingly depend on these devices for critical tasks, it becomes imperative to systematically address these challenges.

Keywords: Digital, Information, Smart Card, Security, Government.

I. Introduction

Smart Cards are the size of credit cards, incorporate embedded technology that establishes an automated connection amid the card and the reader equipment of transit providers. These cards facilitate swift data transfer, crucial for transit providers to gather their fees. The adoption of Smart Cards in lieu of conventional transportation permits or signs leads to reduced currency management, decreased apparatus protection, and improved precautions measures. Furthermore, Smart Cards clasp the potential to enhance accessibility for passengers, enhance the assembly of ridership files, modernize the image of transit systems, and open doors for creative fare structures and marketing strategies. Smart Cards serve multiple purposes for transit providers. They can streamline passenger processing and Billing, managing financial and accounting matters effectively for both travelers and providers. Smart Cards enable travelers to pre-purchase travel, with each journey resulting in a deduction or subsequent Billing. Additionally, transit employees use Smart Cards, primarily for security reasons, granting access and monitoring movements, especially for those with government interests related to safety or security-sensitive functions.

Moreover, Smart Cards act as a valuable source of data for marketing and planning purposes. If they contain demographic data of traveller, transit providers can analyze traffic flows, plan infrastructure development, and tailor advertising to boost ridership. This data may reside on the card's embedded chip or within a remote electronic database. Finally, by incorporating personal and potentially biometric data, Smart Cards can enhance transit system security. They allow for the identification and restriction of passengers who pose security threats or have been involved in security incidents. Biometrics, such as impressions, handwriting, face scans, can be employed to positively identify individuals. Cameras equipped with facial recognition software can track suspects through passenger terminals based on their facial characteristics.

II. Concept of Smart Cards

A Smart Card can be described as a compact, credit-card-sized mechanism equipped with a programmable computer chip capable of storing encoded information. An example of a "key card," often employed for secure access to locked doors, falls within the category of Smart Cards. Smart Cards belong to the family of automatic identification (Auto-ID) systems, alongside technologies such as barcodes, optical character recognition systems, and radio frequency identification (RFID) systems. For instance, a can of soup may feature a Universal Product Code (UPC) barcode on its label, scannable using a laser at the grocery store checkout. Additionally, the packaging containing these cans may also incorporate a UPC barcode, facilitating efficient inventory management across the supply chain, from manufacturing and transportation to wholesale and retail levels.

Smart Cards offer the capability to grant individuals access to buildings while monitoring their entry points, potentially restricting access to certain areas. This system tracks the time and location of their entries. Unlike barcodes, which require direct laser scanning, and basic magnetic stripe cards that necessitate physical contact with a reader, Smart Cards with embedded RFID systems can be identified from a distance. RFID systems offer greater convenience and speed, causing less wear and tear on both the card and scanning hardware due to the absence of physical contact. Smart Cards, typically credit-card-sized and suitable for wallet insertion, contain embedded electronics, such as an integrated circuit chip (ICC) or microchip, as well as a transponder. Unlike basic magnetic stripe cards, which only function upon physical contact with a reader, Smart Cards, especially the newer contactless variants, feature built-in radio frequency antennae. To transmit and receive data, these cards need to be in proximity to a radio transmitter, often referred to as a remote contactless radio frequency interface.

III. The Development of Smart Card Technology

Smart Card technology was initially conceived in 1969, although it has only recently become both technologically advanced and economically feasible. Initially, the banking sector introduced Smart Cards with the aim of reducing losses due to card fraud and enhancing card security. Subsequently, the telecommunications industry embraced them, employing Smart Cards as pay phone cards. In Europe, governments have adopted Smart Cards for various purposes, including the creation of portable personal files (e.g., for health records or to manage eligibility for benefits).

These versatile cards find applications in a multitude of areas, encompassing the secure entry and exit of employees from secured facilities, trusted traveller programs for airlines, travel visas, and immigration services. Notably, the U.S. Department of Transportation (DOT) has implemented an Intelligent Vehicle Highway System, akin to Smart Cards, to monitor individual travel patterns. Smart Cards serve diverse roles, functioning as secure identity tools (e.g., employee ID badges, citizen ID documents, electronic passports, driver's licenses, and online authentication devices), healthcare aids (including citizen health ID cards, physician ID cards, and portable medical records cards), payment methods (both contact and contactless credit/debit cards, as well as transit payment cards), and telecommunications assets (such as Global System for Mobile Communication (GSM) Subscriber Identity Modules and pay telephone payment cards).

IV. Advantages of Smart Cards

Smart cards offer a number of advantages:

(i) **Usual Advantages-** Smart Cards offer numerous advantages, including cost-effectiveness, efficiency, time-saving, and convenience. They play a crucial role in combating identity theft and misuse of governmental benefits while facilitating quicker access to secure areas and international borders. In the broader context, they streamline access to government services. Smart Cards are equipped with computer chips that can: - Act as access keys for buildings. Enable electronic money storage, eliminating the need for cash transactions. Safely store personal identification or biometric data, such as photographs, eye patterns, or fingerprints. Allow the collection of data regarding the time, location, and frequency of card usage. Store personal information, including medical records, DNA data, religion, age, and address, along with personal identification numbers.

(ii) **Unusual Advantages-** Drivers' licenses, credit cards, insurance and health cards, and birth certificates provides for alternative forms of identification. Identification of the specific person is essential so that the administration can look after assistance, such as health caution and public support, to qualified heritors. Individualistic documentation can be enriched through biometric documentation processes, such as a thumb, hand, or retinal scan. Biometric documentation requires a three-step process: (1) enrollment, (2) templates, and (3) matching. Enrollment is the process by which the individual provides biometric data. The enrollment template stores the individual's biometric information. Matching is the process whereby the individual's template is correlated with the individual's biometric measure taken "on the spot." The result is a high level of confidence that the individual present is the individual identified on the template. The template, in turn, can be correlated with other information about the individual to determine whether he or she poses a security concern.

V. Smart Cards: Challenges and Concerns

The use of Smart Cards poses a number of concerns as follows:

(i) Privacy Concerns

The integration of RFID technology within Smart Cards can give rise to heightened privacy issues compared to conventional key cards. These concerns pertain to the collection, storage, access, and potential use of user information. In the realm of transportation, Smart Cards form part of a broader spectrum of emergent expertise known as Intelligent Transportation Systems (ITS). ITS technologies propose an exceptional capability for real-immediate tracking of both individuals and vehicles. Furthermore, these technologies can record and maintain historical travel patterns, encompassing details such as travel destinations, times, and frequencies. They also have the capacity to merge and correlate this data with a wide array of personal information, spanning categories like gender, race, religion, political affiliation, birthplace, residence, employment history, law enforcement records, credit history, income, and more, spanning an individual's entire life.

The aggregation and correlation of such extensive data can reveal significant insights into an individual's behavioral patterns. As databases expand and interconnect with other databases, individual privacy diminishes. Additionally, as technology continues to evolve, it allows for more robust real-time monitoring of individuals, the accumulation of more precise information, and the growth of comprehensive records. Entities may remain oblivious of the depth of knowledge accessible about their actions and may not realize that their whereabouts are under surveillance. The connection involving confidentiality and ITS is interdependent. Privacy is influenced by ITS, but ITS is also influenced by privacy concerns. This complex relationship is further complicated by the fact that both privacy and ITS are dynamic and multifaceted. Smart Card knowledge and its functions are not static; they continually evolve. Factors such as processing speed, database size, and international networks create vast opportunities for data collection. For instance, microprocessors within Smart Cards can store biometric identification data, ensuring that the cardholder is the legal possessor, and that the knowledge aligns precisely with that individual. Smart Cards and the corresponding PC systems can hold precise details for unique biometric identifiers like thumbprints, irises, or retinas. Furthermore, although current technology limits Smart Card readings to a few inches, future developments could extend this range and potentially link GPS technology with Smart Cards for real-time monitoring of individuals' locations beyond ground-based scans. Even before this, the reading radius of Smart Cards may expand to include transit locations and automobiles. In such scenarios, transportation providers could access actual-time statistics about individual passenger locations, enhancing law enforcement efforts in tracking down individuals like graffiti artists, criminals, or potential threats, when coupled with video surveillance.

(ii) Issues and Challenges

Apart from privacy as the major concern, following other challenges also exist with the use of Smart Cards:

a. **Usability Issues:** Electronic payment systems often demand extensive information from users or complicate transactions with intricate website boundaries. For instance, composing credit card payments via a internet site can be cumbersome, as it requires users to input significant individual information and communication elements into web forms.

b. **Safety Concerns:** Electronic payment systems on the cyberspace are vulnerable to theft of capital and individual communication. Consumers are obliged to share credit card details, expense account information, and other particular data online, at times in an unsecured manner. Requiring this information through email or over the cell phone also presents safety hazards.

c. **E-Currency Challenges:** E-cash faces challenges related to its universal acceptance, as it relies on commercial establishments accepting it as a valid payment method. Additionally, e-cash payments require both the customer and the merchant to have accounts with the same bank issuing e-cash, limiting its usability outside of this specific banking network.

d. **Trust Deficiency:** Electronic payment systems have a history of fraud, misuse, and perceived unreliability, which hinders their adoption. Prospective customers often cite this risk as a primary reason for their lack of trust in payment services, leading them to refrain from making online purchases.

e. **User Acceptance:** User acceptance plays a crucial role in determining the success or failure of electronic payment systems. Users' attitudes and human factors significantly impact the effectiveness of these systems. The lack of trust, security concerns, and the evolving business landscape pose challenges to user acceptance. To succeed, electronic payment systems must address the needs and preferences of users and the market.

f. **Lack of Awareness:** Online payments can be daunting, even for educated individuals, leading them to prefer traditional shopping methods over online options. Technical issues, such as server problems during online payment attempts, can discourage users from adopting this payment method.

g. **Rural Accessibility:** Rural areas often lack the technological literacy and access to computers required for online payments. The population in these areas may be less inclined to adopt online payment systems due to limited awareness and familiarity with technological innovations.

h. **Cost and Time Implications:** Electronic payment systems entail significant costs, including setup expenses, equipment costs, and management expenses. Additionally, these systems may take more time to process transactions compared to traditional in-person payment methods.

VI. Addressing Challenges and Issues

This part of the paper explores the means to deal with concerns with regard to Smart Cards:

(i) **Encryption-** Electronic shop is incredibly vulnerable to the belief that e-commerce is unsafe, specifically when it appears to electronic spending. Most electronic expense methods use an encoding technique to combine protection to the communication of peculiar and expense information. There are several encoding methods in use to inhibit frauds of electronic payments.

(ii) **Digital Signatures-** Participants engaged in electronic payment dealings should make utilize of digital signs to safeguard the validation of these dealings, thereby enhancing their security.

(iii) **Firewalls-** It is a cohesive assortment of safety procedures premeditated to prevent unauthorized electronic access to a networked computer system to protect private network and individuals' machines from the dangers of the greater internet, a firewall can be employed to filter incoming or outgoing traffic based on a predefined set of rules called firewalls policies. There are 3 policy actions of firewalls: Accepted: Permitted through the firewall.

(iv) **A further important thing to consider is to confirm the issuing nation and the Billing address-** Make sure the issuing country and Billing address country are the same. This is especially important because minor banks may not have rigorous identification procedures.

(v) **Verification with the Credit Card Issuing Bank:** If online merchants have any suspicions regarding an order and need to verify its details, they can contact the issuing bank to confirm general account information. This precaution helps ascertain that the credit card is not stolen. The contact number for the issuing bank is often based on the first six digits of the credit card number, known as the Bank Identification Number (BIN).

(vi) **Enhanced Identification Procedures:** While preserving customer privacy and facilitating quick online ordering is essential, it's equally important to collect adequate customer identity information during the ordering process. Gathering more than just the customer's name, credit card number, and expiry date is advisable. Merchants may consider verifying customers via phone or requesting a photo ID to be faxed if there are any doubts about the transaction's legitimacy.

VII. Conclusions

Electronic payment, in contrast to physical cash or paper checks, encompasses various methods like debit cards, credit cards, smart cards, and e-wallets. E-commerce's evolution, particularly its online payment methods, is central to our discussion here. The risks associated with online payments primarily involve payment data theft, personal data breaches, and fraudulent customer rejections. Consequently, until electronic signatures become widespread, we must utilize available technology to maintain a reasonable level of network security. While no payment method is flawless, each has its unique advantages. For those prioritizing privacy, options like E-cash or Net Bill Checks offer greater confidentiality, whereas Smart Cards are optimal for security-focused individuals. Both consumers and service providers stand to gain from e-payment systems, ultimately enhancing national competitiveness. The successful adoption of electronic payment systems hinges on effectively managing security and privacy concerns for both buyers and sellers, thereby bolstering market confidence in the system.

References

1   Bhasker, Bharat (2013). Electronic Commerce, Framework, Technologies and Applications. McGraw Hill Education (India) Private Limited., p.9.2-9.16.

2   Chhabra, T.N., Suri, R.K., Verma, Sanjiv (2006). E-Commerce. Dhanpat Rai and Co. (P) Ltd. p.306-328.

3   Dennis, Abrazhevich (2004). Electronic Payment Systems: A User Centred Perspective and Interaction design. Eindhoven: Technical Universiteit Eindhoven. p.1to12.

4   Jing, Yang. (2009). Online Payment and Security of E-Commerce. International symposium on web Information system and application (wise '009). p 1to5

5   Madan, Sushila (2013). E-Commerce, Mayur Paperbacks. P.4.4-4.35.

6   Noor, Aaihan Ab Hamid, Aw Yoke Kheng. A Risk Perception analysis on the Use of Electronic Payment Systems by Young Adult, Wseas Transaction Information Science and Applications.p.1to3 and 8.

7   Rafael Medran Vioque.Alumno De Dectorad.2012-13.

8   Sistemsas de pago.Alfredo Lozano, Josemanuel Agudo, http:/www.icemd. com

9   Whiteley, David, (2007). e-Commerce, Strategy, Technologies and Applications. Tata McGraw-Hill Publishing Company Limited. P.200-201.

10   http:/www.fraudlabs.com 11. http:/www.uaipit.com

Chapter 8

Expunging of Abortion Data of Women to Protect Health Privacy in India

Sanghapriya Ray
PhD Research Scholar
SSOL, Sharda University
Greater Noida, UP, India

Prof. (Dr.) Komal Vig
Dean
SSOL, Sharda University
Greater Noida, UP, India

Arrat Banday
B.A.L.L.B. (Hons.) Student
SSOL, Sharda University
Greater Noida, UP, India

Abstract

The US Supreme Court upheld Roe vs. Wade on June 24, 2022. The judgment will outlaw abortion in nearly half the State, but the effective date may vary. The change will go live in weeks. Google erased user input data from counselling, fertility, addiction, weight reduction, and plastic surgery clinics. Location history tracking is disabled by default and can be erased at any time. In a The New York-based privacy group Surveillance Technology Oversight Project (S.T.O.P.) anti-abortion campaigners monitor pregnant women to coerce them into exercising their reproductive rights. With the verdict overturned, computerised surveillance of pregnant women may increase. The global developments have affected the need for strong protection of health records in India. Abortion and sexual and reproductive health necessitate personal the data privacy legislation. Consensual relationships outside of marriage and potential pregnancies still have a limited societal acceptance in India. Service providers blackmail unmarried persons by alerting their parents or other family members about their pregnancy. Ultrasound technicians and other abortion clinic workers have extorted unmarried abortion seekers for money, sexual favours, and other things. The paper takes a look at rights to abortion vis-à-vis privacy in India.

Keywords- U.S., Data, surveillance, abortion, tracking, Google.

I. Introduction

The Indian Personal Data Protection Bill 2021 was withdrawn from Parliament. The administration wants to replace the "complete framework" and "current digital privacy regulations." India's landmark privacy judgement, *Justice K.S. Puttaswamy (Retd) v. Union of India (2017)*, states that privacy is a fundamental and inalienable right under the Constitution.

It contends that a woman's right to privacy includes the ability to make reproductive decisions, including abortion. It also examines many legal issues. Constitutional principles like individual liberty, bodily integrity, and dignity. One of the tricky issues with is right, though is that there is no recognized or practical concept of privacy in the Indian medical environment, nor are there any effective remedies for privacy violations.

Neha Dixit's 2013 report illuminates such women's experiences. Unmarried women, cohabiting couples, marginalized people, queer individuals, opposite-sex people, polyamorists, and other ethical non-monogamy practitioners can get abortions. It's tough. Non-family discrimination. India does not safeguard non-traditional sex life and sexual and romantic partnerships from abortion prejudice. A 2019 CEHAT study indicated most rape victims were denied safe abortion. Married women seeking abortions without spouse consent have been denied. Their families and communities cannot force children. Without intimidation, a documented and digitized record can lead to "banishment" or humiliation and discrimination for having a pregnancy or a sexual connection outside of marriage. According to the Hidden Pocket Collective's experience as SRHR space practitioners, this dread can induce psychological damage. Abortion seekers also view life-threatening and medically risky abortion as their sole option. Thus, the right to privacy covers married, unmarried, and polyamorous abortion seekers. Before seeing a doctor, they prioritize confidentiality when seeking abortion services.

India's sexual and reproductive health regulations do not address privacy. In 2020, the Hidden Pockets Collective (HPC) mapped north Indian government-owned Primary Health Centres (PHCs). HPC found that doctors required abortion seekers to produce a copy of their Aadhaar number (India's national biometric ID) as evidence of pre-procedure identification. Aadhaar-less participants were denied safe abortion. The Medical Termination of Pregnancy (update) Act 2021, an amendment to the 1971 law, allows PHCs and hospitals to record abortions in confidential clinical documents. Regulation 5 of the 2003 Medical Termination of Pregnancy Regulations requires these data to be kept safe and sealed for five years from the last entry, whereas regulation 4(5) restricts abortion information to the State-appointed Chief Medical Officer. The 2021 Amendment made progressive legislation reforms.

The MTP (Amendment) Act 2021 unmarried women, marital rape victims, non-cis-gender women, and those in "non-traditional family structures" similar abortion rights, according to a September 2022 Supreme Court ruling. It further said that spouses, families, and third parties do not need to consent to abortion. However, the State controls abortion in India. A pregnant woman cannot request an abortion—the doctor decides. Abortion seekers are apprehensive of recording their personal and abortion data due to social shame. Abortion-seekers face numerous forms of discrimination. Abortion service seekers struggle to communicate because sexual and reproductive health issues are stigmatized. We believe the data protection law/framework recognizes that a data breach could have serious, far-reaching, and negative consequences for them.

II. The Toll of a Privacy Violation

The Amendment to the Medical Termination of Pregnancy Act,2021 gave unmarried Indians the same abortion rights as married people for the first time. This was noteworthy since service providers usually denied abortion services to unmarried persons due to legal ambiguity and criminal responsibility under two other laws (one against child sexual abuse and one against female genital mutilation).

Feticide Service providers who reject abortions are unpunished. Doctors and service providers rarely informed unmarried abortion seekers about confidentiality. However, they believed abortion was illegal. If "informational privacy" is not developed in our communities and digital and physical infrastructures for unmarried and non-traditional couples, having a right to privacy is not realistic.

"Patient privacy encompasses a number of aspects, including personal space (physical privacy), personal data (informational privacy), personal choices including cultural and religious affiliations (decisional privacy), and personal relationships with family members and other intimates (associational privacy)," according to the AMA Code of Medical Ethics.

The Indian medical profession lacks an AMA-like privacy definition. MCI Code of Ethics Regulations, doctors "shall not disclose the secrets of a patient that have been learnt in the exercise of his/her profession except – i) in a court of law under orders of the Presiding Judge; ii) in circumstances where there is a serious and identified risk to a specific person and/or community; and iii) notifiable diseases."

Disclosed abortions can lead to slut-shaming, societal degradation, forced marriages, assault, or death. In certain cultures, unmarried pregnancy is shameful and grounds for honor killings. Abortion-seekers may also be stigmatized or assaulted due to family and community opinions about pregnancy and abortion. Due of stigma, young people who have pre-marital sex are occasionally pushed to marry. In marital sex. Abortion-seekers' mental health and well-being are at risk, an issue not thoroughly explored in India.

III. Increased Digitization and Lack of Privacy

Storing sealed paper records of abortion service information is insufficient to ensure the privacy of abortion seekers. Every Aadhaar-linked Ayushman Bharat Health Account (ABHA) will be issued under the Digital Health Mission. Since ABHA numbers are opt-out, everyone gets them. Opting out of the ABHA number system is challenging due to information and computer proficiency gaps.

Name, cellphone number, biometrics (iris scan and fingerprints), home address, date of birth, and other identifying information facts are linked to their Aadhaar number. All medical records including abortion records to a central database. This includes digitising sealed, protected, and revealed paper records only by governmental direction.

Digitizing abortion data ignores the fact that India's digital health infrastructure is among the most attacked, resulting in data leaks and privacy violations. Aadhaar, the world's largest biometric ID database, has been breached multiple times, exposing over a Billion people's data to exploitation. These problems only enhance privacy and confidentiality concerns unless the future data protection legislation/framework can satisfy abortion service seekers.

Companies must secure consumer data under the Information Technology (Amendment) Act, 2008 and Reasonable Security Practices and Procedures and Sensitive Personal Data or Information) Rules, 2011.However, without fines, it is difficult to hold violators accountable and motivate corporations to adopt strong security standards.

Medical record confidentiality, doctor-patient privacy, and no spousal permission can improve safe abortion access. More people may use services created for them if privacy is better defined. The now-withdrawn Personal Data Protection Bill 2021 and its predecessors classified "health data" as "sensitive personal data" and required approval. No way clear means to protect non-personal and anonymized data from being deanonymized or utilised to breach sensitive personal information.

We want the new draught personal data protection framework:

(i) Establish a clear purpose-public interest link for collecting, accessing, and storing healthcare data. This is especially pertinent given conflicts between law and bans determining a foetus's sex (to avoid female-specific abortions) and the right to abortion.

(ii) Define the circumstances under which the state may summon sensitive health data, such as abortion records, together with its necessity, purpose, and scope, and the entities allowed to do so.

(iii) Access such data, how long such records can be maintained, when they must be removed, and the penalties for violating these norms.

(iv) Clarify privacy and confidentiality concerns, the need to link Aadhaar to the ABHA number, and the need for a central database to gather, link, and store all health records with civil society.

(v) Provide technical and non-technical safeguards.

These measures would promote trust in laws and structure that safeguard privacy and personal decisions like abortion.

IV. Flaws in India's Health Data Policies

Indian health data regulation has two issues. First, there is no consensus on what "health data" is under numerous laws and policies. The Information Technology (IT) Rules 2011 only deem health data sensitive and require adequate security practices. These restrictions do not cover health data. Laws and policies have tried to clarify the topic, yet disagreements remain. Lack of standardization makes it unclear how these frameworks interact. "Personal health data" comprises health problems and treatments, according to the National Digital Health Mission (NDHM) strategic overview. The draught Digital Information Security in Healthcare Act 2018 (DISHA) provides a more detailed definition of digital health data, including information about a person's physical and mental health, health services, donations of body parts or substances, and test or examination results. The Data Protection Bill 2021 (DPB) limits health data to "data related to the physical or mental health of the data principal." This fragmentation can leave consumers vulnerable to danger. Second, most of these models do not account for today's health data collecting or surveillance harms. Many policies view it as a resource for a national digital health ecosystem. They prioritize its collection, storage, and sharing to promote research and public health. Thus, they prioritize hospital, clinical, and insurance data. Today, non-clinical health data is collected. Fitness trackers and smart watches track heart rate, sleep, and exercise. Mobile apps capture sensitive menstruation and mental health data. DISHA and DPB draught regulations aim to protect clinical health data privacy, but it is uncertain if they can do so.

As more people utilize digital services, a legal and policy vacuum could allow private firms to abuse health data. It can further complicate surveillance harms from government or law enforcement access to health data from apps or IoT devices. After COVID-19, the federal and state agencies employed contact tracing apps to track illness. Some estimates put the number of central and state-level apps at 19. These apps raised privacy issues. Tamil Nadu's CoBuddy app employed facial recognition to track users' health and location, while the Arogya Setu app acquired more health data than needed for contact tracking. This could enable central and state-level pandemic-resistant surveillance architecture.

Wearable devices allow law enforcement to access health data. Fitbit's privacy policy authorizes government requests for data. US scholars are considering strategies to regulate law enforcement access to sensitive data.

V. Regulation requires Nuance

India requires a clear health data definition. The Data Protection Bill might contain examples of health data and protect it regardless of its source. This may influence government draught laws like DISHA. It defines digital health data but excludes data from applications, the IoT, and non-clinical situations. Addressing this issue will strengthen health data privacy rights.

Health data regulations like DISHA could add a legal layer by generating subgroups and establishing a sensitive approach. Some data is more sensitive than others, and data subjects should have complete control over such personal data. They may have an unconditional "right to erasure" over sensitive mental health information, but their height and weight rights may be subject to constraints and state exceptions. This method clarifies regulations for health data services and gives subjects control over their most private data.

Finally, India's health data regulations presume data exchange and interoperability would improve results. However, this strategy may be risky. After Roe vs. Wade, US researchers have noted how regulations that increase medical information flow hurt abortion-seeking women. This should inspire India to protect privacy while promoting social welfare with its health data legislation.

VI. Conclusions

Politico's publication of documents showing US Supreme Court judges voted to overturn Roe vs. Wade has revived global abortion debate. Aadhaar ID cards, errors, data breaches, and stigma are leaving women seeking abortions in India in a murky area. People in need, especially transsexual women. Behan Box, a gender, equality, and policy outlet, uses interviews, studies, and surveys to address stigma and how it affects decision-making, the law versus practice faced by abortion seekers, and the health effects of Aadhaar ID requirements and service denials. As Behan Box's first article after the U.S. According to the Supreme Court, medical professionals make the final choice. According to a 2021 National Law School of India University study, these practitioners feel shame, guilt, and morality about the practice. The Medical Termination of Pregnancy Act 1971 (MTP) of India was revised in 2003 and 2021 to legalize abortion in all contraceptive failure scenarios. Indian health insurance covers procedures.

References

1   WHO. *The prevention and management of unsafe abortion. Report of a Technical Working Group*. Geneva; 2002.

2   Chhabra R, Nuna S. *Abortion in India: An Overview*. New Delhi: Veerendra Printers; 1993.

3   Government of India Ministry of Health and Family Welfare. *Family Welfare Programme in India Yearbook 2001*. New Delhi: Government of India Ministry of Health and Family Welfare; 2003.

4   Duggal R. *The political economy of abortion in India: cost and expenditure patterns*. Reprod Health Matters. 2004

5   Hirve S. *Abortion law, policy and services in India: a critical review*. Reprod Health Matters. 2004

6   Grimes DA, Benson J, Singh S, Romero M, Ganatra B, Okonofua FE, Shah IH. *Unsafe abortion: the preventable pandemic. Lancet*. 2006

7   Karkal M. *Abortion laws and the abortion situation in India*. Issues in Reprod Genet Eng. 1991; 4:223–230.

Chapter 9

Article 21 with Respect to Digital Data Protection

Varsha Jayan

B.B.A.L.L.B. Student

SSOL, Sharda University

Greater Noida, U.P., India

Vaibhavi Mishra

B.B.A.L.L.B. Student

SSOL, Sharda University

Greater Noida, U.P., India

Ankita Sharma

B.B.A.L.L.B. Student

SSOL, Sharda University

Greater Noida, U.P., India

Abstract

The Constitution of India is based on the principles of dignity, equality, and liberty of which privacy is the underlying principle but protecting privacy has emerged as a new problem in these developing times in terms of technology. With the evolution of technology, methods of criminal acts and ways to infringe the rights of citizens have also evolved. Right to privacy and data protection has become a new problem for individuals. In the pre-personal data protection Bill era, there were legislations of IT acts and the Indian Penal Code that was acting as statutes for data protection. It was not before 2017 that after Justice K.S. Puttaswamy v Union of India case and Justice B N Srikrishna Committee that the right to privacy became a fundamental right under article 21 and a different set of statutes came into existence for data protection known as data protection Bill. Data protection and data protection laws are relatively newfound concepts and therefore it is requisite to comprehend between already existing statutes i.e., the IT laws and the new Bill which was introduced in Parliament as a data protection Bill. This paper set one's sight on how the advancement of technology has led to the infringement of rights of individuals' privacy, it also deals with the evolution of the right to privacy as a fundamental right and the formulation of a data protection Bill, lastly, it forms a comparative analysis between IT act and data protection Bill.

Keywords: Data Protection, Right to Privacy, Technology, Constitution, Covid.

I. Introduction

Life without technology in the 21st century is practically impossible. Consequently, it is not surprising that 'the world is awash with data'. The COVID-19 pandemic further accelerated the reliance on data and related industries as various sectors, including education, IT, and law enforcement, shifted online. As a result, the volume of data available online has reached staggering levels and continues to grow exponentially.

In the modern era driven by digital technology, a common morning ritual for many individuals involves reaching for their smartphones to check their social media accounts. According to IDC (International

Data Corporation) research[1], a staggering 80% of individuals immediately reach for their phones to check their social media accounts. However, while we engage in this seemingly harmless routine, we often fail to realize the invasion of our privacy and the potential misuse of our personal data.

Unfortunately, individuals often fail to realize the extent to which they expose themselves to the online world when they provide their personal data. For example, after browsing through an e-commerce website, individuals may notice that the products they viewed appear as advertisements on platforms like YouTube or Facebook. This is because e-commerce platforms rely on users' browsing history and download records. Such practices, where data is used without proper consent, are just one of the examples of privacy invasion. Moreover, even providing consent for data usage has become as simple as a single click, further compromising privacy. This erosion of privacy is a consequence of technological advancements that have shrunk the world but at the expense of our personal privacy.

The right to privacy was established as a fundamental right under Article 21[2] of the Indian Constitution following the landmark K.S. Puttaswamy judgment[3]. However, the increased use and misuse of data have resulted in constant invasions of privacy, undermining this fundamental right. To address these concerns, the Personal Data Protection Bill was introduced in the Lok Sabha on December 11, 2019, by the Indian government. This Bill was the result of recommendations made by Justice B.N. Srikrishna Committee, which was established after the Puttaswamy judgment. The primary objective of the Bill is to safeguard the right to privacy.

Understanding the right to privacy in the context of the Data Protection Bill is crucial. It is essential to recognize the interdependence between privacy and data protection in the current era of extensive data collection and potential misuse. As data collection continues to rise, it is imperative to establish robust laws that protect individuals' privacy rights and regulate data usage effectively.

II. Evolution of Digital Personal Data Protection Bill

The advancement of the Digital Personal Data Protection Bill marks a major milestone in India's efforts to establish comprehensive data protection. The need for such legislation became apparent due to the rapid expansion of internet and digital technologies, leading to an increased exchange of personal data and growing concerns about privacy breaches. This section will delve into the various stages of the Bill's development, highlighting noteworthy events and significant changes that occurred during its evolution.

The Supreme Court of India delivered a historic ruling in 2017 in the case of Justice K.S. Puttaswamy (Retd) vs. Union of India, establishing the right to privacy as a fundamental right under Article 21 of the Indian Constitution. This groundbreaking judgment served as the cornerstone for recognizing and protecting personal data as an essential component of the right to privacy.

The Government of India established an Expert Committee in July 2017 to address the new issues around data protection and privacy. The committee was headed by Justice B. N. Srikrishna, the committee's primary objective was to investigate matters concerning data protection in the nation and propose suggestions for a strong legislative framework.

In July 2018, the committee presented the draft Digital Personal Data Protection Bill, 2018. This preliminary rendition of the Bill sought to institute extensive measures for safeguarding personal data and regulating its processing. It integrated principles like consent, purpose limitation, data minimization, and accountability, aligning with international data protection standards and best practices.

On December 4, 2019, the Indian cabinet ministry approved the drafted legislation, naming it the Digital Personal Data Protection Bill 2019, after extensive discussions and stakeholder interviews. The Bill suggested the establishment of a Digital Data Protection Authority (DPA) with the responsibility of overseeing and enforcing data protection regulations, promoting awareness, and ensuring compliance. Moreover, the Bill included several provisions aimed at empowering individuals, including the right to access, modify, and delete their personal information.

On December 12, 2019, the Bill was referred to a Joint Committee of both Houses of Parliament (JPC) for review and recommendations in recognition of the significance of various viewpoints and professional insights. The JPC, presided over by Member of Parliament Shri P.P. Chaudhary, was tasked with assessing the Bill, taking stakeholder input into account, and making revisions as required.

The Joint Parliamentary Committee presented its report on the Digital Personal Data Protection Bill and the modified Bill to both Houses of Parliament on December 16, 2021, following significant debates and discussions. The legislation's finalization and addressing the changing issues relating to data privacy in the digital era took a big stride forward as a result of this.

The Digital Personal Data Protection (DPDP) Bill, which outlines procedures for handling personal data and ensuring data protection, was withdrawn in August 2020. The draught Bill was made available for public comment in November 2022, and extensive consultations were held to solicit opinions and recommendations from a range of stakeholders, including business leaders and members of civil society. The Bill was revised and altered after the consultation process based on the input received. The revised DPDP Bill was then approved by the Union Cabinet, clearing the way for its introduction in the Parliament during the Monsoon Session, which would begin on July 20, 2023. To secure the protection of personal data of Indian residents, the Bill attempts to address issues with data breaches, consent-based data acquisition, and sanctions for infractions.

III. Legal Fundamentalism

The right to digital data protection has emerged as a prominent topic in legal and political discussions worldwide. As advancements in technology continue to shape our lives, the pertinent question arises: Is data protection truly a fundamental right? This section explores the legal foundation of data protection and the recognition it has received at both the European and Indian levels.

In the European context, the right to data protection has been expressly recognized as a fundamental right. This right is protected by the European Union's Charter of Fundamental Rights which recognizes the need of preserving individuals' data. Furthermore, the European Court of Human Rights (ECHR has held that the right to data protection falls, within the purview of Article 8[4] of the European Convention on Human Rights (ECHR). These legal provisions emphasize the significance given to data protection as a fundamental right in the European Union.

IV. Is Data Protection Really a Fundamental Right?

In India, the Supreme Court's landmark judgment of 2017 has led to the recognition of data protection as a Fundamental right for India. The Supreme Court of India has ruled that the Right to Privacy is a fundamental right guaranteed by Article 21 of the Indian Constitution in the case of Justice K.S. Puttaswamy (Retd) vs Union of India[5]. It has been recognized by the court that privacy is of fundamental importance for an individual's rights to life and liberty. This groundbreaking judgment has laid down a strong basis for considering the protection of personal data to be an essential aspect of privacy.

The recognition of data protection as a fundamental right provides a sound legal basis for its protection and regulation. Legal frameworks can be created to ensure that people have control over their personal data and that it is treated in a way that respects their privacy and other fundamental rights by acknowledging its fundamental nature. The mere designation of data protection as a basic right, however, does not ensure that it will be properly implemented and upheld.

To effectively safeguard the right to data protection, it is important to establish comprehensive legislation that defines the extent to which digital data protection rights are guaranteed, lays out the duties of data controllers and processors, and offers ways for individuals to exercise their rights.

There are difficulties in putting data protection as a fundamental right into practice. Digital Data Protection regulations must negotiate a challenging environment brought on by technological improvements and an increased reliance on data for many different objectives. It is a complex challenge to strike a balance between the legitimate objectives of data-driven enterprises, governmental monitoring obligations, and individual privacy rights. These cases have played a pivotal role in making privacy a part of Article 21 hence they play a key role in formation of digital data protection Bill.

(i) In R. Rajagopal v. State of Tamil Nadu (1994)[6]:The Supreme Court declared that Article 21's guarantee of the right to life and personal liberty includes the right to privacy as a crucial component. The court determined that people have a right to privacy in matters that are personal and intimate to them and that this right should be protected from needless interference.

(ii) State of Karnataka v. Selvi and Co. (2010)[7]: The right against self-incrimination and the admissibility of evidence gathered from Narco analysis, brain mapping, and lie detector tests were discussed in this case. The Supreme Court concluded that involuntary administration of these tests breaches a fundamental component of Article 21's right to life and personal liberty, which prohibits self-incrimination.

(iii) Canara Bank v. District Registrar and Collector (2005)[8] : The Supreme Court emphasized that people have the right to prevent unauthorized publication of their personal information while upholding the right to privacy in this case. The court determined that although the right to privacy is not addressed expressly in the Indian Constitution, it nevertheless qualifies as a fundamental freedom under Article 21's guarantee of the right to life and personal liberty.

These examples emphasize the significance of Article 21's recognition of privacy as a basic right and the necessity of strong data protection regulations to protect people's personal information. In order to address the issues raised in these cases and provide a comprehensive legal framework to safeguard personal data and privacy rights in the digital age, the Digital Personal Data Protection Bill has been introduced. A trustworthy and secure digital environment for all people is promoted by the Bill, which aims to give individuals control over their personal data and impose strict requirements on data processors to uphold data protection standards.

V. Justice B.N. Sri Krishna Committee

In 2017 the Indian Government established a Justice B.N. Sri Krishna Committee to investigate data protection issues and provide advice on establishing strong data protection legislation in India. The report of the Committee, delivered in July 2018, has significantly influenced the draft Personal Data Protection Law that aims at providing for comprehensive personal data protection measures while safeguarding the right to privacy under Article 21 of the Indian Constitution.

The fact that Article 21 is recognized as a basic right of life and liberty in relation to data protection means that it has significant importance. The Supreme Court of India unambiguously confirmed the right to privacy as a fundamental component of Article 21 in historic decisions like Justice K.S. Puttaswamy (Retd.) v. Union of India (2017). These basic principles were recognized in the Sri Krishna Committee Report, which sought to safeguard people's personal information and guarantee their Inalienable Rights of Privacy.

Article 21 is a fundamental pillar of the Personal Data Protection Bill. The Bill ensures that data protection regulations are in harmony with the fundamental constitutional principles of individual liberty and privacy. The committee acknowledged that in today's digital and data-centric society, protecting personal data is crucial to protect individuals from potential abuse, surveillance, and infringement of their privacy rights.

The committee's report highlighted a number of crucial aspects that emphasize the significance of Article 21 within the Personal Data Protection Bill:

(i) Informed Consent: The need of obtaining informed and express consent from data subjects before collecting and processing their personal information. This allows people to have control over their personal information and is consistent with the idea of autonomy stated in Article 21.

(ii) Purpose Limitation: The report emphasized the necessity for data processors to identify the purposes behind the collection of data. This principle ensures that data is not used beyond its designated purpose, safeguarding individuals from unauthorized and arbitrary processing, as guaranteed by Article 21.

(iii) Data Localization: The report discussed the significance of data localization in safeguarding data sovereignty and citizens' privacy rights. The concept of data localization is consistent with the State's obligation to protect its citizens' life and personal liberty, as enshrined in Article 21

(iv) Data Protection Authority: The committee recommended the establishment of an independent Data Protection Authority to oversee and enforce data protection laws. This authority acts as a guardian of privacy rights, ensuring compliance with data protection regulations, thereby upholding the essence of Article 21.

(v) Right to be Forgotten: The report recognized the right to be forgotten, allowing individuals to erase or anonymize their personal data under certain circumstances. This right aligns with an individual's autonomy and control over their personal information as guaranteed by Article 21.

(vi) Data Breach Notification: The report recommended a robust data breach notification mechanism to promptly inform individuals about data security incidents. This promotes transparency and empowers individuals to take necessary steps to protect their personal data, ensuring that their right to privacy is safeguarded.

The Justice B.N. Sri Krishna Committee's report played a pivotal role in shaping the Personal Data Protection Bill, with Article 21 serving as a guiding principle throughout the process. The report's emphasis on informed consent, purpose limitation, data localization, and the establishment of a Data Protection Authority, among other recommendations, underscores the importance of protecting individuals' privacy rights and ensuring that data protection laws align with the fundamental right to life and personal liberty as guaranteed by Article 21 of the Indian Constitution. The Personal Data Protection Bill, based on the committee's recommendations, seeks to establish a comprehensive framework for data protection while preserving individuals' right to privacy in the digital age.

VI. Digital Personal Data Protection Bill

A comprehensive piece of legislation called the "Digital Personal Data Protection Bill" is intended to safeguard the security and privacy of digital personal data in India. The need for effective data protection measures has never been higher in a time when technology is an essential part of daily life and data-driven innovations have become the norm. While ensuring that people's fundamental right to privacy, as protected by Article 21 of the Indian Constitution, is protected, this Bill aims to address the issues brought on by the widespread collecting, processing, and sharing of digital personal data by both public and commercial institutions.

The responsibility of data fiduciaries, which are organizations or people who choose the reason for and methods for processing digital personal data, are described in Chapter 2 of the Bill. These requirements include having legal justifications for processing data, giving data principals (those whose data is being processed) clear and transparent notices, and gaining explicit consent before processing any data. The legislation recognizes that consent should be freely given, explicit, and informed, enabling data subjects to make decisions about their personal data with knowledge.

The proposed legislation introduces the concept of "deemed consent" for situations in which express consent cannot be obtained. Presumed consent may be appropriate where data is needed to carry out a contract, to adhere to a legal requirement, to protect the vital interests of a data principle, or for the data fiduciary to pursue a legitimate interest. The use of presumptive consent is subject to certain limitations under the Bill in order to ensure that data processing is still fair and transparent.

Additionally, data processors have a general duty to maintain the security and accuracy of the data they handle. To prevent unauthorized access, alteration, disclosure, or destruction of personal data, they must take the necessary precautions. The implementation of data protection measures by data fiduciaries must also follow industry standards and best practices.

The measure sets additional requirements for processing children's personal data in recognition of their unique vulnerability in the digital world. To protect children's data, data fiduciaries must take special care and have parental or guardian approval. This clause tries to create a balance between protecting children's privacy and online safety while enabling them to take advantage of digital services.

According to the decision of the Government, significant data guardians will be subject to additional requirements under this law. The fact that they often collect a large amount of personally identifiable data or have important influence on the lives of citizens, makes it mandatory for these organizations to comply with stricter data protection laws. The Law intends to make key data fiduciaries subject to more rigorous standards of liability and data protection so as to minimize the risks related to their processing activities.

The rights and obligations of data principals are the subject of Chapter 3 of the law. Data fiduciaries are required to provide data principles with access to information about their personal data. They have control over the accuracy and relevancy of their information because they can ask that obsolete or inaccurate data be removed or corrected. Data principals also have the right to file complaints about how their data is processed, giving them a way to express their concerns and demand remedy.

Also included in the law is the ability for data principals to appoint representatives to represent them. When data principals could have trouble independently exercising their rights to data protection, this clause is especially pertinent.

The transfer of personal data outside of India is covered in Chapter 4. Under some circumstances, such as when contracts provide an acceptable degree of data protection in the recipient nation or when the Data Protection Board of India has given its approval, data fiduciaries are permitted to transmit data outside the country.

For special objectives such those related to journalism, literature, the arts, or research, the Bill allows exclusions from several sections. Such exemptions must, however, strike a reasonable balance between the right to privacy and the interests of the data principals.

Chapter 5 of the Act establishes a framework for compliance to ensure adherence to its provisions. The regulatory body, known as the Data Protection Board of India, is responsible for overseeing compliance, investigating complaints, and conducting audits. This Board plays a crucial role in enforcing data protection standards and ensuring that data fiduciaries comply with the Bill's provisions.

The Bill includes a review and appeal mechanism to help with compliance. Data fiduciaries have the right to request reviews of Board decisions, guaranteeing a fair assessment of enforcement measures.

In case of disputes, the Bill offers alternate dispute resolution procedures, encouraging the speedy resolution of disputes involving data protection.

Furthermore, the Bill includes provisions for data fiduciaries to make voluntary commitments. These initiatives give data fiduciaries the chance to be proactive about compliance and show their dedication to data protection.

In situations where non-compliance is established, the Bill empowers the Board to fine data fiduciaries financially where non-compliance is proven. These sanctions are meant to discourage data fiduciaries from breaking data protection laws and emphasize how crucial it is to abide by the Act's rules.

Chapter 6, it encompasses miscellaneous provisions, including the authority to make rules to facilitate the implementation of the Act. The Central Government holds the power to amend schedules, if necessary, to keep the legislation up-to-date and relevant in the face of evolving technological advancements.

A smooth transition to the new data protection framework is ensured by the Bill's measures for handling implementation challenges. A smooth transition to the new data protection framework is ensured by the Bill's measures for handling implementation challenges.

To minimize conflicts and redundancies, it is crucial to guarantee that the Act is consistent with other applicable laws. To do this, the Bill stipulates that the Act must be such.

Thus, the "Digital Personal Data Protection Bill" is a significant stride towards safeguarding individuals' privacy rights in the digital age. By imposing obligations on data fiduciaries and granting powers to the Data Protection Board, the Bill seeks to enhance data protection and privacy measures.

VII. The Imperative of Data Protection in the Digital Era

The rate at which the human race is creating and amassing data on the virtual world in today's quickly evolving digital era is nothing short of astounding. By 2025, it is expected that the entire planet will have amassed a staggering 175 zettabytes of data. A wealth of personal information, from fundamental demographic data to intimate details about people's lives, can be found within this enormous database. Technology's emergence has brought with it a previously unheard-of level of connectivity and comfort, but it has also given rise to grave worries about the right to privacy.

As we traverse the digital landscape, our online activities inevitably expose us to the gathering of personal data. From browsing through online shopping platforms to playing games on the internet, every interaction leaves behind a digital footprint. However, the darker side of this data collection comes to light when we discover that our information is being harvested without our explicit consent or knowledge. This realization raises pertinent questions about the responsible and ethical use of personal data in the virtual realm, calling into question the trustworthiness of technology and the platforms through which data is collected and shared.

A clear reminder of the consequences of data exploitation is the Facebook-Cambridge Analytical data controversy. Millions of Facebook users had their personal data illegally stolen in this well-publicized case for the purpose of political advertising. The consequences of this breach showed the risks involved in relying on technology to protect our personal information as well as the right of individuals to privacy. These occurrences undermine public confidence in the digital ecosystem by raising questions about the motives and methods used by data collectors.

The large amount of personal information that businesses are storing in various databases has exacerbated privacy concerns. The rise of technology leaves a lot of sensitive data points open for attack, including social security numbers, credit card numbers, medical histories, and family histories. The difficulty lies in striking a balance between securing people from potential data breaches and unauthorized access and leveraging data-driven technologies' capacity for the common benefit. To establish clear guidelines for the collecting, storage, and use of data and boost public trust in technology, a comprehensive privacy regulation is necessary.

Fostering confidence in technology requires openness and informed consent. A strict privacy law would oblige data fiduciaries to be open and upfront about their data practices and to inform individuals of the precise purposes for which their data will be used. Additionally, by obtaining their express agreement before collecting their data, individuals are ensured that they retain control over their personal information. When used in conjunction with stringent data security measures, this technique encourages a climate where individuals may believe that their data is being handled ethically and responsibly.

Trusting technology with our right to privacy has far-reaching effects that affect many facets of our existence. Personal data is continuously being gathered and analyzed from the moment we wake up and check our smartphones to the moment we go to bed with wearable devices monitoring our sleep. There is a tone of opportunities for innovation, efficiency, and comfort in today's data-driven environment. However, worries about potential abuse and invasions of personal privacy are also raised by the sheer size and complexity of data collecting.

Data is frequently referred to as the new currency in the digital world, as people unknowingly trade their personal information to gain access to internet services, social networking, and other conveniences. Whether intentionally or unintentionally, the data we provide offers insightful information about our preferences, actions, and even our weaknesses. Technology businesses frequently use this wealth of personal information to customize content, recommendations, and adverts, resulting in a more tailored user experience. While increasing user satisfaction, this level of personalization also raises moral concerns about how much of our personal information is being used and shared without our knowledge or agreement.

In the context of data privacy, the idea of permission assumes special significance. In order to access their platforms, many online services require customers to agree to lengthy terms and conditions agreements. However, the legalese used in these agreements makes it challenging for consumers to completely understand the ramifications of giving consent. In essence, there aren't many options left for consumers; they can either agree to the terms and conditions without fully comprehending them or decide not to use the service at all. The confidence that people have in technology businesses and their dedication to protecting user privacy is undercut by this absence of meaningful permission. Data security and data breaches are major worries as the number of personal data collected keeps increasing dramatically. Sensitive personal data of millions of people has been made public due to high-profile data breaches, such those that big firms and governmental organizations have faced. The effects of such breaches go beyond monetary losses; people may also experience mental pain, identity theft, and possible reputational harm. When there are data breaches, people start to doubt whether their personal information is actually secure in the digital world, which seriously undermines their trust in technology.

Furthermore, as technology continues to advance, it brings forth new challenges to the protection of privacy rights. The progress in artificial intelligence, machine learning, and data analytics has led to the creation of sophisticated algorithms capable of predicting and influencing human behavior. However, these algorithms often rely on extensive data mining and profiling, raising concerns about potential discrimination and bias in decision-making processes. For example, algorithms used in hiring procedures might unintentionally perpetuate existing biases, resulting in discriminatory outcomes.

The significance of trust in technology concerning the right to privacy cannot be overstated. People need to feel secure knowing their personal information is handled with the highest care and respect and that they have real influence over how it is used and shared. A guiding concept for data-driven technologies should be the ethical and appropriate use of personal data. We can create a digital ecosystem where technology and privacy coexist peacefully and where trust in technology is increased by passing comprehensive privacy regulations, encouraging openness, and equipping people with true consent. By doing this, we can fully utilize technology while protecting everyone's fundamental right to privacy in the modern world.

VIII. Conclusions

The Digital Personal Data Protection Bill represents a vital measure in tackling privacy issues in the digital age. As technology becomes more ingrained in our daily lives, finding a balance between its advantages and safeguarding individuals' right to privacy becomes essential. The Bill's development, influenced by the Justice B.N. Sri Krishna Committee's suggestions, underscores the acknowledgment of privacy as a fundamental right protected under Article 21 of the Indian Constitution.

The primary objective of the Bill is to empower individuals by providing them with control over their personal data. This is achieved through measures like informed consent, purpose limitation, and data localization. The Bill also sets up a Data Protection Board responsible for monitoring compliance, conducting audits, and addressing complaints, thereby ensuring the effective implementation of data protection standards.

While the measure is an important step forward, digital data security success depends on data fiduciaries adhering to its correct implementation. In order to guarantee that digital personal data is managed responsibly and ethically, transparency, accountability, and ongoing monitoring will be essential.

Recognizing that digital data privacy is not a barrier to innovation but rather a fundamental basis for fostering trust and confidence in technology is crucial as we embrace the advantages of a data-driven world. The secret to building a safe and respectful digital ecosystem for all users is striking a perfect balance between data-driven progress and individual privacy rights. We can embrace technical breakthroughs while preserving the privacy and dignity of every person in the digital era by following the principles of the Digital Personal Data Protection Bill and establishing a culture of ethical data handling.

Endnotes

1   https://www.cbsnews.com/philadelphia/news/study-80-of-people-grab-smartphone-within-15-minutes-of-waking/
2   art 21 Protection of life and personal liberty No person shall be deprived of his life or personal liberty except according to procedure established by law.
3   K.S. Puttaswamy and Anr. vs. Union of India ((2017) 10 SCC 1)
4   Art 8. ECHR Everyone has the right to respect for his private and family life, his home and his correspondence.
5   AIR 2017 SC 4161.
6   AIR 1995 SC 264.
7   AIR 2010 SC 1974.
8   AIR 2005 SC 186.

Chapter 10

Personal Data Protection Law in India: Need and Challenges

Akansha Rai

Ph.D. Research Scholar

Shree Guru Gobind Singh Tricentenary University

Gurugram, Haryana, India

Abstract

Security has been a well-established idea spreading over hundreds of years, with lawful specialists endeavouring to characterize it in light of their skill. In the present interconnected world, where data unreservedly circles on the web and is available to everybody, understanding the importance of security has become critical. We are presently in a time where information has turned into a significant item, defenseless to double-dealing without any legitimate guidelines. Security, in its vastest sense, can be portrayed as the "assumption for fitting utilization of individual information." Lately, information has earned worldwide consideration, with individuals turning out to be more mindful of its importance and financial worth. Numerous Western purviews have recognized people's freedoms over their own information and the unfortunate behavior of key part in the data innovation industry, who tried to benefit from it. Luckily, these uncalled for rehearses have confronted lawful repercussions lately, with organizations being considered answerable for undermining their clients' security. Be that as it may, there is as yet a need to lay out a far-reaching framework for tending to responsibility in instances of information breaks, as this viewpoint is still during the time spent improvement. The public authority's honorable endeavors to handle these worries through the PDP Bill ought to be recognized. Regardless, certain basic issues stay annoying and should be tended to before the Bill is sanctioned. The establishments proposed under the PDP Bill have critical blemishes that might actually sabotage the whole motivation behind the regulation.

Keywords: Bill, Data Innovation, Personal Data Protection, Security.

I. Introduction

protection has been a well-established idea spreading over hundreds of years, with lawful specialists endeavoring to characterize it in light of their skill. In the present interconnected world, where data unreservedly courses on the web and is available to everybody, understanding the importance of security has become pivotal. We are as of now in a period where information has turned into a significant item, powerless to double-dealing without legitimate guidelines. Security, in its vastest sense, can be depicted as the "assumption for suitable use of individual information."

As of late, information has accumulated worldwide consideration, with individuals turning out to be more mindful of its importance and financial worth. Numerous On September 2, 2020, the Service of Gadgets and Data Innovation (MeitY) of the Indian government practiced its position under area 69A of the Data Innovation Act, alongside the significant arrangements of the Data Innovation (System and Protections for Hindering of Access of Data via Public) Rules 2009. MeitY made a move by obstructing 118 portable applications in light of arising dangers. As per MeitY's notice, these applications were engaged with exercises that imperiled India's sway, trustworthiness, protection, and public request. Different protests were gotten by MeitY from different sources, including various reports of the abuse of explicit portable applications on Android and iOS stages. It was found that these applications were illegally taking and communicating clients' information to servers situated beyond India. The assortment, mining, and profiling of this information by elements antagonistic to India's public safety and safeguard raised quick worries about the nation's power and honesty, requiring brief activity. MeitY's goal was to shield the interests of millions of Indian portable and web clients, guaranteeing the security, wellbeing, and sway of India's the internet.

Yet again MeitY's choice has focused on the squeezing need for hearty Information Assurance Regulations in India. In the computerized time, people readily or unwittingly share delicate individual data on different advanced stages, for example, web-based business sites, versatile applications, online class stages, web banking, and e-wallets. Clients frequently award authorization for their own information to be utilized by essentially tolerating the agreements without completely perusing the Security Strategy. This security weakness uncovered great many clients to likely dangers from programmers and other noxious entertainers.

II. Need of Data Protection

The right to security is a central qualification cherished in Article 21 of the Indian Constitution, which frames our key privileges. This rule was confirmed by a nine-judge seat of the High Court in the milestone instance of Equity K.S. Puttaswamy versus Association of India on August 24, 2017. In that judgment, the court pronounced the "right to protection" as an essential part of Part III of the Indian Constitution.

The justification for why the issue of whether the right to security, a central right was brought under the steady gaze of a nine-judge seat can be addressed. In 2017, a seat of five appointed authorities in the High Court was hearing the Aadhaar Card and the right to security situation when they communicated the requirement for a nine-judge seat to decide the essential idea of protection prior to administering on the fundamental Aadhaar case. The Head legal officer contended in the Aadhaar case that albeit past High Court decisions had perceived the right to security, it had not been acknowledged as a key squarely in the Kharak Singh judgment (by a six-judge seat in 1960) and the M P Sharma judgment (by an eight-judge Constitution seat in 1954). In this manner, it was urgent to meet a nine-judge seat to choose the centrality of the right to protection.

The far-reaching understanding of the right to security by the High Court provoked the public authority to make strides towards ordering Individual Information Insurance regulations. Numerous nations have put forth attempts to guarantee the security of their residents' information somehow. While information assurance regulations are exceptionally cutting-edge in the ongoing general set of laws, many non-industrial countries are yet to perceive the requirement for such regulations. Legislators face various difficulties in managing the immense information market, and an absence of mindfulness about arising advancements hampers the improvement of information security statute. Purviews like the EU, UK, and Canada are driving the way in propelling information regulations and actually resolving different issues looked by individuals around the world.

In India, critical headway has been made lately through different decisions and regulation, with legal activism assuming an essential part in propelling the reason for information security. As of now, information insurance on the web is managed under the Data Innovation Act, 2000

(IT Act). Be that as it may, the deficiency of these arrangements has required the requirement for independent regulation. The Parliament is currently fostering the Individual Information Security Bill, 2019 (PDP Bill). Whenever established, this Bill will engage residents to protect their information and furnish them with legitimate cures in instances of information security infringement. The draft Bill incorporates arrangements for pay to impacted people and presents misdeed risk in information assurance.

The idea of misdeed obligation in information security started in the US during the nineteenth century when the press and media extended essentially. Regulations were established to safeguard people from the nosy way of behaving of the press, forcing sensible limitations on the right to speak freely of discourse. At first, these regulations basically designated people of note who were bound to be covered by the news media. Be that as it may, with the development of the web, anybody turned into an expected wellspring of data, and organizations began taking advantage of this information by encroaching on individuals' security. This constrained legislator to widen the extent of protection regulations to cover each resident. Be that as it may, whether or not organizations ought to be considered responsible for repaying each impacted individual stayed unanswered. Courts were isolated on this issue because of the enormous number of people included and their absence of mindfulness about their misfortunes, making it trying for the courts to decide the meaning of pay. In the new instance of Lloyd v. Google LLC, the High Court of the Assembled Realm collectively decided for Google, denying the solicitation for pay to 4,000,000 Apple clients who were illegitimately followed by Google and had their information sold. The Court accentuated the trouble of finding out the people who endured.

III. The Notable Elements of the Bill

Following the restriction on Chinese applications, people normally have worries about the security of their own information drifting around. They may seek information about the safeguards and norms imposed by the Bill on the collection, processing, and cross-border transfer of such data. Here are the key points outlined in the Bill:

(i) Application of the Act to processing of Personal Data:

a. The Bill governs the processing of personal data collected, disclosed, shared, or processed within India by the government, Indian companies, Indian citizens, or entities created under Indian law.

b. It also applies to data fiduciaries or processors operating outside India if the processing is related to business conducted in India or involves profiling of individuals within India.

c. However, the Act does not apply to anonymized data, which refers to personal data that has undergone an irreversible process of transformation to a form that cannot identify the data principal.

(ii) Categories of personal data:

a. The Bill classifies data into three categories: Personal Data, Sensitive Personal Data, and Critical Personal Data.

b. Personal data includes information that can identify an individual, collected online or offline.

c. Sensitive personal data includes financial information, biometric data, caste, religious or political beliefs, or any other category specified by the government in consultation with the Authority and sectoral regulators.

d. Critical Personal Data refers to personal data notified by the Central Government as essential.

(iii) Obligations of data fiduciary:

a. A data fiduciary, which includes individuals, organizations, or the State, determines the purpose and manner of processing personal data.

b. Personal data can only be processed for specific, clear, and lawful purposes.

c. The processing must be fair, reasonable, and ensure the privacy of the data principal.

d. The data fiduciary should process personal data based on the consent given by the data principal or for purposes incidental or linked to the consented purpose.

(iv) Notice requirements for collection or processing of personal data:

a. Data fiduciaries must provide a notice to the data principal containing information such as the purpose of processing, categories of personal data collected, contact details of the data fiduciary and data protection officer, withdrawal of consent rights, consequences of failure to provide data, and cross-border data transfer details, if applicable.

b. The notice should be given at the time of collection or as soon as reasonably possible if the data is not collected directly from the data principal.

(v) Quality and retention of personal data:

a. Data fiduciaries must ensure that the processed personal data is complete, accurate, not misleading, and updated according to the purpose for which it is processed.

b. Personal data should not be retained beyond the necessary period to fulfill the processing purpose and should be deleted afterward.

(vi) Accountability of data fiduciary:

a. Data fiduciaries are responsible for complying with the provisions of the Act in all their processing activities.

b. Consent for processing of personal data:

Personal data cannot be processed without the consent of the data principal, who can withdraw consent at any time. The burden of proof for obtaining consent lies with the data fiduciary.

These measures aim to protect individuals' personal data and ensure responsible handling and processing in accordance with the Bill's provisions.

IV. Remedies Available in Personal Data Protection Bill

In India, massive changes have occurred because of legal mediations, particularly the earth shattering High Court judgment in K.S. Puttaswamy v. Association of India. This administering has prepared for the advancement of regulations like the Individual Information Assurance (PDP) Bill, which was formed in view of the proposals of the B.N. Srikrishna Panel. The PDP Bill addresses the underlying exertion by administrators to direct information and guarantee security assurance. Whenever passed, this Bill will supplement the ongoing IT Act, 2000, which by and by administers information on the web. The arrangements of the IT Act were lacking and neglected to address numerous significant parts of online information security. Interestingly, the PDP Bill is a thorough regulation that characterizes information, orders it into various sorts, and lays out requirement councils.

This conversation explicitly centers around the common cures gave in the draft PDP Bill, 2019. As indicated by Segment 64(1) of the Bill, any individual (alluded to as an information head) who endures hurt because of an infringement of the Demonstration or its guidelines and guidelines by an information trustee or information processor has the option to look for remuneration from the party in question. The Bill characterizes an information head as the person to whom the information relates (Segment 3(14)). Pay can be guaranteed from information processors and information guardians, as characterized in Segments 3(15) and 3(13) separately. These definitions envelop people, the State, associations, or juristic elements.

Pushing ahead, the Bill lays out establishments to deal with remuneration claims. Under Area 62, the Focal Government will name a settling official to deal with claims under Segment 64. In the event that the petitioner is disappointed with the choice, they can speak to the Redrafting Court, as framed in Segment 67. Area 67 additionally guides the Focal Government to lay out the Redrafting Court to hear and determine requests from the sets of the arbitrating official.

The arbitrating official, referenced in Area 62, will be named by the Focal Government to deal with pay claims under Segment 64. This official has the ability to gather applicable people and request vital archives as proof. In the event of rebelliousness with the arrangements, the official can force suitable punishments.

The Re-appraising Council assumes a critical part in the PDP Bill, as it has huge institutional independence. As per Area 67(1), the Council has expansive investigative ward over the choices of the mediating official. The Court will be going by a director named by the Focal Government and will practice different powers and useful independence. For instance, the Executive can comprise seats and disseminate liabilities among them. The Administrator likewise has the position to move cases starting with one seat then onto the next when important. Individuals from the Council will be selected by the Focal Government as per Segment 70.

A distressed party can interest the Council in no less than thirty days from the receipt of the request being pursued against. Nonetheless, the Council might consider requests even after the predetermined time has lapsed whenever happy with the explanations behind not documenting inside the given period. Quite, the Council has similar abilities as a common court yet is not limited by the methods laid out under the Code of Common Methodology, 1908. The PDP Bill concedes the Investigative Council procedural independence, with the necessity of sticking to the standards of regular equity. This independence is fundamental for compelling settlement in the quickly advancing mechanical scene and guaranteeing the conveyance of equity.

V. Emerging Issues and challenges with respect to the Bill

The Public authority's admirable exertion in resolving different issues through the PDP Bill has demonstrated successful. In any case, there are still a crucial worry inside the Bill that should be settled before it tends to be established. The foundations laid out under the PDP Bill have remarkable weaknesses that might actually sabotage the general target of the law. There, right off the bat, is a troubling degree of government impact over these bodies. Area 68(2) specifies that the Focal Government might decide the arrangement interaction, terms of office, compensations, and different circumstances for the Director and individuals from the Redrafting Court. Since the Bill as of now incorporates the State inside the meaning of information processor and information guardian under Segment 3, this makes a reasonable infringement of the standards of regular equity in cases including the State. To guarantee equity and decency, it is important to diminish the impact of the State in the Redrafting Council, taking into account it is a semi legal substance. Furthermore, the broad procedural independence allowed to these organizations subverts their responsibility. Giving them complete opportunity to lay out their working could bring about an absence of governing rules in the tasks of the Court. While the Council has abilities comparable to a common court without being limited by the Common Methodology Code, 1908, this independence enjoys benefits, for example, expanded effectiveness and the capacity to adjust to quickly advancing innovative turns of events. In any case, it is urgent to keep a feeling of obligation, and forfeiting it for some other explanation may not be judicious.

To address these worries inside the Bill, models from purviews with further developed information security law, like the European Association's Overall Information Assurance Guideline (GDPR), can be thought of. Article 82 of the GDPR gives arrangements to remuneration, expressing that people who experience material or non-material harm because of an infringement of the Guideline reserve the option to look for pay from the regulator or processor. Also, Article 79 of the GDPR manages the implementation of the arrangements, guaranteeing the right to a powerful legal solution for people who accept their privileges under the Guideline have been encroached. This article takes into consideration legal objections to be brought under the steady gaze of courts that work freely of the State, guaranteeing a better quality of reasonableness. In India, the current common court framework ought to be viewed as an option for petitioners to guarantee equity and give people the opportunity to pick.

VI. Conclusions

The proactive moves made by the Indian legal executive and government impart trust for a promising future in the field of records security statute. Nonetheless, it is critical to stay careful and perceive that there is as yet a lot of work to be finished. The dubious arrangements found in the PDP Bill have raised worries as they award the public authority uncontrolled position to encroach upon residents' protection under ambiguously characterized exemptions possibly. Indeed, even Equity B.N. Srikrishna, the top of the Board, has censured the public authority's one-sided translation of the Council's proposals. It is of most extreme significance for the public authority to focus on resolving these issues to forestall the stagnation of information security regulation improvement in India.

As expressed in its preface, the target of the Bill is to defend people's security with respect to their own information, direct the stream and use of individual information, lay out trust among people and elements engaged with handling individual information, safeguard the freedoms of people whose information is handled, make a system for hierarchical and specialized measures in information handling, characterize norms for online entertainment delegates and cross-line moves, address unapproved and hurtful information handling, and lay out the Information Insurance Authority of India. The central standards fundamental the Bill intently line up with those of the Overall Information Security Guideline (GDPR).

Given the ongoing Coronavirus circumstance, where social separating measures and remote work have altogether expanded dependence on computerized stages, India actually must facilitate the endorsement of the new Private Information Security Bill. The sanctioning and requirement of the Bill, alongside its effect, will decide the destiny of information security for a huge number of Indian residents.

References

1   Information Technology Act, 2000.
2   https://www.meity.gov.in/cyber-security.
3   https://prsindia.org/policy/report-summaries/non-personal-data-governance-framework.

4   https://prsindia.org/policy/report-summaries/free-and-fair-digital-economy.

5   Digital Personal Data Protection Bill, 2022.

6   https://blog.mygov.in/editorial/indias-digital-protection-Bill-a-milestone-in-data-protection-and-global-competitiveness/.

7   https://www.hrw.org/news/2022/12/23/india-data-protection-Bill-fosters-state-surveillance.

8   https://www.meity.gov.in/writereaddata/files/Data_Protection_Committee_Report.pdf.

9   https://www.meity.gov.in/data-protection-framework.

10 GDPR and India by Aditi Chaturvedi, The Centre for Internet and Society, India , available at https://cis-india.org/internet-governance/blog/gdpr-and-india-a-comparative-analysis.

11 The Constitution of India.

12 EU General Data Protection Regulation, available at https://gdpr-info.eu/.

Chapter 11

Abraham Maslow's Privacy Pyramid: Understanding Human Needs in a Digital Age

Renu Yadav
B.A.L.L.B. (Hons.) Student
SSOL, Sharda University
Greater Noida, UP, India

Archi Agarwal
B.A.L.L.B. (Hons.) Student
SSOL, Sharda University
Greater Noida, UP, India

Abstract

The theory of Maslow's Hierarchy of Needs provides that human beings are fulfill their needs in a hierarchical order. They do not fulfill their wish at the highest level. According to the hierarchy of needs, they take a step of success of fulfilment of needs one by one. Privacy has been seen with the context of this theory as a tool or an approach which supports the human beings to be satisfied of various needs. Dumbwaiters were invented this concept so the servants couldn't interrupt the "private" life of the family, whereas previously the presence of a large wait staff had social cache. This concept of privacy has evolved and will began to. Humans still have the desire to keep some information to ourselves and to be private. We might want to present an idea of ourselves to the public but do not honestly want the public to know who we really are and what we are going to do because we want our needs should be fulfill and the privacy of our income or wealth to be maintained. The present paper gives us the idea about the need of privacy among the human beings and how it is relevant in modern times.

Keywords – Hierarchy, Fulfilment, Needs, Privacy, Wealth.

I. Introduction

Abraham Maslow was a social psychologist gave the theory of hierarchy of needs in 1943 because he was more interested in the psychological needs of people than the in psychological problems of the people. Maslow's Hierarchy of Needs is a physiological concept-based theory which describes that the motivation to fulfill the needs of human arranged in a hierarchical pyramid. We have also seen this in a biological basis for why people don't always want to be surrounded by each other. Consider animals that want to keep competitors away during the breeding season. They require privacy when mating like (most) humans. We assume that our need not to broadcast our sexual activity is related to "privacy", when in fact it may have evolved as a necessary practice to ensure that an unwanted competitor does not interrupt our own mating ritual. The theory has described in the pyramid arranging five different levels of human psychological and physical needs in order of importance.

II. Maslow's Hierarchy of Needs- Examples of Security in real life

Many aspects of hierarchy are directly related to security problems and the hierarchy needed to solve these problems today. In the world of cyber space, various attacks fit into the original hierarchy of needs created by Maslow nearly 80 years ago. While this may be a metaphorical connection, the value for security professionals is to understand the psychological motivation behind a breach. Additionally, many of the examples focus on individual attacks, but it is important to remember that an exploited individual can direct attackers directly into an enterprise to exploit.

Although often motivated by money, we are starting to see an increase in non-monetary attacks and the violation of Privacy is considered as a non- monetary attack. As these different motivators become more common, security professionals who are aware of the different end goals of an attack can better prepare their defenses.

The hierarchy of needs through Maslow theory –

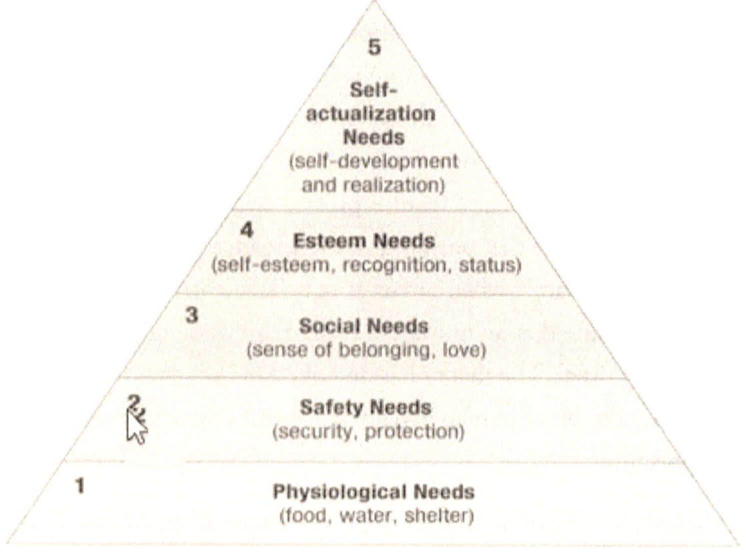

(i) Physiological Needs

Physiological needs consist of the necessary things for human survival such as food, water and shelter. If an individual has unable to fulfill these needs, the deficiency in their life will be quickly and readily apparent. For Maslow, he considered physiological needs as critically important and non-Fulfilment of these needs' shows poverty in the life of human. Attackers utilize this psychological understanding with attacks such as phishing attempts that promise a monetary reward. While not a direct connection to food and shelter, this monetary focused attack can jeopardize both for impacted individuals.

(ii)   Safety Needs

Once the physiological need of individual fulfilled then they can then focus on their physical needs for security, safety, and shelter. They feel to the need of a secure environment for their well- being. Therefore, they opt those things who would help to fulfill this in their day-to-day lives. Attackers utilize this knowledge by deploying scare tactics on users. These exploits often target human beings over the phone or email, informing them something tragic occurred to threaten their online security. The attackers will present themselves as the hero of the tragic situation and happily offer assistance in exchange for money or access to the individual's computer.

A good example here comes from when these types of attacks center around geopolitical affairs. Consider the attack on the Iranian Nuclear Power Plant in 2018 which shut down a nuclear facility. This attack shows how cybersecurity, or lack thereof, can quickly become a national security concern- jeopardizing the security of an entire nation. These first two levels are important to the physical survival of the person. Once individuals have basic nutrition, shelter, and safety, they seek to fulfill higher-level needs.

(iii)   Belongingness and love needs

After the fulfilment of safety and security needs, the next level of need in Maslow's hierarchy of needs is love and belonging. Humans lives in a social relationship in a society, so they also have the desire for things like friendship, closeness, acceptance, love and affection which arises itself. Attackers, again the most common phishing attempt, try to exploit these own desires. Impersonating others to create and exploit false relationships is a common example of this hierarchical class. If the attackers have the knowledge of this needs, it can become a huge problem to fulfill this need. As it can be seen that the world shrinks due to the spread of the internet and social media, it seems that the hackers focus on the love and affiliation need of human has to be increased. This need gave a motivation to maintain a privacy even in online sphere to fulfill the need of love.

(iv)   Esteem needs

Recognition of success or status represents the normal human desire to be valued and accepted by others. This level also includes self-esteem, which means respect and self-acceptance. An imbalance at this level can lead to low self-esteem or an inferiority complex. People who suffer from low self-esteem may find that external validation from others—fame, honor, accolades, etc. fulfills their needs at this level only partially or temporarily.

(v)   Self-Realization

The fifth level of the pyramid is self-actualization. At this stage, people feel that they have reached their full potential and are doing their best. Self-actualization is rarely a permanent feeling or state. Rather, it refers to the constant need for personal growth and discovery that people have throughout their lives. Self-actualization can occur after achieving an important goal or overcoming a specific challenge and can be characterized by a new sense of self-confidence or satisfaction.

III. Analysis of the Needs

Maslow's hierarchy of needs is divided into two ways which helps easy to understand. Starting at the bottom of the pyramid in which basic needs have to be fulfilled. These needs are very important for the survival of people include food, safety and shelter.

At the top of the pyramid live growth needs, which consist of more complex desires, such as psychological needs. After fulfilment of basic needs people move towards their highest level of need. Looking at each category, scarcity needs begin with the basics of survival, which relate to the physical needs of food, water, Shelter. Another level of hierarchy is safety and security. Moving up the pyramid, needs become more complex and psychological needs form the next category. They consist of things like belonging to family and friends. Next, esteem needs include, among others, a sense of self-respect. Finally, at the top of the pyramid is self-actualization, or the desire to achieve one's goals.

IV. Significance of Privacy as a Human Need in Technology Driven Society

Is there any requirement of this in the relation of Privacy of humans? A well understanding of the importance of privacy and why it needs protection is a key. This is important for fulfill our needs because it motivates us to do something different, but our ability has explained the importance of privacy to others in your organization is also invaluable. This will encourage others to protect their personal data because they will better understand why it is important.

(i) Capacity limitation

Privacy is a tool through which the power of state and private company has limited. Personal data is used to make very important decisions in our lives, and it must be protected for the fulfilment of our other needs

Personal data may be used to influence our reputation in both negative and positive sense. It can be used to influence our decisions through research. . It can be used as a tool to track our personal information and it can be used by the wrong hands and evil eyes through which personal data can be used to give us a great harm and become a dangerous thing for us.

(ii)   Respect for individuals

Privacy is about respecting people and also those which closely related to people. If a person has a reasonable desire to keep something private, it is very disrespectful behavior to ignore their wishes. It is obvious that the desire for privacy can conflict with important values, so privacy may not always win the balance. Sometimes people's privacy wishes are simply put aside because the resulting harm is deemed insignificant. Although this does not cause much harm, it shows a lack of respect for the person concede. In a way, it says, "I look out for my interests, but I don't look out for yours."

(iii)   Reputation management

Privacy allows people to control their reputation. It can work both in negative and positive sense. How others judge us affects our opportunities, friendships, and overall well-being. Although we cannot completely control our reputation, we must have some ability to protect our reputation from unfair damage. Protecting one's reputation depends not only on protection against lies, but also on protection against certain truths. Knowing private details about people's lives does not necessarily lead to a more accurate assessment of people. People judge badly, judge hastily, out of context, judge without hearing the whole story and understand more hypocritically. Privacy helps people protect themselves from these uncomfortable decisions.

(iv)   Maintenance of social boundaries

People set boundaries for others in society. These limits are both physical and cognitive. We need places of solitude to retreat, places where we are free from the gaze of others, to relax and feel at home. We also create boundaries of knowledge and have a complex of these boundaries for our many different relationships. Privacy helps people manage these limits. Violating these boundaries can create awkward social situations and damage our relationships. Privacy also helps reduce the social frictions we face in life. Most people don't want everyone to know everything about them – hence the expression "none of your business". And sometimes we don't want to know everything about other people – hence the expression "too much information".

(v)   Confidence

In a relationship, whether it is personal, professional, administrative or business, we depend on trust in the other party. A breach of confidentiality is a breach of that trust. In professional relationships, such as our relationships with doctors and lawyers, this trust is key to maintaining integrity in the relationship. We also trust the other people we interact with and the companies we do business with. When trust is broken in one relationship, it can make us reluctant to trust other relationships.

(vi)   Life management

Personal data is necessary to make so many decisions that affect us, from getting a loan, a business license or a job, to our personal and professional reputation. Personal data is used to determine if we have been investigated by an authority, if we have been searched at an airport or if we have been refused the ability to fly. Personal data affects almost everything, including the messages and content we see online. Without knowledge of what data is being used, how it is being used, without the ability to correct and change it, we are practically helpless in today's world. In addition, we are helpless without the ability to influence how our data is used or the ability to object and hear legal complaints if the use of the data could harm us. One of the hallmarks of freedom is autonomy and control over our lives, and we cannot have that when so many important decisions about us are made in secret without our knowledge or participation.

(vii) **Freedom of thought and speech**

Privacy is the key to freedom of thought. Looking closely at everything we read, or watch can convince us that we are looking for ideas outside the mainstream. Privacy is also key to protecting speech from unpopular messages. And privacy doesn't just protect assistive features. We may want to criticize people we know to others, but we don't share the criticism with the world. A person may want to explore ideas that their family, friends or co-workers don't like.

(viii) **Freedom of Social and Political Activities**

Privacy helps protect our ability to communicate with others and participate in political activities. A key component of freedom of political association is the ability to do so in private if desired. We protect voting privacy because we worry that if we don't, people's true consciences will be cold when they vote. The privacy of the relationships and activities leading up to the polling station is also important, because this is how we form and discuss our political beliefs. A careful eye can interfere with these functions and affect them unnecessarily.

(ix) **Ability to Change and Have Second Chances**

Many people evolve and develop throughout their lives; they are not static. Having a second opportunity, being able to get past a mistake, and being able to reinvent oneself all have significant value. This talent is fostered by privacy. It enables individuals to develop and mature free from being constrained by all the foolish things they may have done in the past. There are certainly some wrongdoings that shouldn't be protected, but there should be some since we want to promote and support development.

(x) Freedom to act as per own will without justifying one's actions.

A major reason privacy is important is not having to explain or justify yourself. We can do many things that may seem strange, clumsy, or worse when viewed from afar by others who lack full knowledge or understanding. It can be a heavy burden to constantly wonder how others see everything we do and be willing to explain.

V. Conclusions

Abraham Maslow's pyramidal "Hierarchy of Needs" model is a highly-influential way of organizing human needs from the most "basic" to the most advanced. Maslow's argument is that the most basic needs must be met before people can move "up" to the more advanced needs. According to Maslow, people should start fulfilling their needs at the bottom of the pyramid, which includes the first level of the hierarchy. People cannot achieve the needs higher up the pyramid until they have taken care of the ones below. Maslow argued that it is only after meeting all five needs that humans can truly thrive. First Level of the hierarchy includes physiological needs, which are basic needs for survival, such as food, water, warmth, sleep, etc., and maintenance of homeostasis. Similarly, they move in a hierarchy of scaling that includes security, love and belonging, esteem and fulfillment. Humans must fulfill their needs in whole or in part through the hierarchy.

Privacy empowers individuals to make choices and decisions without external interference. It allows people to control their personal information, thoughts, and actions, promoting autonomy and freedom. Maintaining privacy fosters trust in technology and the digital ecosystem. When individuals trust that their personal data is secure, they are more likely to engage with technology and share information, leading to a more robust and innovative digital society.

References

1   https://www.cobalt.io/blog/mapping-cyberattacks-maslows-hierarchy-of-needs

2   https://courses.lumenlearning.com/wm-introductiontobusiness/chapter/maslows-hierarchy-of-needs/

3   https://teachprivacy.com/10-reasons-privacy-matters/

4   https://unacademy.com/content/cbse-class-12/study-material/business-studies/maslows-hierarchy-of-needs/

5   https://www.cobalt.io/blog/mapping-cyberattacks-maslows-hierarchy-of-needs

6   https://courses.lumenlearning.com/wm-introductiontobusiness/chapter/maslows-hierarchy-of-needs/

7   https://teachprivacy.com/10-reasons-privacy-matters/

8   https://canadacollege.edu/dreamers/docs/Maslows-Hierarchy-of-Needs.pdf

9   Maslow A.H. A theory of human motivation. *Psychol. Rev.* 1943; 50(4):370–396.

Chapter 12

Analyzing Data Protection Laws in the context of WhatsApp Privacy Policy Updates

Sangeeta Chakravarty
Ph.D Research Scholar
Bennett University
Greater Noida, UP, India

Abstract

Everyday technological developments are taking place, and more and more data is being shared virtually. Over-the top platforms are very popular in currents times and more threat lies in data sharing via Over-the-top messaging apps through smartphones. In K.S. Puttaswamy case, the Supreme Court recognized a fundamental right to privacy of every individual guaranteed by the Constitution, within Article 21 in particular and Part-III on the whole and also laid down a positive duty on the Government to legislate or enact laws on privacy issues. WhatsApp is an OTT platform for messaging apps through smartphones. After being sold to Facebook there has been drastic changes in its privacy policy. WhatsApp is a free of cost service-based app that let users share text messages, videos and images and also used for business communications. In order to generate revenue, its parent company Facebook shares WhatsApp data to third party advertising companies.

In Europe GDPR i.e., General Data Protection Regulations govern these issues pertaining to data protection. In India, in the absence of Data Protection Law WhatsApp has not implemented its new privacy policy till the time data protection laws are enforced. This is a doctrinal method of research to study and analyse the data protection laws of Europe specially GDPR and how it dealt with the data sharing of users' data by WhatsApp to its parent Company Facebook. Also draw legal implications by comparing it with Indian Data Protection Bill 20019.

Keywords: Data, privacy, fundamental right, Article 21, India.

I. Introduction

While the enactment of Personal Data Protection Bill 2019 is still awaited, WhatsApp an Over-the-top messaging app through smartphones[1] updated its privacy policy. Meta Inc.[2] is Parent Company of WhatsApp LLC located in California U.S.A.

In February 2014, Facebook Inc. paid a staggering $19 Billion to acquire WhatsApp LLC. For its customers, WhatsApp offered a free, secure messaging platform via smartphones. WhatsApp's revised privacy policy also reflects its intention to monetize its model for commercial and "advertisement-targeting purposes." This quote perfectly captures the state of affairs now. You are the product if you don't pay for it.[3]

The concern over WhatsApp's privacy policies and data sharing practices with Facebook began as early as 2016. WhatsApp made a change to its privacy policy in that year, allowing it to share some user data with 'Facebook' for a variety of uses, including targeted advertising. Users of WhatsApp had thirty days to choose not to employ this sharing mechanism; nevertheless, beyond that time, data sharing would still take place. The Delhi High Court heard a case challenging the policy because it violated the privacy of citizens.

The Delhi High Court did not strike down the policy but set a deadline (September 25, 2016) and ordered that all data gathered prior to that date be erased and not shared with Facebook. Whatsapp complied with the court's directive. By way of a complaint alleging abuse of dominance, the privacy policy was also contested before the Competition Commission of India (CCI). The CCI ruled in support of WhatsApp's.

In 2021 there were more changes in the privacy policy. Both data regulatory authorities and privacy experts had raised concerns about the extensive data collecting and sharing methods introduced by this update. The most recent modification to WhatsApp's policies, however, has resulted in a mass exodus because users are concerned that Facebook would have access to their private communications. Despite the fact that most users of digital sites accept privacy and cookie agreements without even reading them, this still holds true. WhatsApp has made it clear that only Business Accounts would have their data collected; nevertheless, private chats will be secured by end-to-end encryption.

II. Various Features of WhatsApp

WhatsApp has the following features:

(i) Business WhatsApp

In order to support small businesses, WhatsApp Business, an enterprise app, was released in January 2018. So, in January of 2018, WhatsApp released WhatsApp Business, a commercial app aimed at helping out small businesses. This Programme was developed to help businesses communicate with their customers or clients. Additionally, WhatsApp Enterprise Solution was created by Facebook for international corporations with a worldwide clientele. They could do ecommerce or offer customer support services online through WhatsApp communication services.

(ii) Using WhatsApp Pay

According to this approach, WhatsApp Payment is an exclusive payment processing service for India. With the help of banking partners, WhatsApp offers its users the opportunity to make payments and conduct other financial operations. In exchange for providing such services, the firm could receive a commission.[4]

(iii) Message Encryption

This is the function that will probably help WhatsApp win the trust of your users. With the use of this feature, users can send encrypted communications that are sent straight to the intended recipient.

People will attempt to obtain the message content and modify it to their advantage if there are any fraudulent activities occurring in the area. WhatsApp has guaranteed that it would not take place.

Without the user's permission, the encryption ensures that messages sent through the app are not exposed or leaked to third parties.

WhatsApp Users Worldwide as of June 2021 in Ten Countries with Maximum Users in Millions

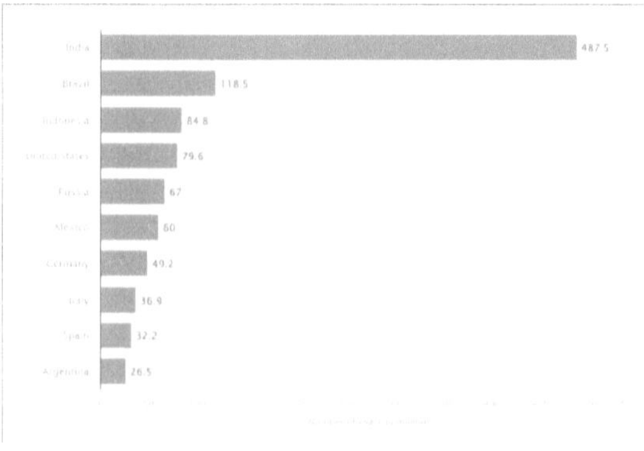

Source: L. Ceci, Global WhatsApp users in Selected Countries, 25.may.2022, https://www.statista.com, (retrieved on 1 Aug.2023)

III. One Sided Bargaining Contracts

When it comes to bargaining power between individuals and enterprises that process personal data, there exists a noticeable inequality as enterprises are at leverage over individuals.

There were allegations that WhatsApp by 'take it or leave it' approach in its privacy policy updates is offering one sided bargaining contract to its users and as consumers no better option is left than accepting the terms and continue using. One sided bargaining contract are against the interest of consumers.

IV. Network Effects

The effect takes place when the advantage that a consumer receives grows in proportion to the increase in the number of other consumers; in other words, the utility grows as the size of the network expands. If we take a messenger app as an example, the apps only valuable if a large number of people utilise it to communicate with one another. On the other side, the application is pointless if nobody is familiar with it. Therefore, the value of the platform will increase in proportion to the growth of the user base of the messaging app. There are two kinds of network effects: direct network effects, in which the value of goods or services rises as the number of users grows, for example, WhatsApp, Uber, etc. and "indirect network effects", in which the value of the network rises as the "number of users" of a good encourages the development of more complementary goods or services. Some of the examples of indirect network effects are OLX, Nykaa, Amazon, etc.

V. Terms of WhatsApp's Updated Privacy Policy

WhatsApp announced an updated version of their privacy policy in January 2021. Existing WhatsApp users received notice, and their sole option was to agree to the revised privacy rules. Users were given deadlines to meet in order to continue using the app. The updated policy made it crystal clear that WhatsApp will compile data namely Usernames , Profile image, Phone numbers saved in the address book, IP address, and user phone numbers, Status details, such as the most recent time a user was online, Transaction and payment data, if enabled, Usage and log information and Device information; Location-related information.[5]

The privacy statement most importantly stated that WhatsApp will permit Meta firms and other third parties to view messages that users share with businesses on the platform.[6] Despite this, WhatsApp will not restrict the operation of the app for users who choose not to agree to the new and modified terms of service and privacy policy agreements.

The current view held by WhatsApp is that the current situation should continue until India's data protection law takes effect. Earlier, WhatsApp made it very clear that any user who does not agree to these terms of service will first receive repeated reminders, after which functionality will start to be banned. This includes limiting access to your WhatsApp chats and finally your calls. While India has been adamantly opposing the new privacy policy, German regulators have subsequently halted its implementation, and the European Union (EU) is also considering doing the same.

WhatsApp has stated that it will not restrict functionality for users who disagree with its new privacy policy but will keep reminding them of the upgrade. This will continue to be in place until the new data protection law is implemented. WhatsApp previously stated that customers who reject the new privacy policy will gradually lose the ability to make and receive calls as well as texts.

The Facebook-owned instant messaging service explains how it handles user data and works with Facebook to make it possible for integrations with other Meta platform apps. In order to provide services and personalize the user experience, according to WhatsApp, user data is obtained.

So here, in the garb of providing better services even the contact details of those who are not using the messaging app but, just because, their phone number is saved on the address book of one of the users are also being shared with the WhatsApp, as user has consented to share others details also of which he never had the right in the first place.

VI. Concerns

The Policy raises a number of concerns -

(i)   WhatsApp will continuously collect the user's precise location information, IP addresses, and other information to estimate their general position and average travel location. WhatsApp will share user data with Meta-owned companies and third parties.

(ii) Will continuously gather the user's specific location information, "IP addresses", and other information in order to estimate their general position as well as their normal travel destination. This information might be utilized for ad targeting on other platforms, such as Instagram.

(iii) It violates the data localization principle by allowing the transmission of personal data outside of the countries.

(iv) Using the information for purposes other than those for which it was gathered with consent. does not adhere to the "principle of purpose limitation."

(v)   The District Court in California is specified as being the jurisdiction for any disputes that may arise.

(vi) If users disagree with the revised privacy policy, they will be forced to cease using WhatsApp or have their functionality restricted (take it or leave it approach). There are hardly any options left for users to reject such a policy.

(vii) WhatsApp is popular and dominant over the top platform for instant messaging through smart phones in India so other options for instant messaging through smart phones are very less. Especially when most of the people are using WhatsApp it becomes difficult for the user to switch to other platforms to share and exchange data, thereby reducing consumer choices in the relevant market.

(viii) WhatsApp by integrating data with Facebook and Instagram might result in anti-competitive agreements causing Appreciable adverse effects in the market by leaving less consumer choices.

(ix) WhatsApp being in dominant position in the market if over the top messaging app through smart phones might abuse its dominant position by sharing more data to advertisers in a concealed way, as non-price competition is involved here.

(x) Mergers of Facebook with WhatsApp and Instagram leaves the apprehension of network effect and consumer harm, and such combinations needs to be checked by Competition Commission of India.

(xi) Here data being shared with other apps is non-price competition element resulting in consumer harm. It also raises privacy concern.

VII. Actions Taken by The Government

The Government of India under the direction of Supreme Court, requested following details regarding data processing from Meta highlighting Puttaswamy judgement:

(i) To retract the most recent terms and privacy policy that the messaging service had put forward for users in India.

(ii) To respect Indian users' data security and personal information privacy. To complete a questionnaire to learn more about its business practices and data-sharing policies.

(iii) Specifics regarding the type of information that WhatsApp collects from users in India.

(iv) Details about the information that WhatsApp in India collects from users.

(v) With regard to the user consent that was sought, the permissions that were asked for, and the importance of each of these elements in light of the operation and the particular service that was provided.

(vi) To describe the distinctions between its privacy policies in India and those in other nations.

(vii) To clarify whether WhatsApp profiles Indian users based on their app usage and the specific type of profiling used.

(viii) Information on WhatsApp's data security, information security, cyber-security, privacy, and encryption policies, as well as information on how WhatsApp's privacy policies differ in India compared to those in other countries; this includes information on how WhatsApp's privacy policies differ in other countries.

In their statement to the High Court of Delhi, the "Ministry of Electronics and Information Technology" of "the Government of India" raised privacy concerns- the categories of sensitive personal data that are being gathered, the fact that there is no way to review or change the information or withdraw consent after the fact, the fact that the information is shared with third parties, including other meta companies, no distinction between personal data or sensitive data collected etc. The Information Technology Act, 2000's Rule 6(4) [7] forbids the disclosure of data acquired by a third party from a body corporate, and the Government of India has called attention to this breach. [8]

VIII. Data Protection and Importance of Right to Privacy

Hondius[9] describes data protection as *"the body of law which secures for every individual, whatever his nationality or residence, respect for his rights and fundamental freedoms and in particular his right to privacy, with regard to automatic processing of personal data relating to him"*. Data protection includes governing and regulating data by Sovereign States. Data protection is challenging for States as data is a borderless asset that can be accessed from anywhere in the world using technology. Data Protection Bill aims to safeguard the fundamental rights and freedoms of a data subject.

Author Lee Bygrave contends that using legal means to protect data is not always necessary. Although it has been associated with a specific form of legislation since the 1970s, it has now attained a level of generality that transcends such legislation. He defines data protection as a set of (legal and/or non-legal) measures intended to protect people from harm caused by the processing of (manual or computerised) information about them and embodying a set of principles on such processing.[10] In the context of this research, the term "data protection" refers to the legal protection afforded to an individual (referred to as the data subject) against the processing of data concerning that individual by a third party or an institution.

Edward Snowden, a former employee of the CIA, revealed to the Washington Post and the Guardian that the National Security Agency (NSA) has been using Microsoft, Google, Apple, Yahoo, Facebook, Skype, and YouTube to spy on users all over the world and gain access to their private information. Snowden made his revelations in interviews with both publications. Spying on foreign communications that transit via computers in the United States is the purpose of the PRISM programme, which has been operational since 2007.

The processing of 'personal data' requires to be done in a fair and lawful manner, according to the fundamental premise of data protection regulations. In this context, fairness also means that individuals are not coerced into providing personal information to a data controller or into agreeing to the controller's intended use of that information. Inferences of fairness implying safeguards against monopoly abuse by data controllers can be drawn from this. The Predominant challenge before the regulatory authorities, would be relating to the jurisdictional conflict of data protection laws around the world and to safeguard fundamental rights. There is a need for digital sovereignty when it comes with data protection of citizens in India.

The term "digital sovereignty" refers to a state's right to control its network in order to further its own national interests, the three most crucial of which being security, privacy, and trade.[11] So from the use of word privacy it could be inferred that how enterprises are dealing with personal data of individuals needs to be governed too.

Table 1: Data Protection Laws in other Countries

S. No.	Territory	Governing Law
1	European Union	The General Data Protection Regulation (GDPR)
2.	California	The California Consumer Privacy Act [CCPA]
3	South Africa	Protection of Personal Information Act [POPI Act]
4	India	The Data Protection Bill,2019 (pending)

Privacy is an old concept. First recognized in 1604 in **Semayne case**.[12] It was recognized that the house of everyone is to him as his castle and fortress. Semayne case sets a good precedent for territorial privacy. In the classic work known as the right to privacy, which was published in 1890, Justice Louis Brandeis and Mr. Samuel Warren, a lawyer from Boston, argued that the law should recognise this right and impose liability on those who violate it. In the article, they claimed that the law should impose liability for any incursions into the private sphere, as preservation of the private sphere is the foundation of individual freedom in the modern day.[13]

Both Article 12 of the Universal Declaration of Human Rights and Article 17 of the International Covenant on Civil and Political Rights explicitly identify the right to privacy as a fundamental human right. The right to privacy was universalized in Article 12 of the UDHR. As a result of the protection provided by the UDHR, a number of nations developed a greater appreciation for personal privacy and codified privacy protections in their own legal systems. Even though the right to privacy was acknowledged internationally as an essential component of human dignity as early as 1948, it was not until the Puttaswamy verdict in 2017 that the term "privacy" was included in Article III of the Constitution.

IX. Right to Privacy in Context of Puttaswamy Judgement

Many nations have enacted data protection legislation in response to the issue of the invasion of an individual's privacy caused by the processing of personal information. It is necessary for the framework for the protection of data to have a relationship between the data principal and the data fiduciary that is founded on the fundamental expectation of trust. In the Puttaswamy judgement, these different types of privacy breaches as well as the principles of decisional autonomy were reinstated.

The Supreme Court of India stated that the concept of privacy, defined as "the right to be let alone," is predicated on the idea that one should be afforded a separate area in which to conduct one's own affairs. The concept of privacy gave each person the power to claim and regulate the aspects of their humanity that were intrinsic to their individual personalities. Privacy, the Court ruled, "protected the individual from the searching glare of publicity in matters which were particular to his or her life" and "recognised the heterogeneity, the right of the individual to be distinct and to stand against the tide of uniformity in creating a zone of solitude.

Privacy has always meant "freedom from governmental intervention" in the context of public law. However, as a result of developments in science and technology as well as other influences, the concept of privacy has taken on a broader connotation in today's society. In 1888, Justice Thomas Cooley used the expression the right to be let alone in general meant Privacy. According to him, "the right to respect for private life was the right to be let alone".

In the case law of **Ram Jethmalani vs. Union of India**[14], The right to privacy, as recognised by the Supreme Court, is an essential component of the right to life, which encompasses those spheres of individual autonomy that are shielded from governmental intrusion except in the case of criminal activity.

In the famous case law of **South Africa Court in National Media Ltd vs. Jooste**, [15]Justice Harms defined *"privacy as an individual condition of life characterized by exclusion from the public and publicity. The condition embraced all those personal facts which a person concerned had determined him to be excluded from the knowledge of outsiders and in respect of which he had the will that they be kept private"*.

So Supreme Court in, **K.S. Puttaswamy (Retd.) vs. Union of India**[16], examined whether the right to privacy is a part of the right to life and personal liberty protected by Article 21 of the Constitution. The Supreme Court has ruled that privacy is a fundamental human right. The protection of one's right to privacy is a guarantee of one's autonomy in making important life decisions. Personal intimacies (marriage, procreation, and family), including sexual orientation, were also cited as being fundamental to a person's worth. According to the document, one's right to privacy is an expression of one's independence, dignity, and uniqueness.

Therefore, in this specific case, the Aadhaar Card Scheme was contested on the grounds that it violates Article 21 of the Indian Constitution's fundamental right to privacy by gathering and combining demographic and biometric data on the population for use in a variety of contexts. Due to divergent legal precedents regarding the constitutional status of the right to privacy, the case was brought to a constitutional bench of (nine) judges. The petitioners argued that the right to privacy is a fundamental freedom that goes hand in hand with personal freedom and dignity.

In contrast, the Union of India argued that the right to privacy is not a constitutionally protected fundamental right. The principal defence of the UOI was that:

(i)   The "right to privacy" would have been specifically stated in the Constitution if the Constitution's creators had meant for it to be a basic right.

(ii)   The idea of privacy is subjective and hard to define. It might be challenging to articulate exactly what is meant by the term privacy. An idea as nebulous as that cannot be elevated to the status of a fundamental right.

(iii)   Existing legal provisions offer persons an adequate level of protection against invasions of privacy; and

(iv)   the right to privacy is a valid claim that is supported by common law, but an individual claim like this cannot be elevated to the status of a fundamental right.

The Supreme Court of India, in deciding whether or not privacy is a fundamental human right, reached the Conclusions unanimously (with a Bench of nine judges). In a unanimous (nine judge bench) decision, the Supreme Court of India ruled that privacy is a fundamental individual right. In the modern era of rapid technological development, threats to privacy can come from either the state or non-state actors; therefore, the right to privacy can be asserted against either. One aspect of the right to privacy recognised by the courts is protection against having one's personal information disclosed without consent. A person's right to privacy and autonomy online may be compromised if their personal data is used without their consent.

In this landmark case on August 24, 2017, a nine-judge panel of India's Supreme Court confirmed the constitutionality of the "right to privacy" under "Article 21" of the "Indian Constitution". Privacy was ruled to be a fundamental human right and should be included in Part III of the Indian Constitution. Additionally, the Court ruled that privacy is not an absolute right, and that any state or non-state actor invading private must meet the following three criteria: Legitimate Aim, Proportionality and Legality.

X. Legal Framework akin to Data Protection

We do not currently have any explicit laws dealing with data protection and privacy. The Information Technology Act of 2000, which has some pertinent provisions pertaining to the protection of data, is however serving the purpose.

(i) **Information Technology (Reasonable Security Practices and Procedures and Sensitive Personal Data or Information) Rules 2011**

On April 11 in 2011, a gazette was published by the Ministry of Communication and Technology. The Information Technology (Reasonable Security Practices and Procedures and Sensitive Personal

Data or Information) Rules were finally authorised for implementation in 2011. The term "sensitive personal data" is given a legal definition under this act. Information such as passwords, biometric medical data, sexual orientation, or specifications of financial instruments such as debit/credit card or bank account information are examples of the types of information that fall under this category.

On the other hand, it does have a provision for an exemption, which stipulates that any material that is either publicly accessible or available in the public domain, or if it is covered by the Right to Information Act of 2005, is not subject to the restriction. If this is the case, the information in question is not considered to be sensitive personal data.

(ii)   Information Technology Act, 2000

Both cybercrime and cybersecurity are addressed under the IT Act. Various parts of the document address cybercrime and data breaches and lay forth procedures that businesses should follow when dealing with their clients' private information.

In addition to defining cybercrime, it also describes the punishments that can be given for offences of this nature, such as Section 72 of the Information Technology Act of 2000. This outlines the consequences that would be incurred in the event of a breach of privacy or confidentiality. It stipulates that anybody in a position of power or authority with access to sensitive data may release such information in the form of an "electronic record", but only with the consent of the individual whose personal data is being compromised.

In accordance with Section 72, the person in charge of such data may face imprisonment of up to two years, a fine up to ten lakh rupees, or both, in case there is a breach.[17]

(iii)   The Personal Data Protection Bill, 2019

Personal Data Protection Bill, 2019 includes 98 sections and fourteen chapters. The primary objective of the Bill is safeguarding of an individual's right to privacy. It covers a wide range of issues, some of which include the restrictions on the transfer of personal data outside of India and the obligations of data fiduciaries.

"Article 51 of the Constitution of India, which is also a part of the Directive Principles of State Policy, requires the state to endeavour to foster respect for international law and treaty obligations in the dealings of organized people with one another".

Article 12 of the Universal Declaration of Human Rights and Article 17 of the International Covenant on Civil and Political Rights both clearly recognize right to privacy as a fundamental human right ("ICCPR"). The ICCPR is recognized as a human rights instrument by the Protection of Human Rights Act of 1993, and it requires nations to take action to ensure the realization of these rights and protection from interference by private parties.

The duty on Government to legislate on privacy issues which was laid down in Puttaswamy Judgement. Data protection Bill was designed towards confirming the right to privacy.[18] There are various forms of privacy violations, and one form is right to information and privacy. Under Data protection Bill one can challenge two sets of constitutional rights one is privacy and other is equality. User rights are involved when it comes to data protection.

This Bill lays down a positive duty on government to legislate on data protection. Privacy has been upheld to the status of Fundamental Right in the Puttaswamy judgement. The objective of Bill is to ensure interest and security of State. This Bill has two main goals: one is to preserve individuals' privacy, and the other is to unleash the potential of the data economy. Data fiduciaries are person who are using the data has a duty to notify that a data breach has taken place. It aims to secure various right like right to be forgotten, right to correction and eraser etc. Right to be forgotten was only regarding publication but now it is stopping someone from processing. This act tries to bring transparency in processing of data.

(iv)   Salient Features of Personal Data Protection Bill 2019

The significant features of Personal Data Protection Bill 2019[19] are as under -

a.   Strives to safeguard personal data. Data regarding or pertaining to a *natural person* who is directly or indirectly identifiable is what it defines as personal data. Data that is not personal is referred to as non-personal data.

b.   Provides for compensation against harmful processing of personal data. Harm has been defined in the form of an exhaustive list. It includes: bodily or mental injury, financial loss, denial of service/benefit, identity theft, discrimination, and unreasonable surveillance[20].

c.   Puts a duty on Data fiduciary to notify Data Protection Authority regarding any data breach.

d.   Exemption to State authorities regarding processing of personal data if expedient or in the interest of national security or public order.

e.   A data principal has the *"right to receive his personal information"* when it has been "processed automatically" and the right to prevent the continued "disclosure of personal" information that is no longer, required for the purposes for which it was obtained.

f.   Selection Committee to make recommendations for appointment of Data Protection Authority.

In the context of, the "International framework for data protection", renowned author Lee A. Bygrave has laid out the basic ideas that go along with it: Single statute legislation to ensure clarity and coherence; independent enforcement body to oversee the implementation of legislation, broad framework of laws to enable smooth adoption of modifications in line with the changing needs of technology and innovation , and advisory body for regulator to aid effective understanding and implementation of laws by the enforcement body.

XI. GDPR and Data Protection

The General Data Protection Rule, also known as GDPR, is a regulation that was created with the intention of enhancing the level of protection afforded to individuals' personal data inside the European Union. Beginning on May 25, 2018, it will be legally binding. GDPR makes it possible to establish rules for the lawful processing of data and requires that, in the event of doubt, the legality be verified, and consent be obtained.

(i) Pre-Requisites

One of three bases for validity must be met before personal data can be processed:

a. A contract's performance necessitates the collection of data (Article 6, section 1b GDPR). As a result, the information is necessary to fulfil, "a contract between the user and the company".
b. Data processing is justified by legitimate interests (Article 6, section 1f DSGVO[21]).
c. Users give their permission for their data to be processed. (DS-GVO, Article 6 section 1a). Objection options (opt-out) are insufficient in this case, hence active permission is required. Because the first two points can be interpreted in different ways, it's best to

choose the route via the user's consent if you're in question. This puts Messenger conversation on a legal footing.

(ii) Six Principles of Privacy

Wrapped up in every article of the GPDR there are the six privacy principles. These principles arrive early in the legislation at Article 5(1) and include: Lawfulness, Fairness, and Transparency; Limitations on Purposes of Collection, Processing, and Storage, Data Minimization, Accuracy of Data, Data Storage Limits, Integrity and Confidentiality.

These privacy principles show what the GDPR is all about. Article 5(1) provides a means to check if you are not sure if your data privacy practices meet the standards set in later articles. The two words that best describe the GDPR are accountability and transparency. The six privacy principles protect and keep both of these in place. The person in charge of the data must make sure that it is:

"Processed lawfully, fairly, and in a transparent manner in relation to the data subject."[22]

XII. Conclusions

So elaborate study and critical analysis of WhatsApp's updated privacy policies reveals that WhatsApp generates revenue through WhatsApp Business, WhatsApp pay and third-party advertising companies. The mode of obtaining consent *prima facie* does not appear to be fair. Specially after takeover by Meta, WhatsApp's privacy policy has tried to gain coerced consent from its consumers. One sided bargaining contract, undue gains from network effects, use of vague and ambiguous terms by such a global enterprise pose a threat to privacy.

The fact that if consumers do not want to agree with such terms can switch to other platform like Signal, WeChat, Telegram etc. is not practically conducive because it does not work with other apps unlike it is in mobile telecommunications. All related users have to be on WhatsApp in order to communicate with each other.

So, there is a need for strict data protection laws in India to safeguard fundamental right to privacy. Data protection laws should be such that it brings transparency in the process of data sharing and thereafter processing of data. Users' data should be used as they have consented and is necessary for the working of the app. There should be same policy in other countries also. WhatsApp should not be allowed to disrespect India users' privacy by releasing different kind of policy in European countries and India. GDPR bring private as well government entities within its ambit for breach of personal data. In India, WhatsApp has agreed to not to implement its policy till data protection laws are enacted.

Endnotes

1   Here in after over-the-top messaging app through smartphones will be referred as messaging app.
2   Earlier known as Facebook Inc. now it has become Meta Inc.
3   Tomas Laurinavicius, 'If you are not paying, you're the product, Life designed, 22.Jan.2021, https://tomaslau.com/blog/you-are-the-product, (retrieved on 5.Aug.2022)
4   Launch of WhatsApp Pay was challenged in the "Competition Commission of India". Where CCI has upheld the launch of WhatsApp pay as not being anti-competitive.
5   Tirthesh Shah, "'WhatsApp privacy policy: The controversy so for alarming the need of data protection law in India'", DevDiscourse, 9. Aug.2021, "https://www.devdiscourse.com/article/technology/1685916-whatsapp-privacy-policy-the-controversy-so-for-alarming-the-need-of-data-protection-law-in-india," (retrieved on 7. Jul.2022)
6   Ibid
7   *"Rule 6: Disclosure of information*

(1) Disclosure of sensitive personal data or information by body corporate to any third party shall require prior permission from the provider of such information, who has provided such information under lawful contract or otherwise, unless such disclosure has been agreed to in the contract between the body corporate and provider of information, or where the disclosure is necessary for compliance of a legal obligation:

(4) The third party receiving the sensitive personal data or information from body corporate or any person on its behalf under sub-rule (1) shall not disclose it further."

8   Richa Banka, 'Government wants WhatsApp privacy policy blocked', "Hindustan Times", 20.Mar.2021, "https://www.hindustantimes.com/india-news/govt-wants-whatsapp-privacy-policy-blocked-101616199308268.html", (retrieved on 7.Jul.2022)

9   Hondius 1983 Neth Int LR 103, (retrieved from Shod Ganga)

10   Data security is not the same as data protection. Data security safeguards the interests of data controllers, processors, and users, whereas data protection protects individuals' privacy against data processing.- Lee Bygrave (Retrieved : Anneliese Roos, The law of data (privacy) protection: a comparative and theoretical study, 2008-2009,"University of South Africa,Institutional Repository,Pretoria, http://hdl.handle.net/10500/1463, (Jul.2022))

11   Akanksha Prakash, What is the purpose of Data Protection Law in India?, Business Today, 9.Jul.2021, https://www.businesstoday.in/opinion/columns/story/what-is-the-purpose-of-data-protection-law-in-india-300925-2021-07-09, (retrieved on 19.Jul. 2022)

12   Semayne v Gresham [1604] Yelverton 29, https://www.dealingwithbailiffs.co.uk/Semayne-v-Gresham1604.htm, (retrieved on 10. Jul.2022)

13   Ibid 46

14   (2011) 8 SCC 1

15   National Media Ltd. and Another v Jooste (335/94) [1996] ZASCA 24; 1996 (3) SA 262 (SCA); [1996] 2 All SA 510 (A); (26 March 1996),South African Supreme Court of Appeal, http://www.saflii.org.za/za/cases/ZASCA/1996/24.html, (retrieved Jul.2022)

16   Justice K. S. Puttaswamy& Others. vs. Union of India & Others, Writ Petition, (Civil) No. 494 of 2012 decided in 2017.

17   Section 72, The Information Technology Act, 2000, Act no. 21 of 2000

18 Lalit Panda, 'Decoding India's Data Protection Bill: Day 2', MediaNama, 19-20. Jan.2022, https://www.youtube.com/watch?v=wzPPhgCfAc4, (retrieved on 5. Jul.2022).

19 Hereinafter referred to as PDPB

20 Personal Data Protection Bill

21 "The DSGVO (*Datenschutz-Grundverordnung*) is a privacy regulation" that came into effect in "May 2018", in Berlin, Germany. "It's called the GDPR (General Data Protection Regulation) in English".

22 Article 5 (1) GDPR.

Chapter 13

Data Protection Dynamics: Insights into India's Evolving Legal Framework

Shailendra Saxena
Senior Manager
HCL Technologies
Noida, UP, India

Dr. Sanchita Ray
Assistant Professor (Political Science)
SSOL, Sharda University
Greater Noida, UP, India

Abstract

Data protection refers to the set of practices, policies, and measures put in place to safeguard the confidentiality, integrity, and availability of data. It involves protecting data from unauthorized access, loss, or theft and ensuring that it is used only for legitimate and authorized purposes. Data protection is essential in both personal and organizational contexts, considering the growing reliance on digital data and the potential risks associated with its mishandling. data protection is essential for preserving individuals' rights, maintaining trust between organizations and their customers, complying with laws and regulations, and safeguarding sensitive information from misuse or unauthorized access. It is a critical aspect of responsible data management in the modern digital world. Data protection regulations vary from country to country, and each nation has its own set of laws and guidelines concerning the handling and protection of data. Each country's data protection laws are influenced by its legal and cultural context, but the overall objective is to protect individuals' privacy rights and promote responsible data management by organizations. As technology continues to advance, many countries are updating and strengthening their data protection frameworks to address new challenges and protect their citizens' data more effectively. This chapter deals with the importance of data protection and India's constant effort towards data protection.

Keywords- Data Protection, GDPR, Data Security, Data Privacy, Information Technology Act of 2000.

I. Introduction

Data protection is the process of protecting sensitive information from damage, loss, or corruption. The importance of data protection increases as the amount of data created and stored continues to grow at unprecedented rates. There is also little tolerance for downtime that can make it impossible to access important information. Consequently, a large part of a data protection strategy is ensuring that data can be restored quickly after any corruption or loss. Protecting data from compromise and ensuring data privacy are other key components of data protection. The coronavirus pandemic caused millions of employees to work from home, resulting in the need for remote data protection. Businesses must adapt to ensure they are protecting data wherever employees are, from a central data centre in the office to laptops at home.

Although some businesses use the terms data protection, data security and data privacy, they have different purposes. ***Data protection*** safeguards information from loss through backup and recovery. An example of data protection is backing up your data, so if data is corrupted or deleted due to a disaster or a cyberattack, it can be recovered or restored. ***Data security*** refers specifically to unauthorized access or use where it could be leaked, deleted or corrupted. An example of data security is the use of encryption to prevent hackers from using your data if it is compromised. ***Data privacy*** refers to concerns about how data is processed, including data sensitivity, regulatory requirements, consent, and notifications. An example of data privacy is the use of a separate, secure database for personally identifiable information (PII).

II. Principles of Data Protection

The key principles of data protection are to safeguard and make available data under all circumstances. The term *data protection* describes both the operational backup of data as well as business continuity/disaster recovery (BCDR). Data protection strategies are evolving along two lines: data availability and data management. Data ensures users have the data they need to conduct business even if the data is damaged or lost. The two key areas of data management used in data protection are data lifecycle management and information lifecycle management. Data lifecycle management is the process of automating the movement of critical data to online and offline storage. Information lifecycle management is a comprehensive strategy for valuing, cataloguing and protecting information assets from application and user errors, malware and virus attacks, machine failure or facility outages and disruptions. More recently, data management has come to include finding ways to unlock business value from otherwise dormant copies of data for reporting, test/dev enablement, analytics and other purposes.

ESG Data Protection Family Tree

This chart from Enterprise Strategy Group describes the complementary, yet distinct, branches of the data protection family tree.

III. The Purpose of Data Protection

Storage technologies for protecting data include a disk or tape backup that copies designated information to a disk-based storage array or a tape cartridge. Tape-based backup is a strong option for data protection against cyber-attacks. Although access to tapes can be slow, they are portable and inherently offline when not loaded in a drive, and thus safe from threats over a network. Organizations can use mirroring to create an exact replica of a website or files so they're available from more than one place. Storage can automatically generate a set of pointers to information stored on tape or disk, enabling faster data recovery, while continuous data protection (CDP) backs up all the data in an enterprise whenever a change is made.

IV. Data protection and Privacy Laws

Data protection and privacy laws and regulations vary from country to country, and even from state to state -- and there's a constant stream of new ones. China's data privacy law went into effect June 1, 2017. The European Union's General Data Protection Regulation (GDPR) went into effect in 2018. In the United States, the California Consumer Privacy Act supports the right for individuals to control their own personally identifiable information. Compliance with any one set of rules is complicated and challenging.

Personal data can consist of anything from a name, a photo, an email address or bank account details to posts on social networking websites, biometric data or the IP address of a person's computer.

Coordinating among all the disparate rules and regulations is a massive task. Being out of compliance can mean steep fines and other penalties, including having to stop doing business in the country or region covered by the law or regulation. For a global organization, experts recommend having a data protection policy that complies with the most stringent set of rules the business faces, while at the same time using a security and compliance framework that covers a broad set of requirements. The guidelines for data protection and privacy apply across the board and include safeguarding data; getting consent from the person whose data is being collected; identifying the regulations that apply to the organization and the data it collects; and ensuring employees are fully trained in the nuances of data privacy and security.

V. Data protection for GDPR compliance

The European Union updated its data privacy laws with a directive that went into effect May 25, 2018. The GDPR replaces the EU Data Protection Directive of 1995 and focuses on making businesses more transparent. It also expands privacy rights with respect to personal data. The GDPR covers all EU citizens' data regardless of where the organization collecting the data is located. It also applies to all people whose data is stored within the European Union, whether they are EU citizens or not.

(i) GDPR compliance requirements include the following:

These are the requirements of compliance under GDPR -

Barring businesses from storing or using an individual's personally identifiable information without that person's express consent.

a. Requiring companies to notify all affected people and the supervising authority within 72 hours of a data breach.
b. For businesses that process or monitor data on a large scale, having a data protection officer who's responsible for data governance and ensuring the company complies with GDPR.

Organizations must comply with GDPR or risk fines as much as 20 million euros or 4% of the previous fiscal year's worldwide turnover, depending on which is larger.

6 GDPR compliance benefits

GDPR can spur enterprises into adopting practices that deliver long-term competitive advantages. The GDPR, in Recital 1, notes that the protection of personal data is a fundamental right. However, in Recital 4, it says that this right must be balanced with other rights.

Recital 1 states: "The protection of natural persons in relation to the processing of personal data is a fundamental right. Article 8(1) of the Charter of Fundamental Rights of the European Union (the 'Charter') and Article 16(1) of the Treaty on the Functioning of the European Union (TFEU) provide that everyone has the right to the protection of personal data concerning him or her."

Recital 4 states: "The processing of personal data should be designed to serve mankind. The right to the protection of personal data is not an absolute right; it must be considered in relation to its function in society and be balanced against other fundamental rights, in accordance with the principle of proportionality. This Regulation respects all fundamental rights and observes the freedoms and principles recognized in the Charter as enshrined in the Treaties, in particular the respect for private and family life, home and communications, the protection of personal data, freedom of thought, conscience and religion, freedom of expression and information, freedom to conduct a business, the right to an effective remedy and to a fair trial, and cultural, religious and linguistic diversity."

VI. Data Protection Laws in India

In India, the rapid digital transformation and increasing reliance on technology and the internet have undeniably brought about substantial changes in various aspects of society and the economy. This digital growth has not only led to the emergence of numerous opportunities and benefits, but it has also given rise to new challenges, particularly in terms of cybersecurity and data protection.

The expansion of the internet and the growing use of technology have made India a prime target for cyberattacks and data breaches. With a large population and an increasing number of internet users, the amount of digital data generated and collected is substantial. This valuable data, if not adequately protected, can be vulnerable to exploitation, leading to privacy breaches, identity theft, financial fraud, and other cybercrimes.

In India, the issue of data breaches is constantly brought up. The largest data breach, according to the WEF World Economic Forum's Global Risk Report 2019, occurred in India. called the "Aadhaar leak case". The Aadhaar leak case, which exposed the personal information of over 1.1 billion Indian people, was a significant wake-up call to the vulnerabilities in India's data protection framework. Such incidents can have severe consequences for individuals, businesses, and the overall trust in digital services.

According to a report by the Internet and Mobile Association of India (IAMAI, 2020) 'In order to meet the rising demand for digital services and to secure data protection, the country's cybersecurity workforce will require an extra 1 million qualified personnel by 2025,'. Thus, the shortage of qualified cybersecurity personnel, as indicated by the report further emphasizes the urgent need to bolster the cybersecurity workforce in the country. As technology continues to advance, skilled professionals are crucial in defending against cyber threats and securing sensitive data.

There are multitude of laws in India which protects certain aspects of data protection and privacy. These laws include the Information Technology Act, 2000; Indian Contract Act, 1872; Copyright Act, 1957; and Indian Penal Code, 1860.

(i)   Information Technology Act, 2000

The Information Technology Act, 2000 (hereinafter referred to as the "IT Act") is an act to provide legal recognition for transactions carried out by means of electronic data interchange and other means of electronic communication, commonly referred to as "electronic commerce", which involve the use of alternative to paper-based methods of communication and storage of information to facilitate electronic filing of documents with the Government agencies.

(ii)   Grounds on which Government can interfere with Data

Under Section 69 of the IT Act, any person, authorized by the Government or any of its officer specially authorised by the Government, if satisfied that it is necessary or expedient so to do in the interest of sovereignty or integrity of India, defence of India, security of the State, friendly relations with foreign States or public order or for preventing incitement to the commission of any cognizable offence relating to above or for investigation of any offence, for reasons to be recorded in writing, by order, can direct any agency of the Government to intercept, monitor or decrypt or cause to be intercepted or monitored or decrypted any information generated, transmitted, received or stored in any computer resource. The scope of section 69 of the IT Act includes both interception and monitoring along with decryption for the purpose of investigation of cyber-crimes. The Government has also notified the *Information Technology (Procedures and Safeguards for Interception, Monitoring and Decryption of Information) Rules, 2009,* under the above section.

The Government has also notified the *Information Technology (Procedures and Safeguards for Blocking for Access of Information) Rules, 2009,* under section 69A of the IT Act, which deals with the blocking of websites. The Government has blocked the access of various websites.

(iii) **Penalty for Damage to Computer, Computer Systems, etc. under the IT Act**

Section 43 of the IT Act, imposes a penalty without prescribing any upper limit, doing any of the following acts:

a. accesses or secures access to such computer, computer system or computer network;

b. downloads, copies or extracts any data, computer data base or information from such computer, computer system or computer network including information or data held or stored in any removable storage medium.

c. introduces or causes to be introduced any computer contaminant or computer virus into any computer, computer system or computer network.

d. damages or causes to be damaged any computer, computer system or computer network, data, computer data base or any other programmes residing in such computer, computer system or computer network.

e. disrupts or causes disruption of any computer, computer system or computer network.

f. denies or causes the denial of access to any person authorised to access any computer, computer system or computer network by any means; (g) provides any assistance to any person to facilitate access to a computer, computer system or computer network in contravention of the provisions of this Act, rules or regulations made thereunder.

g. charges the services availed of by a person to the account of another person by tampering with or manipulating any computer, computer system, or computer network, he shall be liable to pay damages by way of compensation to the person so affected.

h. destroys, deletes or alters any information residing in a computer resource or diminishes its value or utility or affects it injuriously by any means.

i. steel, conceals, destroys or alters or causes any person to steal, conceal, destroy or alter any computer source code used for a computer resource with an intention to cause damage.

(iv) Tampering with Computer Source Documents as provided for under the IT Act, 2000

Section 65 of the IT Act lays down that whoever knowingly or intentionally conceals, destroys, or alters any computer source code used for a computer, computer programme, computer system or computer network, when the computer source code is required to be kept or maintained by law for the time being in force, shall be punishable with imprisonment up to three years, or with fine which may extend up to Rs 2,00,000 (approx. US$3,000), or with both.

(v) Computer related offences

Section 66 provides that if any person, dishonestly or fraudulently does any act referred to in section 43, he shall be punishable with imprisonment for a term which may extend to three years or with fine which may extend to Rs 5,00,000 (approx. US$ 8,000)) or with both.

(vi) Penalty for Breach of Confidentiality and Privacy

Section 72 of the IT Act provides for penalty for breach of confidentiality and privacy. The Section provides that any person who, in pursuance of any of the powers conferred under the IT Act Rules or Regulations made thereunder, has secured access to any electronic record, book, register, correspondence, information, document or other material without the consent of the person concerned, discloses such material to any other person, shall be punishable with imprisonment for a term which may extend to two years, or with fine which may extend to Rs 1,00,000, (approx. US$ 3,000) or with both.

VII. Amendments as introduced by the IT Amendment Act, 2008

Section 10A was inserted in the IT Act which deals with the validity of contracts formed through electronic means which lays down that contract formed through electronic means "shall not be deemed to be unenforceable solely on the ground that such electronic form or means was used for that purpose".

The following important sections have been substituted and inserted by the IT Amendment Act, 2008: Section 43A – Compensation for failure to protect data, Section 66 – Computer Related Offences, Section 66A – Punishment for sending offensive messages through communication service, etc. (This provision had been struck down by the Hon'ble Supreme Court as unconstitutional on 24th March 2015 in Shreya Singhal vs. Union of India), Section 66B – Punishment for dishonestly receiving stolen computer resource or communication device, Section 66C – Punishment for identity theft, Section 66D – Punishment for cheating by personation by using computer resource, Section 66E – Punishment for violation for privacy, Section 66F – Punishment for cyber terrorism, Section 67 – Punishment for publishing or transmitting obscene material in electronic form, Section 67A – Punishment for publishing or transmitting of material containing sexually explicit act, etc, in electronic form, Section 67B – Punishment for publishing or transmitting of material depicting children in sexually explicit act, etc. in electronic form, Section 67C – Preservation and Retention of information by intermediaries, Section 69 – Powers to issue directions for interception or monitoring or decryption of any information through any computer resource, Section 69A – Power to issue directions for blocking for public access of any information through any computer resource, Section 69B – Power to authorize to monitor and collect traffic data or information through any computer resource for cyber security, Section 72A – Punishment for disclosure of information in breach of lawful contract, Section 79 – Exemption from liability of intermediary in certain cases, Section 84A –Modes or methods for encryption, Section 84B –Punishment for abetment of offences and Section 84C –Punishment for attempt to commit offences.

The Information Technology (IT) Act of 2000, although updated in 2008, is seen as insufficient to address the complexities of the modern digital environment and the growing challenges in data protection and cybersecurity. It lacks the comprehensive provisions and mechanisms required to safeguard data and handle emerging cybersecurity threats adequately.

The Personal Data Protection Bill, 2019 was introduced with the aim of establishing a comprehensive data protection framework in India, addressing concerns related to data privacy and security. This Bill was intended to provide individuals with more control over their personal data and establish rules and regulations for the collection, processing, and storage of personal data by various entities, including government and private organizations. However, the Personal Data Protection Bill 2019, which aimed to enhance data protection and privacy for Indian residents, has faced delays in becoming law. The introduction of the proposed Digital Personal Data Protection Bill, 2022, shows that the government recognizes the need for data protection but also highlights the challenges in reaching a consensus on its final form.

The Digital Personal Data Protection Bill, 2022 introduces a new framework for personal data protection, making it paramount to comprehend and understand its applicability and functions. The Government of India sees this released Bill as one of the parts of its larger vision of a digital economy, this vision will include a comprehensive digital India act that would in course of time replace the existing The Information Technology (IT) Act of 2000.

The delay in passing comprehensive data protection legislation leaves a gap in India's data protection framework and may hinder the country's ability to cope with the evolving cybersecurity landscape effectively. Strengthening data protection laws is essential not only to protect individual rights but also to foster trust in digital services, support the growth of the digital economy, and enhance India's position in the global digital landscape.

As the digital landscape evolves, it's crucial for policymakers, industry stakeholders, and the government to work together to develop and implement effective data protection laws that address the challenges posed by the digital era and safeguard the interests of all stakeholders involved.

VIII. Conclusions

Data protection has become an increasingly critical concern in today's digital age, where vast amounts of personal data are generated and processed. The implementation of data protection measures is essential for upholding individuals' privacy rights and fostering trust between organizations and their customers or users. Data protection laws and regulations, such as the GDPR in the European Union and the upcoming Personal Data Protection Bill in India, play a vital role in setting the standards for data handling practices and empowering individuals to have more control over their personal information.

For organizations, investing in robust data protection measures is not just a legal obligation but also a strategic imperative. By prioritizing data privacy and security, businesses can build strong relationships with their customers, enhance their reputation, and reduce the risk of data breaches or misuse that could lead to financial and reputational damages. Moreover, as technology continues to advance and data becomes even more pervasive, the importance of data protection will only grow. It is crucial for organizations to stay updated with the evolving data protection landscape and proactively adapt their practices to comply with relevant regulations and protect their customers' data.

In the digital era, data protection is not just a compliance checkbox; it is a fundamental aspect of ethical and responsible business conduct. Embracing a privacy-first approach not only benefits individuals and businesses but also contributes to creating a safer and more secure online environment for everyone involved. Data protection in India is an evolving and crucial aspect of the country's regulatory landscape in the digital age. The introduction of the PDPB reflects India's commitment to ensuring data privacy and security while also fostering a conducive environment for digital innovation and economic growth. By empowering individuals with greater control over their personal data and establishing accountability for organizations handling such data, the PDPB seeks to build trust between businesses and their customers.

As technology continues to advance and data becomes a valuable asset, the need for robust data protection measures becomes more critical. India's efforts to enact data protection legislation align the country with global best practices and demonstrate its commitment to protecting citizens' data in line with international standards. While the PDPB is a significant step forward, its implementation and enforcement will be essential to achieve the intended goals of data protection. Regulatory authorities and businesses will need to collaborate to ensure compliance with the new law and build a culture of data privacy and security. Moreover, as India's digital ecosystem continues to evolve, it is essential for the data protection framework to remain agile and adaptable to emerging challenges and technological advancements. Overall, data protection in India is a multi-faceted process that requires ongoing efforts from all stakeholders - government, businesses, individuals, and civil society - to create a safe and secure digital environment where personal data is treated responsibly and ethically. By fostering a privacy-conscious approach and upholding data protection principles, India can bolster its position in the global digital economy while protecting the rights and interests of its citizens in the digital era.

References

1　Duggal Pavan (2016) Data Protection Law in India, Universal law publishing co, Delhi 2016.

2　Duggal Pavan (2015), Indian National Cyber Security Policy-A Legal Analysis, Saakshar Law Publications

3　Eu General Data Protection Regulation (GDPR): An Implementation and Compliance Guide

4　Singh, Shatakshi (2018) Data Protection – Protection of What, Protection from Whom and Protection for Whom – An Analysis of the Legal and Judicial Provisions in India and Abroad. NLIU Law Review, 7, 79-117.

5　Subramaniam, Aditi. (2016). The Privacy, Data Protection and Cybersecurity Law Review. The Law Review. Retrieved from - https://thelawreviews.co.uk/edition/1001264/the-privacy-data-protection-andcybersecurity-law-review-edition-5.

6　Subramaniam, Hari. (May 15, 2017). Data Protection 2017. ICLG. Retrieved from https://iclg.com/practiceareas/data-protection/data-protection-2017/india

7　The Information Technology Act, 2000, No. 2, Acts of Parliament, 2000 (India).Jhingan, Seema and Yadav, Neha. (April 4, 2018). Worldwide: An Overview of Data Protection Laws in India and European Union. Retrieved from http://www.mondaq.com/india/x/687750/data+protection/An+Overview+Of+Data+Protection+Laws+In+India+And+European+Union.

8　Justice K S Puttaswamy v Union of India and Ors, Supreme Court of India, Writ Petition (Civil) No. 494. https://sci.gov.in/supremecourt/2012/35071/35071_2012_Judgement_26-Sep-2018.pdf.

9   KPMG Assets. (2018). Indian Data Protection Regime – Close to Reality? Retrieved from https://assets.kpmg/content/dam/kpmg/in/pdf/2018/08/personal_data_protection_Bill.pdf.

10   Nadeau, Michael. (May 29, 2019). General Data Protection Regulation (GDPR): What you need to know to stay compliant? Retrieved from https://www.csoonline.com/article/3202771/general-data-protectionregulation-gdpr-requirements-deadlines-and-facts.html

11   The new EU Regulation on the protection of personal data: what does it mean for patients? Can be accessed at https://www.eu-patient.eu/globalassets/policy/data-protection/data-protection-guide-for-patients-organisations.pdf

12   https://unctad.org/page/data-protection-and-privacy-legislation-worldwide

13   https://gdpr-info.eu/.

14   Internet Adoption in India ICUBE 2020 June 2021.

Chapter 14

Privacy in an Information-Driven World

Dr. Sandeep Kumar
Associate Professor
University Institute of Legal Studies &
Chairman, Dept. of Buddhist Studies
Himachal Pradesh University
Shimla, Himachal Pradesh, India

Jefrin Johny
B.B.A.L.L.B. Student
SSOL, Sharda University
Greater Noida, UP, India

Tushar Sikarwar
B.A.L.L.B. (Hons.) Student
S.S.O.L, Sharda University
Greater Noida, U.P., India

Abstract

Information flows freely in today's internet driven world. Free flows of information and privacy are contradicted concepts which are required to be balanced well. The rampant information should not breach one's privacy. Sometimes a false information or sensitive information may lead to severe consequences like rumors, social tensions, misperceptions etc. The question arise who is responsible for this false information spread- individual, some organizations or government? The tension between individuals, government and organizations is escalating due to difference in needs, which are often contradictory. Privacy protection through law inevitably conflicts with other important values and imposes costs both financial and otherwise. The variety of potential uses and users of data frustrate any holistic attempt to protect data privacy. A paradigm shift from data maximization to privacy by design may help individuals to protect privacy. Information technology can be used for both to protect and disrupt privacy of an individual. This is very sensitive matter of concern, and it demands high efficiency and intellect while dealing with this. The paper shall analyze the impact of information technology vis-à-vis informational privacy.

Keywords – Technology, privacy, information, data, impact.

I. Introduction

We are going through the era in which the universe of data doubles every two years and quintillions of bytes of data are generated every day[1]. Now as Billions of smartphones and other devices collect and transmit data over high-speed global networks, store data in ever-larger data centers, and analyze it using increasingly powerful and sophisticated software. With the advancement of Artificial Intelligence, technological improvements, advance networking, rapidly growing storage capacity and eventually quantum computing deploy, this data explosion will grow even faster and bigger.

Facial recognition systems offer a preview of the privacy issues that emerge. With the benefit of rich databases of digital photographs available via social media, websites, driver's license registries, surveillance cameras, and many other sources, machine recognition of faces has progressed rapidly from fuzzy images of cats to rapid (though still imperfect) recognition of individual humans. A number of applications require access to various data of individual for effective working of application. This access works in two folds as while it helps to work the particular application efficiently at one hand it has access to personal data of individual.

As artificial intelligence evolves, it magnifies the ability to use personal information in ways that can intrude on privacy interests by raising analysis of personal information to new levels of power and speed.

II. Concept of Informational Privacy[2]

Paul Schwartz conceives of informational privacy as a constitutive value[3] and Daniel Solove sees informational privacy as a right to have one's information "treated thoughtfully," to understand the disclosures of one's personal data, and to participate meaningfully in the use of that data. Informational privacy is the "claim of individuals, groups, or institutions to determine for themselves when, how, and to what extent information about them is communicated to others."[4] Informational privacy could refer at least to the following: information inalienability, disclosure control, and dissemination control.[5] It is distortion to describe every instance of the dissemination of information about an individual as a loss of privacy.[6] Any definition of informational privacy seems, at best, debatable and, at worst, arbitrary.[7] As personal information becomes more important and is accessible to a greater number of people and institutions, the need to protect such private information intensifies. Privacy for information that society views as truly shameful or criminal cannot be expected.[8]

The protection of informational privacy is not solely about protecting personal information from being lost but extends to the gathering of excessive personal information and processing personal information unfairly.[9]

There exist three policy approaches to informational privacy which have been reproduced in **Table here**. It becomes clear that preventive, corrective and recovery mechanism may be adopted to provide protection to information.

Table - 1 Policy Approaches to Informational Privacy

Nature	Ex ante regulation	Information disclosure	Ex post liability
	Prevention mechanism	Corrective mechanism	Recovery mechanism
Principle	Enforcing minimum standard of care should reduce the probability of a harmful event from occurring.	Disclosure of potential harm becomes a light-handed incentive device. Threat of internalizing costs creates incentive for firms to improve their practices. Awareness of potential harm enables consumers to take action to prevent future loss.	Compensate injured parties for their loss. Court holds injurers liable for loss, forcing them to internalize costs of engaging in harmful activity.

Adapted from Sasha Romanosky & Alessandro Acquisti, *"Privacy Costs and Personal Data Protection: Economic and Legal Perspectives"* 24(3) Berkeley Technology Law Journal 1061(2009) at 1068.

(i)   Privacy as Control

Privacy is the capacity to negotiate social relationships by controlling access to personal information. [10] The variations of the standard version of the control theory of privacy can be found in Fried.[11] The right to privacy can be regarded as a distinctive sort of right in virtue of the special kind of interest it protects.[12] Privacy is about Control and a privacy failure is a control failure.[13] Privacy-control seeks to achieve informational self-determination through individual stewardship of personal data and by keeping information isolated from access; encourages a property approach to personal information that transforms data into a commodity; and supports a move to an intellectual property regime for privacy.[14] Arguments based on the "privacy as control" perspective tend to concentrate on the extent of what must be controlled for privacy to be maintained. Theorists supporting the access-based definition of privacy have offered explicit explanations for their analysis, based on the ability of that analysis to explain situations that cannot be explained by a control-based theory.[15]

(ii)   Privacy as Restricted Access

An individual or group has privacy in a situation if and only if in that situation the individual or group or information related to the individual or group is protected from intrusion, observation, and surveillance by others.[16] There are two kinds of private situations-naturally private and normatively private. Naturally private situations are situations in which people, because of the circumstances of the situation, are naturally protected from intrusion or information-gathering by others. In normatively private situations, some people (the outsiders) are morally or legally forbidden from intruding or gathering information about others (the insiders) who are allowed in the situation. The hard core of concerns and interests covered by the label 'privacy' is related to the interest of individuals in limiting access to themselves.[17]

(iii)   Place of Consent in Privacy

The dominant approach to addressing these concerns and achieving privacy online is a combination of transparency and choice.[18] Personal data use in cyberspace increasingly is structured around an empty process of consent that takes both formal and informal variants.[19] Opt-in is the standard whereby the entity that gathers information from individuals assumes that it cannot disclose it or use it for secondary purposes without first getting permission from those individuals. Opt-out is the situation where the information-gathering entity can further use and disclose the information by default until such time as the individual says "no."

(iv)   Transparency vis-a-vis Privacy

The right to privacy and the right to information are both essential human rights in the modern information society. For the most part, these two rights complement each other in holding governments accountable to individuals. But there is a potential conflict between these rights, when there is a demand for access to personal information held by government bodies. Where the two rights overlap, States need to develop mechanisms for identifying core issues to limit conflicts and for balancing the rights. In many countries, the t-wo rights are intertwined constitutionally. Both are focused on ensuring the accountability of powerful institutions to individuals in the information age.[20] They also mutually enhance each other: privacy laws are used to obtain policy information in the absence of a right to information law and right to information laws are used to enhance privacy by revealing abuses.[21] Every national right to information law has an exemption for personal privacy.

III. Inherent Risks to Privacy

"Antibug" technologies will arise, but the resulting surveillance arms race can hardly favor the "little guy."[22] The rich, the powerful, police agencies, and technologically skilled elite will always have an advantage.[23] The collection of personal information by impersonal government and private-sector roadway surveillance agencies has political, as well as psychological and practical dimensions.[24]

Among the key characteristics of technology in the surveillance realm are amplification, routinization, and sublimation.[25] Amplification refers to the ability of technology to extend the ability to gather information and intrude into private life. Examples of amplification are linked directly to the sensory abilities. Routinization is the process of making intrusion into private life an ongoing process. Sublimation is the means by which a technique for privacy invasion becomes increasingly difficult to detect.

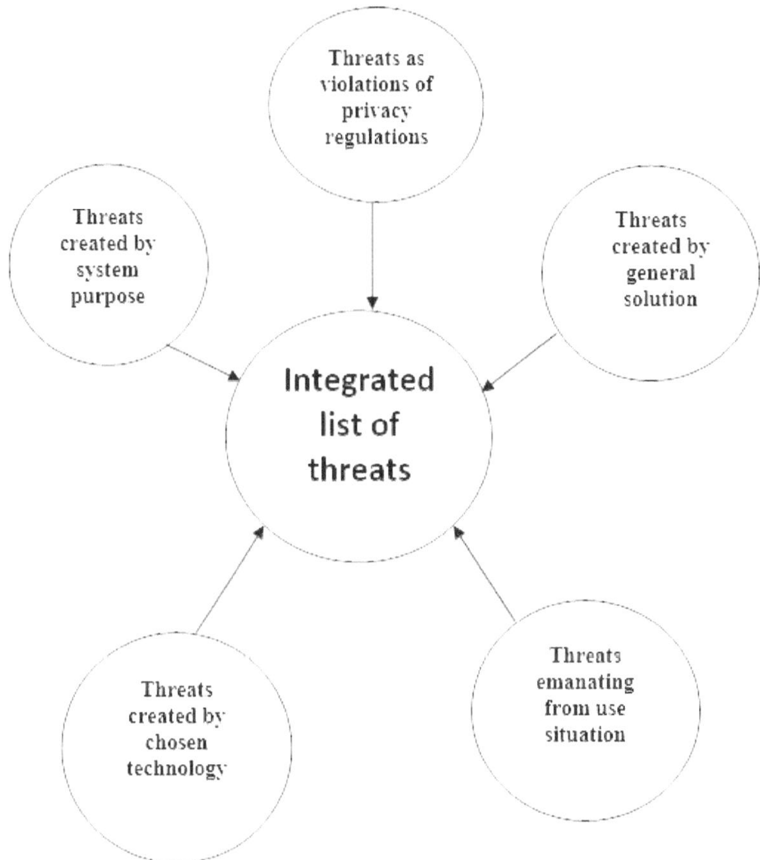

Figure -1: Five-Pronged Approach to Privacy Threats[26]

Figure 1 enlists threats surrounding privacy- beginning with violations of privacy regulations, created by system purpose, chosen by technology, created by general solution and emanating from use situation. It makes clear that threat to privacy may come from any of these sides.

Table – 2

Some Techniques of Data Surveillance

Sr.No.	Technique	Feature
1.	Front- End Verification	Cross-checking of data in an application form, against data from other personal data systems, in order to facilitate the processing of a transaction.
2.	Computer Matching	Expropriation of data maintained by two or more personal data systems, in order to merge previously separate data about large numbers of individuals
3.	Profiling	A set of characteristics of a particular class of person is inferred from past experience, and data-holdings are then searched for individuals with a close fit to that set of characteristics

Data Surveillance or Dataveillance is the systematic use of personal data systems in the investigation or monitoring of the actions or communications of one or more persons. Some techniques of Data Surveillance have been discussed in **Table 2**. As surveillance became a central, constitutive component of modernity, so it became increasingly a social ordering device on a steadily greater scale.[27] Two broad types of surveillance can be distinguished- mass surveillance and targeted surveillance. Mass surveillance is also known as "passive" or "undirected" surveillance. CCTV and databases are examples of mass surveillance. Targeted surveillance is surveillance directed at particular individuals and can involve the use of specific powers by authorized public agencies. Targeted surveillance can be carried out overtly or covertly and can involve human agents. Body surveillance transforms the body into a source of information, and in this way, biotechnology becomes a form of 'information' technology, which raises a number of questions. The practice of self-surveillance by use various existing and emerging technologies, such as GPRS enabled smartphones will decrease information privacy in troubling ways.

IV. Analogy Between Privacy and Technology

Technology can both improve and impair privacy. Privacy concerns which may have arisen because of certain uses of earlier information and communications technologies, but which are now also inextricably associated with and exacerbated by the Internet, can be analyzed under two general headings- dataveillance and data gathering, and data exchange and data mining. Still another effect of new information technologies is the erosion of privacy protection once provided through obscurity or the passage of time. The technology boosts our privacy in the present (we do not have to meet people face to face), but it threatens the privacy of our past. Encryption technologies, anonymous remailers, multinational access, and other features of the Internet regulate access as well as distort the application of laws and facilitate their evasion also provide important tools for protecting vital interests. Digital technologies offer individuals enormous privacy protection and the ability to access information with disclosing anything about themselves.

The design of privacy enhancing technologies depend on the extent of the cooperation, and goodwill, of the different participants: some assume that service providers can be trusted to respect the rights and preferences of users; other scenarios assume a very hostile environment where an adversary may exploit all available means to breach privacy.[28]

Table – 3

Various Privacy Invasive Methods

Sr. No.	Method	How Method Works
1.	General	Invading privacy by using technologies that are not primarily designed and implemented for invading privacy.
2.	Technological flaws	Covers implementation and design flaws that cause to a data breach. Functional limitations of PET also belong to this category
3.	Technical infrastructure	The design of systems and technical infrastructure and the nature of systems lead to data breach. Some examples are cookies, handheld data storage devices, cloud computing. Invasive: This refers to systems that are particularly designed for invading privacy. These systems are used for Surveillance and profile building.
4.	Surveillance	The primary intention of using surveillance technologies is to provide protection.
5.	Profile building	The primary purpose of these technologies is building users profiles for commercial purposes such as target advertising. Additionally, built profiles are used for surveillance and surveillance technologies can also be used for building profiles. Becon applications, online tracking tools, packet inspections, collecting all search terms are some of these tools.
6.	Malicious programs	These are programs designed and used for malicious activities. For example, computer viruses that delete records containing PI.

Adapted from Rasika Dayarathna, "Taxonomy for Information Privacy Metrics", 6 (4)
Journal of International Commercial Law and Technology 194 (2011) at 202.

(i)   Privacy Invasive Technologies

Privacy-destroying technologies can be divided into two categories: those that facilitate the acquisition of raw data and those that allow one to process and collate that data in interesting ways. In practice, just a few mistakes can turn a system designed to protect personal information into one that destroys our secrets.

Table 3 presents a summary of various technologies used for invading privacy.

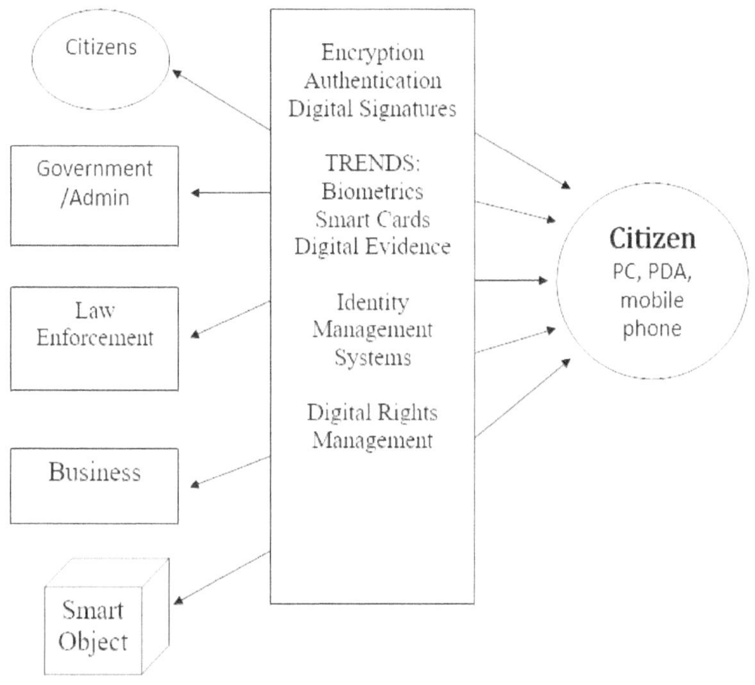

Figure – 1 User controlled processes[29]

Figure 1 illustrates that in today's developing Information Society, the citizen interacts in a "user-controlled" manner with different actors through Internet and mobile communication technologies and devices (ICTs). In these user-controlled interactions, the citizen is generally aware of the interaction processes taking place and of the data exchanged and is probably in control of the use of the underlying security and privacy mechanisms such as the use of a password or the privacy policy of a website.

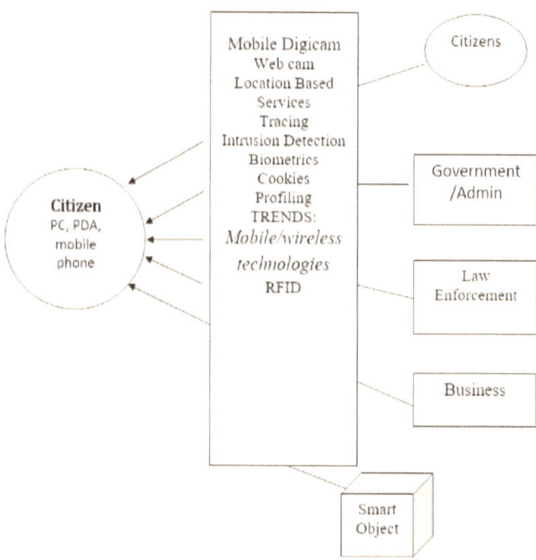

Figure – 2 Non- user controlled processes[30]

An increasing number of situations in which the user is either not aware that data is collected, or his system unlawfully accessed, or, if he is aware, he is not able to control this process can be observed from Figure **2**.

(ii)   Privacy Enhancing Technologies[31]

Privacy can be engineered into systems design, systems can be built without much thought about privacy, or they can be constructed in ways intentionally designed to destroy it, in order to capture consumer information or create audit trails for security purposes. Some of the technologies place control over personal data in the hands of the data subjects themselves, others limit the amount of data that can be collected and used for profiling purposes.

(iii)   Opacity Tools and Transparency-Enhancing Tools (TETs)

Transparency-Enhancing Tools and Privacy Enhancing Technologies are complementary technologies to reduce information asymmetries between interacting partners. Opacity tools "hide" personal data in accordance with the principle of data minimization. Transparency enhancing tools (TETs) provide users with information about privacy policies or granting them online access to their personal data.[32] Transparency-enhancing technologies provide tools to the end users, or their proxies acting on behalf of the user's interests (such as data protection commissioners), for making personal data processing more transparent to them.

(iv) Classification of Privacy Enhancing Technologies (PETs) on the Basis of Makers

Three overlapping types of Privacy Enhancing Technologies are distinguishable. Systemic instruments are ones that arise from the decisions (intended or unintended) of the engineers who design networks, machinery or computer code, and from the technical standards and protocols that are then devised to enhance network efficiency. Collective instruments are those created as a result of government policy and are authoritative policy applications where government or business makes explicit decisions to 'build privacy into' the technical systems for the provision of services and goods. [33] Instruments of individual empowerment, itself an 'umbrella' term, require explicit choices by the consumer or citizen. In this instance, the PET is only activated when individuals choose a measure of privacy-enhancement in their transactions.[34] Of the last-mentioned, the most prominent initiative is the Platform for Privacy Preferences (P3P) initiative, constructed by the World Wide Web Consortium (W3C). Privacy enhancing technologies may make it possible to reach equilibria where data holders can still analyze aggregate and anonymized data, while subjects" individual information stays protected.

(v) Platform for Privacy Preferences (P3P)

P3P helps the user screen information requests and gives the user control over the delivery of requested information, including negotiation of privacy terms between the user and the service provider. It operates as a kind of digital analog to caller ID and caller ID blocking, whereby an answering party wishes to know who is phoning but may be denied this information if the calling party blocks the request. It increases the explicitness with which privacy policies are expressed, allowing the user and the service provider to specify the terms of use for each data item.

(vi) Privacy by Design[35]

Privacy by design' refers to a solution in which privacy is part of the architecture, rather than an add-on. Data minimization principle, one of the most important sub-principles and probably the most adequate for privacy by design, is expected to play an important role in defining which personal data appears to be necessary to use by whom.

(vii) Privacy-Friendly Identity Management

Privacy-Friendly Identity Management give users control of their own identifying information and minimize the personal information required by service providers. Examples of "User-Centric" identity management systems include Microsoft's CardSpace, IBM's Idemix and the 6th Framework Programme Privacy and Identity Management for Europe (PRIME) project prototypes. Vendor Relationship Management (VRM) is an associated concept that supports individuals in managing their relationships and personal data exchanges with businesses rather than the other way around, as is common with Customer Relationship Management systems. It allows users to store data on their own systems are more privacy- protective than those which keep data on central servers.

V. Conclusions

Privacy of individual is crucial for a healthy society. Data privacy is also of same importance. A change in "personal data" is taking place in India. India needs to strengthen its governance systems for managing data if it is to take the lead in the data economy. It is a positive thing that both the central and state governments in India are developing platforms to digitize documents and provide online services to citizens. Establishing these platforms will improve governance efficiency but results in the formation of digital footprints. India must therefore find a way to balance the flow and widespread usage of data with upholding privacy, security, safety, and ethical norms in order to fully realize its potential. It is necessary to evaluate the options for general regulatory, legislative, self-regulatory and voluntary steps that can enhance privacy in order to ensure effectiveness. Transparency and security are another major concern to be kept on priority so that the object must be fulfilled.

References

1   Bernard Marr, "How Much Data Do We Create Every Day? The Mind-Blowing Stats Everyone Should Read," Forbes, May 21, 2018. A quintillion is a 1 followed 30 zeroes.

2   It is useful to use the term 'information privacy' to refer to the combination of communications privacy (data about individuals should not be automatically available to other individuals and organizations, and that, even where data is possessed by another party, the individual must be able to exercise a substantial degree of control over that data and its use) and data privacy (an interest in being able to communicate among themselves, using various media, without routine monitoring of their communications by other persons or organizations. (See generally Roger Clarke, "Introduction to Dataveillance and Information Privacy, and Definitions of Terms" http://www.rogerclarke.com/DV/Intro.html).

3   Constitutive privacy views limit on access by the State and community to personal information as a necessary means of restricting these entities' sovereignty.

4   Alan F. Westin, Privacy and Freedom 7 (Athenum, New York, 1967).

5   C. Edwin Baker, "*Autonomy and Informational Privacy, or Gossip: The Central Meaning of the First Amendment*", 21 Social Philosophy and Policy 2004 (2) 215 at 217.

6   Raymond Wacks , Privacy: A Very Short Introduction 42, (Oxford University Press, New York, 2010).

7   Elbert Lin, "*Prioritizing Privacy: A Constitutional Response to the Internet*", 17 Berkeley Technology Law Journal 1085 (2002) at 1092.

8   Stewart Baker, Skating on Stilts: Why We Aren't Stopping Tomorrow's Terrorism 317, (Hoover Institution Press, Stanford,2010).

9   Information Commissioner's Office (ICO) (U.K.), The Privacy Dividend 6 (2008) available at http://www.ico.gov.uk/news/current_topics/ privacy_dividend.aspx (accessed on 17th August,2012).

10   Philip E. Agre & Marc Rotenberg (eds.), Technology and Privacy: The New Landscape, Technology and Privacy: The New Landscape, (MIT Press, Cambridge,1998).

11   See C. Fried, "Privacy: A Rational Context", *Ian Kerr and Jane Bailey* (eds.) Anatomy of Values (Cambridge University Press, New York ,1970).

12   James Rachels, "Moral Philosophy", 4(4) Philosophy and Public Affairs 323 (1975) at 333.

13   Bruce Schneier, "Schneier on Security: Privacy and Control", Journal of Privacy and Confidentiality 2 (2010) at 3.

14   Paul M. Schwartz, "Internet Privacy and the State", 32 Connecticut Law Review 815 (2000) at 820.

15   James Waldo, Herbert S. Lin, and Lynette I. Millett (eds.), Engaging Privacy and Information Technology in a Digital Age 61 (2007).

16   James H. Moor, "The Ethics of Privacy Protection", 39 (1-2) Library Trends 69 (1991) at 76.

17   Ruth Gavison, The New Encyclopedia of the Social Sciences, http:// www.gavison.com/a2612-privacy

18   Helen Nissenbaum, "A Contextual Approach to Privacy Online" 140 (4) Daedalus Fall (2011) 32 at 34.

19 Paul M. Schwartz, "Privacy and Democracy in Cyberspace", 52 Vanderbilt Law Review 1609 (1999) at 1684.

20 David Banisar, "The Right to Information and Privacy: Balancing Rights and Managing Conflicts" 9 (The World Bank Institute,2011).

21 Id. 10.

22 David Brin, The Transparent Society: Will Technology Force us to Choose Between Privacy and Freedom? (Basic Books, New York ,1998) (accessed on 14th September,2012).

23 Ibid.

24 See Dorothy J. Glancy, "Privacy on the Open Road", 30 Ohio Northern Law Review 295 (2004). The realization that such centralized tracking. is possible impresses a profound sense of powerlessness upon an individual and affects his choices about where, and where not, to go.

25 Marc Rotenberg, "Preserving Privacy in the Information Society", (EPIC,2000), available at http://www.unesco.org/webworld/points_of_views/rotenberg_1.html (accessed on 20th August,2012)

26 G.W. van Blarkom, J.J. Borking & J.G.E. Olk, Handbook of Privacy and Privacy-Enhancing Technologies The case of Intelligent Software Agents 24 (PISA Consortium, Hague,2003).

27 David Lyon, "Everyday Surveillance: Personal data and social classifications", 5(2) Information, Communication & Society 242 (2002) at 245.

28 *Better use of personal information: opportunities and risks,* 25 (Council for Science and Technology, London, 2005) available at http://www.bis.gov.uk/assets/cst/docs/files/cst-reports/05-2177-better-use-personal-information (accessed on 30th August,2012). Past here signifies information which is electronically stored or exchanges even a second ago.

29 IPTS (Institute for Prospective Technological Studies) Report to the European Parliament Committee on Citizens Freedoms and Rights, Justice and Home Affairs (LIBE) on Security and Privacy for the Citizen in the Post-September 11 Digital Age: A Prospective Overview, (2003).

30 Ibid.

31 The Dutch and Canadian Data Protection Authorities together with the Dutch TNO jointly published the study 'Privacy-Enhancing Technologies- A Path to Anonymity'[vRGB+95] in 1995. It proved that technology could be used as a means of protecting individuals against

misuse of their personal data by reducing the processing of personal data. Protecting one's information also comes at a cost: money spent for an anonymizing service and privacy enhancing technologies, time spent learning to use the protecting technology, or hassles incurred when changing one's behaviour and habits.

32 "Online Privacy: Towards Informational Self-Determination on the Internet" 2, Manifesto from Dagstuhl Perspectives Workshop 11061 (2011) available at http://drops.dagstuhl.de/opus/volltexte/2011/3205/pdf/dagman_v001_i001_p001_11061.pdf (accessed on 2nd September,2012).

33 Prominent examples are the attempts to develop public-key infrastructures for government service-delivery.

34 Encryption instruments, devices for anonymity and pseudonymity, filtering instruments and privacy-management protocols are members of this family, offered through a variety of proprietary devices for online use.

35 It was coined by Ann Cavoukian.

Chapter 15

Comprehending the Subtle Difference of Data Protection in Lieu of Privacy Rights

Mayank Chamoli
B.A.L.L.B. (Hons.) Student
SSOL, Sharda University
Greater Noida, UP, India

Mudit Sharma
B.A.L.L.B. (Hons.) Student
SSOL, Sharda University
Greater Noida, UP, India

Jyotish Gupta
Assistant Professor
Galgotias University
Greater Noida, UP, India

Abstract

In an era of rapid technological advancement and ubiquitous data collection, the concepts of data protection and privacy rights have become increasingly intertwined, yet their subtle distinctions hold paramount significance in today's digital landscape. This research paper delves into the intricate relationship between data protection and privacy rights, aiming to shed light on their nuanced differences, while emphasizing the necessity of a comprehensive understanding of both in safeguarding individuals' personal information. The paper commences by delineating the fundamental principles underpinning data protection and privacy rights, offering a conceptual framework that distinguishes these two domains. It then proceeds to dissect the multifaceted legal and ethical dimensions of data protection and privacy rights, highlighting the global regulatory landscape and its evolution over time. Furthermore, this research explores real-world case studies and their implications, showcasing instances where the delineation between data protection and privacy rights can become blurred. The analysis underscores the need for harmonizing these concepts to strike a balance between personal data usage and individual autonomy. In conclusion, this paper underscores the urgency of comprehending the subtle difference between data protection and privacy rights, emphasizing their unique roles in preserving the sanctity of personal information in the digital age. A deeper grasp of these distinctions is pivotal in shaping robust policies and legal frameworks that safeguard both individual rights and the responsible use of data.

Keywords: Privacy, Rights, Digital, India, Difference.

I. Introduction

In a period overwhelmed by digital relations and technical progress, The conception of data protection and privacy rights have occurred as essential references in the current terrain. While frequently used interchangeably, these phrases encapsulate separate elements of protecting people's private knowledge and maintaining their liberation. This composition direct delves into the modest yet influential distinctions between data protection and privacy rights, emphasizing their individual preference in a connected world.

(i) Safeguarding Data and Ensuring Privacy Entitlements[1]

Data protection refers to the measures and mechanisms implemented to ensure the security, confidentiality, and integrity of personal data. This involves employing technical and organizational strategies to prevent unauthorized access, data breaches, and misuse of sensitive information. In contrast, privacy rights encompass the legal and ethical principles that grant individuals control over the collection, usage, and sharing of their data. Solitude rights support people's autonomy, authorising them to determine how their data is operated and guarding them from meddling oversight.

(ii) The Convergence and Divergence

Data protection and privacy rights are connected, they have separate directions and objectives. Data security mainly commands the performance of protection efforts to protect data from violations and illegal access. This contains encryption, key management, and frequent protection audits. Privacy privileges, on the different hand, involve themselves with explaining the permitted and moral limitations of data accumulation, use, and allocation. These privileges enable people to deliver knowledgeable support before their data is accumulated and operated.

(iii) The Function of Technology in Society[2]

In the age of the digital era, technology recreates a key function in both data security and privacy ownership. Algorithms progressive encoding, protected data depository, and multi-factor verification grant data safeguard measures. These standards confirm that confidential data stays personal and cannot be readily compromised. Similarly, innovations like blockchain deliver creative explanations for tying data via regionalized and tamper-proof archives.

Technology and innovations also impact the terrain of privacy privileges. With the addition of corresponding gadgets and the Internet of Items, there are better options for data display. Privacy privileges need technical resolutions such as robust support managing venues and data anonymization strategies to assign people in their data's usage handling.

(iv) Laws and Rules for Technology Implementation

Privacy ownership is usually consecrated in lawful functions, such as data security rules and statutes like the European Union's California Consumer Privacy Act (CCPA) and General Data Protection Regulation (GDPR). These regulations authorise someone with privileges to access, accurately, and obliterate their data, improving their authority over confidential data. Data security laws, on different indicators, concentrate on the obligations of communities in defending data, and defining damages for data violations and non-compliance.

II. Striking Harmony between Advancement and Confidentiality

As technology persists to progress, the fragile equilibrium between creation and solitude evolves increasingly critical. Communities and states must promote surroundings where data-driven creation succeeds while appreciating people's solitude requests. Hitting this ratio necessitates translucent facts procedures, robust protection standards, and available media for people to ply their privacy rights.

(i) Data Protection: Safeguarding the Digital Age

In an enhancing connected planet where data is the banknotes of the digital domain, the idea of a data shield has emerged as a crucial need. From confidential attributes to prudent initiative data, the exact extent of digital understanding application needs robust steps to deliver safety and innocence. Data protection, more valuable than just a specialised span, is an important compass that shapes the trust people and associations set in the digital landscape.

(ii)   Defining Data Protection

Data security guides to the bunch of rules, procedures, and actions strived at saving data from unauthorized credentials, usage, exposure, alteration, or obliteration. It contains a comprehensive scope of techniques that span specialised, administrative, and lawful fields. At its core, data security aims to provide that data stays intimate, and objective, and functions just for those with appropriate permission.

(iii)   The Significance of Data Protection[3]

Data Protection is significant due to the following reasons -

Privacy Protection: In a period where private data is always contained and examined, data security evolves as interchangeable with privacy protection. People share an overload of confidential details online, from monetary data to healthiness forms. Robust data defence steps guarantee that this data remains secret and is not misused for unauthorized goals.

a.   **Confidence and Prominence:** Data infringements can carry far-reaching effects for both people and associations. Data breach can guide to essence stealing, financial loss, and a decline of faith. For companies, a data infringement can spoil their standing and influence in consumer decay. Data security, thus, acts as the basis for setting and retaining faith.

b.   **Regulatory Submission:** Data security statutes and limitations, such as the GDPR (General Data Protection Regulation) in Europe and the CCPA (California Consumer Privacy Act) in the United States, cause institutions to attach to rigid data safeguard measures. Non-compliance can direct to weighty damages and lawful repercussions.

III. Elements Of Data Security

Data Security is a combination of encryption, access management, common auditing and monitoring, data depreciation etc. [4]

(i) **Encryption:** Encryption is a cornerstone of data defense. It concerns transforming data into a coded structure that can superior be solved by privileged groups containing the decryption code. Actually, if data is precluded, it stays incredible and ineffective to unauthorized people.

(ii) **Access Management:** Restricting entry to data established on parts and approvals guarantees that only qualified people can consider, alter, or exploit exposed details. This contains inner violations and unauthorized data transferring.

(iii) **Common Auditing and Monitoring:** Ongoing monitoring and recurring audits benefit deciding any particular actions or violations in absolute time. Rapid detection permits for speedy steps to mitigate the influence of possible violations.

(iv) **Data Depreciation:** Including exclusively the needed data and keeping it for the most temporary needed procedure reduces the potential risk exposure. Substitute or obsolete data should be disposed of.

IV. Challenges and Future Trends

As technology advancements, so do the methods utilised by adversarial actors. From cosmopolitan cyberattacks to sociable engineering, the hazard terrain is ever evolving, necessitating constant transformation of data security procedures.

(i) **Global Data Governance:** With data surpassing geographical limitations, transnational data management and harmonization of limitations become necessary. Institutions must guide various legal frameworks to provide observation and support data security regulations.

(ii) The emergence of Privacy-Enhancing Technologies: Privacy-enhancing technologies, homomorphic such as encryption and zero-knowledge evidence, offer innovative methods to save data while even permitting significant research. These technologies show assurance in keeping the proportion between data utilization and privacy.

(iii) Safeguarding Personal Freedom: The Essence of Privacy Rights

In an age marked by the expansion of digital technologies and interconnected, the idea of privacy privileges has accepted epicenter stage in conversations covering individual time and the digital era. Privacy requests are the essential protections that give people rule over their personal data, actions, and associations. This report delves into the importance of privacy ownership, their development, challenges in the modern earth, and possible methods to protect them.

(iv) The Essence of Privacy Rights

Privacy requests are embedded in the basic regulation that people should keep the space to decide how their confidential data is accumulated, used, and transferred. [5] These requests are usually glorified in constitutions, regulations, and multinational compacts as a mechanism to protect people from unnecessary intrusion by both states and confidential entities. At its essence, privacy licenses people to represent themselves, desire knowledge and encounter movements without suspicion of leadership or determination.

(v) Digital Era Challenges

The source of digital technology has reshaped the privacy terrain. With the immense supply and research of unique data, people face unusual challenges in keeping their privacy. Sociable media, online shopping, and intelligent machines always render data, showing apprehensions about management, data infringements, and the decline of obscurity.

(vi) Surveillance vs. Security

The pressure between management for safety objectives and the protection of privacy is a tender recognition. States usually plead that improved management is essential to control offence and terrorism. Nevertheless, this can have the potential to violate personal liberties. Smashing the proper proportion needs fine restrictions, management mechanisms, and reviews and credits to provide that privacy privileges are not overly compromised.

(vii) Corporate Data Practices

Private companies even recreate an important function in shaping privacy requests. The comprehensive user data array for targeted promotion and other goals has submitted inquiries around support, data request, and the prospect for manipulation. Industries like the General Data Protection Regulation (GDPR) in the European Union and the California Consumer Privacy Act (CCPA) desire to provide users with better authority over their data.

V. Conclusions

In the era of digital age, the quick improvement of technology in has led to a remark on the able quantity of personal information being able to be shared. As be Result of the visions on data security and privacy rights have achieved substantial importance. Understanding the subtle disparities between different amount to *protecting rights* and maintaining the delicate between technological creation and personal safeguards. Data protection, revolves around the steps taken to ensure that personal information is gathered, processed, and stored securely and ethically. This encompasses a wide range of practices, including encryption, access controls, and regular security audits. The main aim of data protection is to prevent unauthorized access, breaches, and the improper use of personal data. Organizations and institutions have a responsibility to implement strong data protection methods to safeguard sensitive information and establish trust with their users. On the flip side, privacy rights constitute a more expansive legal and ethical framework that governs an individual's authority over their personal data. Privacy rights empower individuals to dictate how their data is collected, utilized, and shared by entities that amass it. This involves granting consent for data processing, ensuring transparent data practices, and offering avenues for individuals to exercise their rights, such as the right to be forgotten or the right to access their information. Privacy rights are deeply rooted in human dignity, personal autonomy, and the entitlement to live without unwarranted intrusion.

Though data protection and privacy rights are closely linked, they exhibit nuanced yet crucial differences. Data protection primarily focuses on the technical and organizational measures to secure data, while privacy rights delve into the legal and ethical aspects of handling personal data. A robust data protection framework contributes to upholding privacy rights, but it's the broader landscape of privacy laws and regulations that furnishes individuals with the legal basis to manage their data.

Additionally, as technological advancement keeps redefining data collection and utilization, the importance of distinguishing between these ideas escalates. The emergence of artificial intelligence, machine learning, and extensive data analysis necessitates meticulous deliberation on how data security measures align with upholding individuals' privacy entitlements. Achieving the right equilibrium guarantees that progress in technology does not occur at the expense of eroding essential human freedoms.

To conclude, understanding the nuanced divergence between data security and privacy entitlements constitutes a fundamental aspect of the digital age. Data security concentrates on the technical safeguards implemented to protect personal information, while privacy entitlements encompass the legal and ethical frameworks that authorize individuals to govern their data. This comprehension is crucial for individuals to make enlightened selections regarding data sharing, for organizations to makeover intricate regulatory terrains, and for society to assure that technological advancement respects and preserves individuals' rights. As we move ahead, it remains imperative for individuals, organizations, and policymakers to continue navigating this dynamic sphere armed with a lucid grasp of these intricate notions.

Endnotes

1   Safeguarding data privacy: Striking the balance: An in-depth analysis of India's Digital Personal Data Protection act 2023 (Legal Service India) <https://www.legalserviceindia.com/legal/article-12940-safeguarding-data-privacy-striking-the-balance-an-in-depth-analysis-of-india-s-digital-personal-data-protection-act-2023.html> accessed 18 August 2023

2   Simplilearn (2023) What is the importance of technology?: Simplilearn, Simplilearn.com: <https://www.simplilearn.com/importance-of-technology-article#:~:text=Technology%20lends%20immense%20support%20in,the%20comfort%20of%20our%20homes>accessed18 August 2023

3   Encrypt, S. (2021) What is privacy protection? [updated for 2021], Choose to Encrypt. <http://choosetoencrypt.com/privacy/what-is-privacy-protection/> accessed 19 August 2023

4   Top 5 methods of protecting data – titanfile (no date) Google. <https://www.google.com/amp/s/www.titanfile.com/blog/5-methods-of-protecting-data/amp/> accessed 21 August 2023

5   What is privacy protection? (2021) Choose to Encrypt. <http://choosetoencrypt.com/privacy/what-is-privacy-protection/>

Chapter 16

The Aadhaar Act, 2016 and Right to Privacy

Palak Aggarwal

B.B.A.L.L.B. (Hons.) Student

SSOL, Sharda University

Greater Noida, UP, India

Prof (Dr) Phool Singh

Principal

Jagannath Vishva Law College

Doiwala, Dehradun, Uttarakhand, India

Abstract

Aadhaar is a 12-digit unique identification number issued to Indian residents based on their biometric and demographic data. The concept of a unique identification system was proposed by Nandan Nilekani, the co-founder of Infosys, in 2006. The idea was to provide a unique identification number to every resident of India to streamline various government services and reduce duplications and fraud. To combat excessive corruption in the administration of various welfare schemes, the Parliament has now enacted the Aadhaar (Targeted Delivery of Financial and Other Subsidies, Benefits and Services) Act, 2016 to provide targeted delivery of benefits and subsidies directly to the people thus eliminating intermediaries. If properly implemented, this scheme can prove a boon to the poor as they can receive actual benefits more effectively. The Aadhaar Act has many loopholes, including open ended definitions resulting into confusion about the scope of the scheme. But the issue of violation of right to privacy is of prime importance. Right to privacy is not mentioned in the chapter dealing with fundamental rights in the Constitution Despite the Supreme Court's decision, Aadhaar continues to be used for various government services and initiatives. Through this paper, an attempt has been made to analyse the present Act by taking into consideration various national and international privacy principles. It aims to achieve a balance between governmental privileges and right to privacy.

Keywords: Aadhaar Act, Constitution, Constitutional bench, fundamental rights, privacy principles, right to privacy.

I. Introduction

On March 11, 2016, the Lok Sabha passed the Aadhaar (Targeted Delivery of Financial and Other Subsidies, Benefits and Services) Bill, 2016, despite the criticism for being introduced as a money Bill. A week later, when the Rajya Sabha returned the Bill to Lok Sabha with amendments, they were rejected as recommendations on the money Bill are not binding on the lower house. On 25th March, it finally received the assent of the President. It is an Act to provide for, good governance, efficient, transparent, and targeted delivery of subsidies, benefits and services, to individuals residing in India through assigning of unique identity numbers to such individuals. To obtain an Aadhaar number, an individual has to submit his biometric information including photograph, fingerprint and iris scan along with his demographic information. There is every possibility of leak and misuse of the information from such a large database the responsibility of the collection of which is given to private agencies. Therefore, the chances of infringement of privacy are quite high.

With the enactment of this legislation, the issue of right to privacy has once again come into the limelight. According to Black's Law Dictionary Right to Privacy means the right to personal autonomy or the right of a person and the person's property to be free from unwarranted public scrutiny or exposure. The Constitution of India does not expressly provide for right to privacy in Part III and therefore there is confusion over whether it is a fundamental right or not. Various attempts have been made by the apex court to clarify the issue but at present there is no settled law. With regard to the same, only last year, a five judge Constitution bench was set up by the Supreme Court which is yet to give its verdict.

The researchers have sought to address the debate as to whether right to privacy is a fundamental right or not, by first referring to the Constituent Assembly Debates and then to various judicial pronouncements. Secondly, an attempt has been made to analyses The Aadhaar (Targeted Delivery of Financial and Other Subsidies, Benefits and Services) Act, 2016 (hereinafter referred to as "Aadhaar Act, 2016"). It begins by referring various reports and international covenants and discusses different privacy principles. Subsequently, it critically examines the provisions of the Act by applying the privacy principles. Thirdly, the paper seeks to achieve a balance between government privilege and right to privacy by following the doctrine of unconstitutional conditions. Finally, the essay concludes with suggestions as to what needs to be done in the present circumstances.

II. Is Right to Privacy a Fundamental Right?

This part discusses the right to privacy in the Indian context.

(i) Constituent Assembly Debates

K.T. Shah, in his A Note on Fundamental Rights considered the right to privacy to be an essential element of the right to liberty.

On April 30, 1947, while discussing Right to Freedom under clause 8, Mr. Somnath Lahir proposed an amendment. "The privacy of correspondence shall be inviolable and may be infringed only in cases provided by law." Members including K.M. Munshi, Harman Singh and Dr. B.R. Ambedkar were of the favor to include right to secrecy of correspondence as one of the fundamental rights while on the other hand B.N. Rau, A.K. Ayer and K.M. Panikkar dissented. The initial draft report of the Sub-Committee included both right to secrecy of correspondence as well as right against unreasonable search and seizure as fundamental rights. However, after several rounds of discussions, it was decided that the right to secrecy of correspondence is not a fundamental right. The final report of the Advisory Committee that was submitted to the Constituent Assembly did not have any mention of the provisions relating to the right to privacy.

(ii) Judicial Interpretation

In 1954, for the first time in the legal history of India, the Supreme Court in M.P. Sharma v. Satish Chandra held that when the Constitution makers have thought fit not to recognize right to privacy as a fundamental right, we have no justification to make it as a fundamental right by some process of "strained construction". This was reiterated in Kharak Singh v. State of UP wherein the Apex Court recognized right to privacy as a common law right and not as a fundamental right guaranteed by Part III of the Constitution. In both of the two cases, the court followed literal interpretation.

The Supreme Court in its subsequent decisions on this issue did not consider itself bound by the ratio of M.P. Sharma's case since the decision was specifically limited to the scope of Article 20(3) and according to settled law as to the binding effect of precedents, was an authority for what it actually decided. And finally, after a decade later the court rejected the first view in Govind v. State of Madhya

Pradesh and accepted right to privacy as a fundamental right. The court speaking through Mathew, J. held that the fundamental rights explicitly guaranteed by the Constitution have penumbral zones and many of the fundamental rights can be described as contributing to the right to privacy. In the most celebrated case of Maneka Gandhi v. Union of India, Bhagwati, J. distinguished between named rights and unnamed rights. He noted,

"It is not enough that a right merely flows from or emanate from a named right, i.e., rights categorically mentioned in the text of the Constitution". Therefore, an unnamed right (rights not mentioned in the text of the Constitution) to be a part of the named right, it must be integral to the named right or must partake of the same basic nature or character of the named right."

In **Rajagopal v. State of Tamil Nadu**, the court for the first time directly linked the right to privacy to Article 21 of the Constitution. The court held that the right to privacy is implicit in the right to life and personal liberty guaranteed to the citizens of the country by Article 21. It is a „right to be let alone". While dealing with telephone tapping the court in PUCL v. Union of India, observed that telephone-tapping is a serious invasion of an individual's privacy and thus violate Article 21 of the Constitution, unless it is permitted under the procedure established by law. Right to privacy is an essential and integral part of the fundamental right to life enshrined under Article 21 but it is not absolute and may be restricted for prevention of crime, disorder or protection of health or morals or protection of rights and freedom of others. More recently, in 2015 the Apex Court in **K.S. Puttaswamy v. Union of India**, held that if the observations made in M.P. Sharma and Kharak Singh are followed and accepted as law, the fundamental rights guaranteed under the Constitution particularly right to life and personal liberty under Article 21 would be denuded of vigour and vitality.

III. Development of the Aadhaar Act, 2016

This Part shall discuss the steps towards the enactment of Aadhaar Act, 2016 -

(i) Shah Committee Report

In 2012, an expert committee was constituted to identify the main privacy issues in various laws in India while keeping in view the international privacy principles. The Shah committee recommended enactment of a general Privacy Act. The objectives and scope of the Act were recommended as to clarify its ambit and definitions, to specify the constitutional basis of the right to privacy, to frame National Privacy Principles which can be used to harmonize legislations, policies, and practices including the use of personal identifiers, the use of personal information by the government and the private sector etc, to establish the office of the Privacy Commissioner at the regional and central levels and creating a system of complaints and redressal for aggrieved individuals, to prescribe safeguards for physical privacy including search and seizure and to enumerate offences, associated remedies, and penalties.

The committee recommended nine Fundamental Privacy Principles to form the foundation of the proposed Privacy Act in India. Various international privacy principles which can be effectively implemented in Indian conditions were taken into consideration The National Privacy Principles recommended by the committee are as follows:

a. **Notice:** A data controller shall give simple-to-understand notice of its information practices to all individuals, in clear and concise language, before any personal information is collected from them.

b. **Choice and Consent:** A data controller shall give individuals choices (opt-in/opt-out) with regard to providing their personal information and take individual consent only after providing notice of its information practices. The data subject shall, at any time have an option to withdraw his consent.

c. **Collection Limitation:** A data controller shall only collect personal information from data subjects as is necessary for the purposes identified for such collection. Such collection shall be through lawful and fair means.

d. **Purpose Limitation:** Personal data collected and processed by data controllers should be adequate and relevant to the purposes for which they are processed.

e. **Access and Correction:** Individuals shall have access to personal information about them held by a data controller; shall be able to seek correction, amendments, or deletion such information.

f. **Disclosure of Information:** A data controller shall not disclose personal information to third parties, except after providing notice and seeking informed consent from the individual for such disclosure.

g. **Security:** A data controller shall secure personal information by reasonable security safeguards against loss, unauthorized access, destruction, use, processing, storage, modification, deanonymization, unauthorized disclosure or other reasonably foreseeable risks.

h. **Openness:** A data controller shall make the privacy policies, practices and systems open, transparent, and accessible to individuals through mechanisms such as providing information in multiple languages and adopting an open standards/accessible format for the disabled.

i. **Accountability:** The data controller shall be accountable for complying with measures which give effect to the privacy principles.

(ii) 52nd Report of the Standing Committee on Information Technology

In 2014, the committee in its 52nd Report on "Cyber Crime, Cyber Security and Right to Privacy" emphasized the importance of right to privacy. It felt that in view of enormous data, very sensitive in nature, being consigned to cyber space each day particularly in the light of Government's UIDAI programme, the Government should not jeopardize the privacy of citizens.

(iii) Report of the Office of the United Nations High Commissioner for Human Rights on the Right to Privacy in the Digital Age

Also known as Navi Pillay's Report gave the following recommendations:

a. The identification scheme/measures must not arbitrarily or unlawfully interfere with an individual's privacy, family, home or correspondence. Governments must take specific measures to ensure protection of the law against such interference.

b. The State must ensure that any interference with the right to privacy, family, home or correspondence is authorized by laws that are publicly accessible; contain provisions that ensure that collection of, access to and use of communications data are tailored to specific legitimate aims; are sufficiently precise, specifying in detail the precise circumstances in which any such interference may be permitted, the procedures for authorizing, and procedures for the use and storage of the data collected; and provide for effective safeguards against abuse.

c. The degree of interference must, however, be assessed against the necessity of the measure to achieve that aim and the actual benefit it yields towards such a purpose.

d. Practices in many States have, however, revealed a lack of adequate national legislation and/or enforcement, weak procedural safeguards, and ineffective oversight, all of which have contributed to a lack of accountability for arbitrary or unlawful interference in the right to privacy.

e. States should review their own national laws, policies and practices to ensure full conformity with international human rights law. Where there are shortcomings, States should take steps to address them, including through the adoption of a clear, precise, accessible, comprehensive and non-discriminatory legislative framework. Steps should be taken to ensure that effective and independent oversight regimes and practices are in place, with attention to the right of victims to an effective remedy.

(iv) Standing Committee on Finance

In the year 2011, the Standing Committee on Finance presented the 42nd Report for the examination of The National Identification of India Bill, 2010. After considering the contradictions and ambiguities within the Government on its implementation as well as implications, the committee categorically conveyed their refusal of the National Identification Authority of India Bill, 2010 in that form.

With the enactment of the Aadhaar Act, 2016, it is essential to relook those recommendations. Various provisions which were absent in the 2010 Bill, a few of them are included in the present Act:

IV. International Human Rights Treaties

Article 12 of Universal Declaration of Human Rights ("UDHR") provides that, "No one shall be subjected to arbitrary interference with his privacy, family, home or correspondence, nor to attacks upon his honour and reputation. Everyone has the right to the protection of the law against such interference or attacks."

Article 17 of International Covenant of Civil and Political Rights ("ICCPR") reiterates the same principle. Article 8 of European Convention on Human Rights ("ECHR") in addition to the above principle provides that, "There shall be no interference by a public authority with the exercise of this right except such as is in accordance with the law and is necessary in a democratic society in the interests of national security, public safety or the economic wellbeing of the country, for the prevention of disorder or crime, for the protection of health or morals, or for the protection of the rights and freedoms of other." Article 16 of Convention on the Rights if Child ("CRC") talks about right to privacy of a child and Article 14 of Protection of the Rights of All Migrant Workers prescribes privacy rights to migrant workers.

V. Important Issues regarding provisions of the Act

The Unique Identification Authority of India (UIDAI) has been conferred a statutory authority as per the recommendations. A separate provision has been included which specifically deals which methodology of collections of data. The structure and functioning of UIDAI has now been incorporated in Chapter IV of the Act. Regarding financial implications as has been emphasized by the committee, Chapter V which deals with Grants, Accounts and Audit and Annual Report has been added. Chapter VI titled as Protection of Information has been included. It deals in a great length the issues like access and misuse of personal information, security and confidentiality of identity information, authentication records of individuals etc. While a few of the recommendations seem to be accepted included in the new Act, some of them still remain unconsidered. Section 2(f) of the Act defines "benefit" as any advantage, gift, reward, relief, or payment, in cash or kind, provided to an individual or a group of individuals and includes such other benefits as may be notified by the Central Government.

Also, as per Section 2(c), "Records of entitlement" means records of benefits, subsidies or services provided to, or availed by, any individual under any programme.

(i) **Issue:** "Benefits" and "subsidy" includes such other benefits or subsidies as may be notified by the Central Government. What all benefits or subsidies the people will receive under the scheme is not provided in the Act which leads to confusion and chances of exploitation. There is still no clarity regarding the scope of the scheme.

(ii) "Biometric information" as per Section 2(g) means photograph, finger print, Iris scan, or such other biological attributes of an individual as may be specified by regulations. Such other biological attributes of an individual may also include information's like DNA and hence the chances of exploitation cannot be ruled out.

(iii) "Demographic information" includes information relating to the name, date of birth, address and other relevant information of an individual, as may be specified by regulations for the purpose of issuing an Aadhaar number, but shall not include race, religion, caste, tribe, ethnicity, language, records of entitlement, income or medical history. This can be regarded as an attempt to prevent the discrimination on the basis of religion, race, caste etc. enshrined in the Constitution. Also, excluding medical history can be seen as preventing the violation of right to privacy.

(iv) According to Section 2(l), "enrolling agency" means an agency appointed by the Authority or a Registrar, as the case may be, for collecting demographic and biometric information of individuals under this Act. The Authority may engage one or more entities to establish and maintain the Central Identities Data Repository. Many private individuals and agencies can be appointed for collecting information and can lead to misuse.

(v) Section 3(1) of the Act entitles every resident to obtain an Aadhaar number. The committee was of the view of restricting the scope of scheme only to the citizens as the country is facing a serious problem of illegal immigrants and infiltration from across the borders. Therefore, the chances of illegal immigrants getting an Aadhaar number are quite high.

(vi) Section 4 (3) provides an Aadhaar number may be accepted as a proof of identity of the Aadhaar number holder for any purpose. Section 57 articulates that "Nothing contained in this Act shall prevent the use of Aadhaar number for establishing the identity of an individual for any purpose, whether by the State or anybody corporate or person, pursuant to any law, for the time being in force, or any contract to this effect." These provisions contradict the aims and objectives of the Act which confines the ambit of the scheme for the purpose of government purposes only.

(vii) Section 8(2) of the Act provides that the requesting entity shall obtain the consent of an individual before collecting his identity information for the purposes of authentication in such manner as may be specified by regulations. The principle of consent made it mandatory for the data controller to collect the information only after the consent of individual. This principle appears to be followed here.

(viii) Section 8(2)(b) ensures that the identity information of an individual is only used for submission to the Central Identities Data Repository for authentication. The provision seems to follow the „Collection Limitation" principle which made a data controller to collect personal information from data subjects as is necessary for the purposes identified for such collection.

(ix) Section 9 very clearly mentions that the Aadhaar number shall not confer any right or proof of citizenship or domicile. This makes this scheme voluntary and not mandatory as not having an Aadhaar number will not amount to his being a non- citizen. Also, this should be regarded as a step to avoid any conflict with the Citizenship Act, 1955.

(x) Sections 32(1) and 54(w) provide the power to UID Authority to maintain authentication records in such manner and for such period as may be specified by regulations. Even during the very initial stage of the Bill, it was pointed out that maintaining authentication records over a long time period may be misused for activities such as profiling an individual's behaviour.

(xi) Section 28(2) provides, "Subject to the provisions of this Act, the Authority shall ensure confidentiality of identity information and authentication records of individuals." Section 33(1) that provides disclosure of information, including identity information or authentication records can be made pursuant to an order of a District Judge. However, no situations are given in the clause under which a District Judge can do so. Hence, it is completely a discretionary power and may lead to abuse.

(xii) Section 33(2) provides disclosure of information can be made in the interest of national security in pursuance of a direction of an officer not below the rank of Joint Secretary to the Government of India specially authorised by the Central Government. The Rajya Sabha suggested amendment in this clause. The members were of the opinion to substitute the words "public emergency and public safety" for "national security" which is defined loosely. However, Mr. Jaitley refuted this contention and said that „security of state is a well-defined concept. It is something to do with integrity of the India, sovereignty of India. But there is no concept of public emergency. He also said while national security is limited, public safety and public emergency are not constitutional phrase rather they are undefined and unjustified and thus the amendment if done will result in larger encroachment of right to privacy.

(xiii) According to Section 47(1), *"No court shall take cognizance of any offence punishable under this Act,* save on a complaint made by the Authority or any officer or person authorised by it." This may present a conflict of interest as Section 28 makes the authority responsible for the security and confidentiality of identity information and authentication records. There may be situations in which members or employees of the authority are responsible for a security breach. Then in that case the individual cannot approach the court directly without the complaint made by the authority and there is no other remedy left to him.

VI. Other Important Issues

The Act punishes various offences with imprisonment or fine or both. But, nowhere in the whole Act there is any mention of „compensation". With the introduction and development of Victimology in the Criminal Justice System, the Act does not prescribe any remedy in the form of compensation. It is quite illogical to even think of a situation where a person's biometric information is accessed and misused; only the convicted person will be punished, and the state will receive the amount as fine.

The Centre and even the Constitutional Bench of the Supreme Court have consistently declared that the scheme is voluntary in nature and not mandatory. The committee pointed out that although the scheme is voluntary; an apprehension is found to have developed in the minds of people that in future, services/ benefits including food entitlements would be denied in case they do not have Aadhaar number. Even the Rajya Sabha in its amendment to the Bill suggested that there is a need for clarification that benefits will not be denied to those who do not have an Aadhaar number.

According to the Standing Committee on Finance in its 42nd Report, there are lessons from the global experience to be learnt before proceeding with the implementation of the UID scheme. For instance, the United Kingdom shelved its Identity Cards Project for a number of reasons, which included: huge cost involved and possible cost overruns, too complex, untested, unreliable and unsafe technology, possibility of risk to the safety and security of citizen, requirement of high standard security measures, which would result in escalating the estimated operational costs.

The committee also referred to the Report of the London School of Economics" on UK"s Identity Project which states that, *"Identity systems may create a range of new and unforeseen problems. The risk of failure in the current proposals is therefore magnified to the point where the scheme should be regarded as a potential danger to the public interest and to the legal rights of individuals".*

It is also argued that to get an Aadhaar number, an individual has to face a lot of hurdles due to weak outsourced infrastructure, fragile network connectivity and biometric issues due to ageing and people engaged in long hour of physical labor. Aadhaar will become an excuse for not delivering the dues and harassing them in the name of authentication.

VII. Balancing Government's Privilege and Right to Privacy

While passing the Aadhaar Act, 2016, Union Finance Minister Arun Jaitley argued that the privacy of a Billion people could be compromised for government privileges including subsidies, benefits and other services. It means since the scheme is not mandatory, a person can choose either his right to privacy or the benefits provided by the government. However, according to the doctrine of unconstitutional conditions, the government cannot do so. At this stage, it becomes very significant to discuss the said doctrine.

(i) Doctrine of Unconstitutional Conditions

The doctrine of unconstitutional conditions prohibits the government from requiring people to waive their fundamental rights. It bars the government form arbitrarily conditioning the grant of a benefit on the surrender of a constitutional right, regardless of the fact that the government appropriately might have refused to grant the benefit at all.

In India, the doctrine was first adopted in the case of Ahmedabad St Xavier's College v. State of Gujarat. The 9-judge bench of the Supreme Court held, *"The doctrine of unconstitutional condition means any stipulation imposed upon the grant of a governmental privilege which in effect requires the recipient of the privilege to relinquish some constitutional right."*

Dr. B.R. Ambedkar, the chief architect of the Constitution of India explained the basic logic of the doctrine. He asked that if an unemployed person is offered a choice between a job with wages and with an interdict on joining a union and the exercise of his right to freedom of association, can there be any doubt as to what his choice will be? Fundamental rights are of no value to them. "The fear of starvation and the fear of losing a house are factors too strong to permit a man to stand out for his fundamental rights. The unemployed are thus compelled to relinquish their fundamental rights for the sake of securing the privilege to work and to subsist."

What essentially the government asking people is to give away their fundamental right to privacy to receive government benefits in exchange. Generally speaking, there is a lack of legal awareness among Indian people. How can one expect the people to know their right to privacy, which is relatively a new concept, when most of them may not know about the fundamental right to life and personal liberty? How can one expect the economically poor people to choose their privacy over the benefits and subsidies? This makes the scheme indirectly mandatory and essentially there is no option for the people and hence no right to privacy. So, in the present scenario economically sound families can have their right to privacy as they can leave the benefits and subsidies provided by the government including subsidy on LPG. But the financially backward class has effectively no choice at all and the government is violating their fundamental rights indirectly.

VIII. Conclusions and Recommendations

With the enactment of The Aadhaar (Targeted Delivery of Financial and Other Subsidies, Benefits and Services) Act, 2016, the six-year-old debate of violation of right to privacy has received an attention of legal scholars, media and public. Five judge Constitutional Bench is yet to give its verdict on the validity of UID scheme and whether right to privacy a fundamental right. Since, it is not specifically written in Part III of the Constitution, there is lack of clarity regarding its nature. After Maneka Gandhi, the scope of Article 21 is wide enough so as to include a number of rights which are not even mentioned anywhere in the Constitution including the right to privacy. Regarding the 2016 Act, there are certain positive as well as negative points.

Recommendations

To deal with various loopholes, the government should come up with rules and regulations.

(i)   Firstly, it should clearly mention what all benefits are to be given under the scheme.

(ii) Secondly, it should make every possible effort to ensure that no information will be leaked by any private or public individuals or agency.

(iii) Thirdly, regarding scope, the best option could have been to restrict it within citizens and not residents. But since the Act clearly entitles every resident then it should be guaranteed that no illegal immigrants shall receive the Aadhaar number.

(iv) Fourthly, the Act should provide to an individual, remedy of approaching the court directly and not through the UID authority.

(v) Fifthly, keeping in mind the victims, a provision regarding compensation should be added so that they may receive justice in case of misuse of their information.

(vi) Sixthly, the scheme is already voluntary in nature, but it is indirectly mandatory for poor people as they cannot leave the benefits. Therefore, the scheme should be implemented in such a manner so as to minimize if not cancel every chance of violation of right to privacy.

(vii) Seventhly, the Aadhaar number should not be the only criteria of receiving benefits and subsidies. The government should develop an alternative for those who do not want to or do not have an Aadhaar number.

(viii) Finally, the government should focus on implementation of the scheme instead of arguing whether right to privacy is a fundamental right or not. It should try to achieve the real aims and objectives of the scheme while protecting right to privacy.

References

1 https://www.thehindu.com/opinion/op-ed/is-aadhaar-a-breach-of privacy/article62113418.ece.

2 https://www.indiatoday.in/india/story/aadhar-privacy-loophole-968012-2017-03-27.

3 https://jsis.washington.edu/news/the-aadhaar-card-cybersecurity-issues-with-indias-biometric-experiment/.

4   https://economictimes.indiatimes.com/news/politics-and-nation/
    fixing-flaws-adhering-to-security-and-privacy-key-to-aadhaars-future/
    articleshow/62464605.cms?from=mdr.

5   https://timesofindia.indiatimes.com/blogs/straight-candid/the-aadhaar-
    amendment-act-loophole-galore/.

6   https://www.ecommercetimes.com/story/privacy-concerns-key-reason-
    buyers-flee-online-retailers-87301.html.

7   https://www.deccanherald.com/india/aadhaar-act-verdict-
    history-693614.html.

8   https://www.dailyo.in/politics/aadhaar-supreme-court-right-to-privacy-
    pan-uidai-19231.

Chapter 17

Data Protection and International Laws: An Overview

Dr. Shalini Kashmeria
Assistant Professor, Dept. of Laws &
Asst. Director, Centre of Canadian Studies
Himachal Pradesh University
Shimla, Himachal Pradesh, India

Shubhanjali Bajpai
Advocate
District and Sessions Court
Kanpur, U.P., India

Prashansa Sharma
B.A.L.L.B. (Hons.) Student
SSOL, Sharda University
Greater Noida, U.P., India

Abstract

The international framework on data protection is continuously evolving to adapt to technological advancements, global data flows, and emerging privacy challenges such as AI ethics, data breaches, and surveillance. Harmonizing data protection laws and fostering international cooperation remain critical objectives to effectively protect individuals' privacy in a globally connected digital landscape. At the international level, European Union, United Nations, etc. have led the movement for a framework on data protection. Convention 108, also known as the Convention for the Protection of Individuals with regard to Automatic Processing of Personal Data is one of the oldest international treaties in the field of data protection. It provides a framework for data protection laws and practices and has been adopted by multiple countries. The GDPR is a European Union Regulation that came into effect in 2018. It governs how personal data of EU citizens is collected, processed, and stored. It grants individuals greater control over their data, imposes strict requirements on data controllers and processors, and introduces severe penalties for non-compliance. Over the years, more emphasis has shifted on this area. Various countries are taking guidelines from the international and regional instruments while enacting the domestic data protection laws.

Keywords - Data Protection, international, European Union, regulation, communication.

I. Introduction

Globalization and borderless electronic communication have brought huge benefits to individuals. In the global information economy, personal data have become the fuel driving much of current online activity. Every day, vast amounts of information are transmitted, stored, and collected across the globe, enabled by massive improvements in computing and communication power.

In developing countries online social, economic and financial activities have been facilitated through mobile phone uptake and greater internet connectivity. As more and more economic and social activities move online, the importance of data protection and privacy increasingly recognized, not least in the context of international trade. At the same time, the current system of data protection is highly fragmented, with diverging global, regional and national regulatory approaches. Privacy is a universally recognized human right. Article 12 of the universal declaration of human rights declares that "there could be no interference in an individual's privacy and each and every person is entitled to avail this right for their protection."[1]

The UDHR has formed the basis of various international instruments including international covenant on civil and political rights to foster the inclusion of right to privacy as a human right. The protection of personal data has to be considered as an integral part of the right to privacy. There are several regional frameworks for the protection of personal data that have been really instrumental in furthering the objective of laying down appropriate guidelines to guide the corporates while collection of data.

II. Global Developments

Data protection is not the subject of a single, global treaty or agreement rather, it is included in a range of international and regional instruments, each of which covers a particular group of countries. The international concern for privacy is the creation of modern society. Today, there are more than 120 countries already engaged in some form of international privacy laws for data protection to ensure that citizens and their data are offered more rigorous protections and controls.

The international privacy laws for data protection generally follow, or are guided by, the five global privacy principles of :

(i) Notice – advising users, visitors, readers and users of the policies in place to protect personal information.
(ii) Choice and consent – providing people with choices and consent around the use, storage, management and collection of personal information.

(iii) Access and participation – ensuring the information is accessed and used by the correct people within the right security protocols.

(iv) Integrity and security – ensuring that the data is secure and that there is no unauthorized access.

(v) Enfoncement – ensuring that the service, site, solution and platform are aligned with some form of regulation that enforces compliance.

In 2018, the General Data Protection Regulation (GDPR) broke ground as the most forward-thinking and extensive legal provision for the protection of personal data and its ongoing security. This law is an international privacy law for data protection that impacted any organization that processed any personal data (including biometrics) from any EU citizen. It set the standard and has shaped the trends that dominate this sector today. As the amount of data being stored and created continues to increase exponentially, increased data protection has become critical, and indispensable.

This has driven international data protection laws, and offers the following benefits. Brand value is inherent and implicit in a robust policy and framework. Good governance improves a company's competitive advantage. Improvements in automation, digitization and innovation due to business process transformation. Deeper understanding of the data, its value, and the benefits it offers. Improved data management and control, resulting in improved innovation and transformation.

The enforcement of GDPR created a seismic global shift in how countries, organizations and individuals viewed data privacy and saw a rapid global move towards more rigorous controls and protections.

III. Countries having Privacy Law Framework

Here are some of the top line regions and countries that currently have international privacy laws for data protection:

(i) Europe

The GDRP law was less a localized layer of security and compliance and more an international privacy law for data protection that impacted any organization that processed any personal data from any EU citizen. After several companies ignored the GDPR and some were hit by extensive fines and organizations sat up and paid attention. The enforcement of GDPR and the hefty fines, and reputational damage that came with them, has meant that organizations are facing a challenging time. They have to be compliant, and they need the right support to achieve it.

(ii) The USA

While the country lacks formal laws at the federal level, there is some federal legislation that protects data on a more general level. With the devolution of power to the state level, several US states have created their own data-related laws. California's legislation is considered among the most forward thinking with the California Consumer Privacy Act (CCPA) providing robust privacy rights and consumer protection. Other states with Bills in place, or in the process of being passed, include Alabama, Connecticut, Florida, New York, Washington, Illinois, Texas and Virginia.

(iii) Brazil

No general privacy or data protection law currently exists in Brazil, but it is a good example of a country that has been attempting to develop draft data protection legislation. A Draft Bill for the Protection of Personal Data was released in January 2015. It is broadly based on the European Data Protection Directive. Brazil's general data protection law Lei Geral de Protecao de Dados (LGPD) came into effect on September 18th last year and creates a legal framework for the use of personal data of individuals in Brazil, regardless of where the data processor is located.

(iv) France

France is a good example of an established data protection regime under the EU Directive. The Data Processing Act 1978 (revised 2004) sets out the main data protection provisions in France. Several other laws contain minor data protection requirements. The National Commission on Computer Science and Freedoms (Commission national de l'informatique et des libertés) (CNIL)[2] is an independent administrative authority protecting privacy and personal data. CNIL is probably one of the most visible and active privacy regulators in the world.

(v) South Africa

South Africa has implemented the Protection of Personal Information Act (POPIA) with equally stringent and rigorous personal data protection controls in place. The Act has undergone several iterations and evolutions since it was first proposed in 2013 and is set to harden the final layers of the Act in July 2021. The privacy laws and protections outlined in POPIA are of as rigorous a standard as those in the GDPR.

(vi) Bahrain

Bahrain has the Data Protection Law that has the honor of being the first of its kind to be introduced in the Middle East and that provides individuals with rights concerning how their data is collected, processed and stored.

(vii) Canada

Canada implemented the Personal Information Protection and Electronic Documents Act (PIPEDA) that is aligned with EU data protection law. The Digital Charter Implementation Act (DCIA) was introduced by the Canadian Minister of Information, Science and Economic Development on 17 November 2020. If this passes, it will replace PIPEDA and introduce several interesting changes to privacy legislation in the country. This includes a private right to action and fines that could exceed those of the GDPR.

(viii) The United Kingdom

In the United Kingdom, the GDPR will apply until 31st July 2021 and thereafter different regulations will apply. However, the Data Protection Act 2018 has already implemented the requirements of the EU's GDPR into UK law from 01 January 2021. The Data Protection, Privacy, and Electronic Communications (DPPEC) Regulations of 2019 changed the DPA 2018 with the GDPR to create a holistic, UK-specific data protection system that applies within the UK context and is known as the UK GDPR.

(ix)   India

India has enacted a data protection Bill called the Personal Data Protection Bill that embeds many of the tenets of GDPR within the country's context. These include requirements for notice and prior consent for the use of individual data, limitations on the purposes for which data can be processed by companies, and restrictions to ensure that only data necessary for providing a service to the individual in question is collected. Now, the Aadhar number can be used for verification, but prevents private companies from collecting the individual's details.

Other international privacy laws for data protection include Australia, Angola, British Virgin Islands, Denmark, Finland, Nigeria and Israel.

IV. International Data Protection Commissioners' Initiatives

The final data protection initiative with a near-global reach is the work of the international Data Protection authorities. Their main role is the regulation of national data protection laws, but because their work involves more international disputes, they have started to involve themselves in the global privacy debate. Their three main initiatives are an annual meeting and conference; a system for cooperating in international and cross-border complaints; and a statement on global privacy principles.

This third initiative is of the greatest interest. At their 2005 meeting, the International Data Protection Commissioners issued a statement titled: The protection of personal data and privacy in a globalized world: a universal right respecting diversities (usually cited as the Montreux Declaration).[3] The Declaration called for the development of an international convention on data protection, and it is one of the most significant efforts to harmonize data protection laws around the globe.

V. Regional Initiatives

At the regional level, following initiatives have taken place with regard to data protection:

(i) The European Union (EU)

The European Economic Area (EEA) has 31 members[4]and accounts for a significant proportion of the world's population and global trade. The EU has a long history of involvement in data protection, and this section briefly summarizes several of their key initiatives.

(ii) EU Data Protection Directive (1995)

The most significant regional development in data protection regulation is the European Union Data Protection Directive in 1995. The Directive covers the member states of the European Union, but it has also had a significant influence on global privacy developments. Its core principles appear in a similar form in numerous national privacy laws outside Europe.[5] In addition, the cross-border data transfer rules contained in the Directive have set the standard for international data flows for two decades. The Directive includes a mechanism for assessing the 'adequacy' of foreign data protection regimes, and this too has proved to be very influential.

(iii)   EU General Data Protection Regulation

After more than twenty years of operation, the European Union is 'upgrading' the Directive. It is to be replaced by the General Data Protection Regulation (GDPR) – a mandatory regulation that ensures a harmonized approach across all EU member states. EU, and will ensure that the rights of individuals are more effectively protected across the continent. Consistency of interpretation of the new rules will be guaranteed. In particular, in cross-border cases where several national data protection authorities are involved, a single supervisory decision will be adopted.[6]

The key provision is Article 25(1), which prohibits EU Member States from allowing the transfer of personal information to countries that do not have adequate protections in place. Importantly, Article 25(6) allows the European Commission to determine that a country has "adequate" privacy protections; EC has approved[7] the following nations as adequate: example the ongoing cooperation between the EU and APEC concerning binding corporate rules); this type of collaboration needs to be developed further.

(iv)   Asian- Pacific Economic Cooperation (APEC)

APEC is composed of 21 member economies that together represent approximately 55 percent of the world's GDP. APEC has developed several recent data protection initiatives. The three key initiatives are the development of a set of common APEC Privacy Principles; the development of a system for coordinating complaints that involve more than one APEC jurisdiction; and the development of the Cross-Border Privacy Rules system (CBPRs).

This third initiative is the most relevant, since it has a direct impact on interoperability and cross-border data transfers. The APEC CBPR system is an innovative self-regulatory mechanism for allowing the transfer of data between APEC members where a company has voluntarily joined the scheme.[8] The scheme is very new and only a few nations and a handful of businesses are involved, but it represents an alternative approach to traditional measures for managing cross-border data transfers.

The APEC CBPR system provides standard data privacy policies that businesses can use in order to comply with the APEC privacy framework. The system is meant to facilitate cross-border data flows by providing a voluntary framework to ensure certainty and minimum privacy protections.

(v)   African Union (AU)

The adoption of the African Union Convention on Cyber-security and Personal Data Protection in June 2014 is potentially very significant. The Convention is unusual in that it aims to establish regional and national legal frameworks for cyber-security, electronic transactions and personal data protection. The African Union has 54 member states. Within Africa, another regional initiative has been developed for the members of the Economic Community of West African States (ECOWAS). The ECOWAS Supplementary Act on data protection is unusual as a binding regional agreement. It specifies the required content of data privacy laws and requires member states to establish a data protection authority. To date, seven countries have enacted legislation in compliance with the agreement.[9]

Another regional framework has been developed for the East African Community (EAC) – the EAC Framework for Cyber laws adopted in 2010.[10] This framework recommends that each member state develops a regulatory regime for data protection but makes no specific recommendations on selection of the law.

(vi) The Commonwealth

The Commonwealth is a grouping of fifty-three nations, supported by a Secretariat. The Commonwealth has contributed to the development of data protection regimes through the influence of Commonwealth model laws on national legislation of member countries.

The two relevant model laws are the Privacy Bill and the Protection of Personal Information Bill. They both address key issues relating to information privacy. The adoption of the model laws by several countries has had a positive impact on harmonization. The principles contained in the model laws are heavily influenced by the OECD privacy guidelines and the EU Directive, although there is considerable flexibility in their implementation.

VI. Data Protection Core Principles

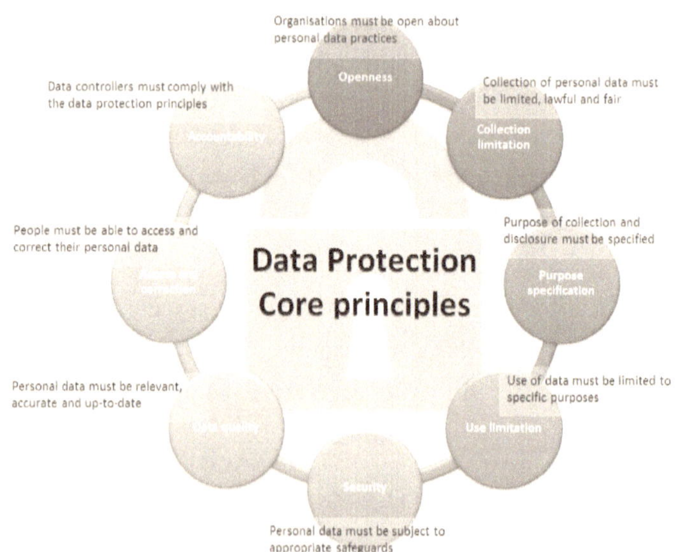

Although numerous attempts have been made to promote global harmonization, there is no single agreed model for data protection law at this stage. However, compatibility is the stated objective of many initiatives (for example, those that have been led by the APEC, the Council of Europe, the EU and the OECD). However, no single initiative has won comprehensive global support. Some individual countries have amended their laws to improve compatibility. New Zealand and the United States have changed their laws to ensure that foreign citizens have data protection and dispute resolution rights (prompted in both cases by a desire to achieve compatibility and to smooth other areas of difference with the EU). This demonstrates that these countries see compatibility as an important objective.

VII. Conclusions

Cooperation and effort are needed to develop practices aimed at ensuring protection for personal data that not only provides necessary protection of sensitive personal data and privacy, but also enables data driven innovations. Work is needed to anchor data protection more firmly in international law. The globalization of society and the pervasiveness of electronic communications make it imperative that data protection rights be applicable and enforceable at an international level. Notably, the processing of appropriately de-identified data would give more flexibility to companies while still maintaining a high level of data protection. Data protection regulation must carefully correspond to the evolving needs and possibilities associated with these changes in order to facilitate potential benefits.

States and international organizations should thus work on dual tracks by both beginning discussions on an international legal framework for data protection, while at the same time finding ways for existing frameworks to interaction.

Endnotes

1   UDHR, Article. 12, 1948.

2   www.cnil.fr.

3   The International Data Protection and Privacy Commissioners, *Montreux Declaration - The protection of personal data and privacy in a globalized world: a universal right respecting diversities*, 2005, https://icdppc.org/wp-content/uploads/2015/02/Montreux-Declaration.pdf.

4   The European Economic Area comprises the 28 European Member States plus Iceland, Liechtenstein and Norway.

7 1995 *Directive on the protection of individuals with regard to the processing of personal data and on the free movement of such data* (95/46/EC).

5   More specific information on the principles governing, and rights afforded by, the Directive can be found in the contribution by the European Commission in Part II.

6   See the contribution by the European Commission in Part II. The contribution contains further details concerning the updates included in the GDPR.

7   http://ec.europa.eu/justice/data-protection/international-transfers/adequacy/index_en.htm.

8   http://www.cbprs.org/default.aspx.

9   See http://unctad.org/en/PublicationsLibrary/dtlstict2015d2_en.pdf

10 *Ibid.*

Chapter 18

Critical Evaluation of the Personal Data Protection Bill, 2019

Vaibhav Kumar
B.B.A.L.L.B. Student
SSOL, Sharda University
Greater Noida, UP, India

Ms. Snigdha Kuriyal
Ph.D. Research Scholar
SSOL, Sharda University
Greater Noida, UP, India

Seelam Saideepak Reddy
B.B.A.L.L.B. Student
SSOL, Sharda University
Greater Noida, UP, India

Abstract

The right to privacy had always been an important aspect of our right to life and personal liberty from the inception and it had always been manifest under Article 21 of the Constitution of India but it was constitutionally recognized on 24th August, 2017 in a historic judgment in the case of Justice K.S. Puttaswamy (Retired) and Another versus

Union of India and Others, Cambridge Analytica scandal later made privacy and data security a hot topic of concern among individuals. After the declaration of the right to privacy as the fundamental right by the Honorable Supreme Court in 2017, the onus to come up with the legislation aiming to secure and advance the protection to individual's data and privacy shifted upon the legislature. Accordingly, the legislature had come up with the Personal Data Protection Bill, 2019. The Bill is prima facie substantially weak on paper and if enacted as it is then it will be procedurally weaker and ineffective. Eventually it will create further menace instead of suppressing the mischief it sought to achieve. The present paper mainly focuses on the shortcomings of this Bill and how it was drafted without strict adherence to the existing international instruments on data protection and privacy.

Keywords: Privacy, right to privacy, data protection, personal data, breach, Bill.

I. Introduction

After the affirmation of right to privacy as a fundamental right under the ambit of Article 21 in Justice K.S. Puttaswamy (Retd.) and Anr. versus Union of India and Ors. the need of a legislation for protection of the data and to maintain individual privacy arose. In 2018, the draft of The Personal Data Protection Bill (hereinafter referred as 'PDP Bill' or 'The Bill') was prepared by the Justice B.N. Srikrishna Committee[1] deriving its basis from European Union's General Data Protection Regulation. The draft by Justice B.N. Srikrishna Committee was revised and then presented in Lok Sabha on 11 December, 2019 which is still pending in the Parliament.[2] The Bill aims to amend Information Technology Act, 2000 by repealing Sections 43A and 87, which are currently addressing the issue of data protection. If approved, it will so replace the current data protection framework established by the Information Technology Act, 2000 (hereinafter referred as 'IT Act') and under Information Technology (Reasonable Security Practices and Procedures and Sensitive Personal Data or Information) Rules, 2011.

II. Legal Framework prior to 2019

Currently, we have two provisions Section 72 and Section 43A of the IT Act, 2000 which are single handedly protecting individual's privacy against the breach and no other provision is available in any statute for penalizing the breach of individual's privacy.

Section 43A puts mandatory duty upon corporates processing, handling or dealing any sensitive data, to practice reasonable security procedures and practices to prevent wrongful gain or wrongful loss to any person and if they fail to do so then they are liable to compensate the aggrieved person.[3] While Section 72 provides that if any person breaches confidentiality and privacy of any other person then the person committing the breach shall be punished with an imprisonment which may extend to 2 years or with fine which may extend to 1lakh rupees or both.[4]

III. Comparative Overview of the Personal Data Protection Bill and the European Union's General Data Protection Regulation

General Data Protection Regulation (hereinafter referred to as 'GDPR') has acted as foundation stone for so many nations to frame their personal data protection legislation. Relying upon the GDPR, India had also prepared its first data protection Bill that is PDP Bill, 2019 following the structure granted by the GDPR but without strict adherence to it.[5]

There is a provision in the PDP Bill for division of critical personal data from personal data, but only if the data is processed on a server that is physically located in India. In contrast, the GDPR does not in any manner recognize data classification under the current guidelines and operating regulations.[6] In addition, the PDP Bill's provision regarding data protection officer designates the data protection authority as having primary responsibility for raising public awareness of pertinent concerns. However, under the GDPR, it is the duty of officer to perform audits, training of staff about data breaches, and raise awareness of the issue.[7]

With the exception of data localization and exports, this Bill does not prescribe higher levels of protection for sensitive personal data, leaving it to Data Protection Authority of India (hereinafter referred as 'DPAI') to determine through regulations any extra protections that may be required.[8] This was also one of the shortcomings in the Srikrishna Bill, and there are considerable differences between the GDPR and these Bills.

IV. Shortcomings of the PDP Bill

Following are the major loopholes of the proposed law:

(i) Restriction upon data principals and their representatives from taking independent actions to enforce the law:

Data principals are unable to get just compliance orders from the DPAI. A data principal may submit a complaint to the DPAI, which may then look into it.[9] Different forms of compliance orders may be issued by the DPAI.[10] A data fiduciary or processor is explicitly given the right to appeal against such a compliance order to the Appellate Tribunal (and eventually to the courts),[11] but the data principal is not explicitly given the same right to appeal. A data principal would have to rely on the very general right of 'any person aggrieved' to appeal against a DPAI judgement.[12] The DPAI's order might not even include reference to a data principal, therefore it is not certain that they are a "person aggrieved." In addition, the exclusion of data principals from

Clause 54(2) it can be seen as denying them right of appeal. Right of data principal to complain will be useless if they are unable to use their rights of appeal to get the DPAI to administer the legislation fairly.

Data Principals are not allowed to approach an Adjudication Officer (hereinafter referred to as 'AO') to impose penalties and investigate for the breach of law. Only the DPAI could initiate an inquiry by an Adjudicating Officer (AO) to determine penalties.[13] Although the injury to data principal may be the cause of the violation and punishment imposed,[14] the data principal is not a "person aggrieved" and is not entitled to appeal against an insufficient penalty[15] because they have no right to be heard.[16]

Currently, the only recourse data principals have for enforcing their rights is to file compensation claims with an AO. Data principal who has been harmed by "any infringement of any provision under this Act."[17] can seek damages by making complaint to an Adjudicating Officer in a prescribed form,[18] and he or she has the right to appeal[19] if they feel that they are still aggrieved. The phrasing of Clauses 64 and 67 implies that compensation proceedings could result from both a violation of a data fiduciary's obligation and a violation of a data principal's right. However, Clause 21 stipulates that data principals must first complain to the data fiduciary regarding rights under Clauses 17 through 19, but nothing is written about complaints regarding obligations. The Bill requires a clear declaration that Clauses 64 also permits damages for breach of obligations. Although "Harm" is generally defined,[20] there is no compensation for violation of Bill per se. An additional challenge for a data principal seeking to enforce the Act is proving injury.

Data principals are not permitted to approach the courts directly for legal enforcement, compliance orders, fines, or compensation. Their access to the courts under this Bill is restricted to two options: first, an appeal to the Appellate Tribunal, which is shown above to have potential for being very limited; second, an appeal against an Appellate Tribunal decision to the Hon'ble Supreme Court, but only in one case that is "on any substantial question of law".[21] Therefore, there is no rationale for asking a court to examine issues that are primarily factual, such as those involving fines or compensation. The Bill seeks to prevent any court from taking cognizance of "any offence" without first receiving a complaint from the DPAI[22] and to stop any investigation into fines without first receiving a complaint from the DPAI.[23] Both clauses may be unconstitutional and is the subject of on-going litigation.[24]

All these defects in PDP Bill are in conflict with the GDPR's vision, which grants the data subjects right to make complaints against the breaches,[25] the right to an effective judicial redressal against supervisory authority for the way it handles the complaints,[26] and the right to compensation lawsuits.[27] India must adopt the GDPR's strategy of maximising data principals' opportunities to ensure that they can take the initiative and pursue enforcement of all aspects of the Act, and that its enforcement is not frustrated by a bureaucratic or under-resourced DPAI, by offering alternative enforcement through the courts.

(ii)   Lack of Independence of DPAI and its AO's

The proposed Data Protection Authority of India (DPAI) was referred as "an independent regulatory agency," by the Srikrishna Committee although in reality, it is not. GDPR required a supervisory authority to "act with complete independence in carrying out its functions and exercising all its powers" and for its members to "remain free from any external influence, whether direct/indirect, and shall neither seek nor take any instructions from anyone".[28]A supervisory body "must act with total independence... and in so doing shall neither seek nor accept any instruction," according to Article 15(5) of the Data Protection Convention 108.

In contrast, Indian Government has the authority to direct the DPAI as it deems appropriate in order to protect a variety of state interests. The

DPAI is obligated to follow written instructions from the government "on the questions of policy," and such directives are (purportedly) not subjected to the judicial review.[29] Additionally, the Bill does not specifically require the DPAI or any of its members to act independently.

There are more provisions in the DPAI's constitution that raise the question of whether the organisation truly possesses the necessary independence. While a body with a majority of part-time members can be viewed as better policy, the DPIA must be made up solely of full-time members,[30] Numerous DPAs exist in Europe and other parts of the world that are made up solely of full-time Commissioners, and their independence is not questioned as a result. Perhaps a more pertinent issue is that, in contrast to the Srikrishna Bill, the selection committee now only consists of bureaucrats from the central government.[31]

(iii) State's power to exempt government agencies

The Government's unjustifiably wide discretion to exempt "any agency of government" from any provision of PDP Bill for any type of processing and for a very wide variety of reasons (Section 35) is unjustified. On the other hand, the Srikrishna Bill restricted such exclusions only on ground of State's security, through legislation passed by the Parliament, and only in cases where it is appropriate for the goal sought to be achieved.[32] As Section 35 does not pass the criteria outlined in Justice K.S. Puttaswamy judgement for legislative interferences with that right, it is also likely that it would be in conflict with India's constitutional right to privacy.

(iv) DPAI's authority to provide new grounds of non-consensual processing

The DPAI has the authority to define "reasonable purposes" as grounds for processing the personal data (including the sensitive data) without the consent in its regulations.[33] DPIA is required by Clause 14(1) to balance interests of fiduciaries, data principals, and the public interest as well as "reasonable expectations" derived from the context and the difficulty of obtaining consent. However, there is no such restriction that the data principal's interests must be clearly outweighed. There are no restrictions on the topics covered by such rules or the goals they are intending to achieve; instead, Clause 14(2) provides an exhaustive range of potential instances. Numerous reasons for non-consensual processing are already included in the Bill.[34]

The result of Clause 14 is that the Bill delegated the DPIA with very few legislative restrictions (and hence little chance for judicial review other than probable Parliamentary disallowance) and did not substantively specify the grounds for legitimate processing. This level of discretion is risky from the standpoint of data principals, who must contend with the lobbying strength of significant commercial organizations.

(v)   Obligation to notify about breach of data

Data fiduciaries are only required to inform DPAI if a data breach "is likely to cause the harm" to the data principal,[35] which is a more liberal requirement than the GDPR requirement to notify the DPA within 72 hours unless breach is "unlikely to result in risk".[36] A lack of certainty about what effect a breach may have should result in notice to DPAI, not in its being withheld. The DPAI will then decide whether a data fiduciary must notify data principals of a data breach.[37]

(vi) Weak rights of data principal

The rights of Data Principal are ignored as follows-

a.   Right of withdrawing consent for processing

The requirement that, unless data principle has a "valid reason" (undefined), they must bear "all legal repercussions" for such withdrawal[38] needlessly and unfairly destroys the value of data principal's right of withdrawing consent for processing.[39] Contrarily, GDPR only stipulates that any processing carried out previous to the withdrawal of consent will remain valid,[40] which is all that is required to safeguard data fiduciaries.

b. Right to access data

It is unclear whether data principle has right to demand from a fiduciary all the personal data that is held (and "being processed"; see Clause 3(31)[41]), rather than only just "any summary thereof".[42] The GDPR, in contrast, only mandates that the controller "must give a copy of the personal data under processing".[43]

(vii) Requirement to define 'harm'

'Harm' is not defined in the Bill, other than by a list of ten things that it includes.[44] It would be much better if the Bill included an objective definition of 'harm', which could then be followed by the list of things that it includes (which are useful, since they are quite broad).[45]

(viii) Exemption from outsourcing of data on foreigners

When there is any contract between an Indian data processor and an overseas data controller, the Government has the authority to exempt certain processing of the personal data of the foreign nationals who are not resident in India.[46] Some groups of Indian data processors will undoubtedly appreciate this "outsourcing exception," but they do not require one because the Act's responsibilities do not generally apply to data processors (as defined in Clause 3(15)[47]), and the few provisions that do are reasonable.

If India wants to be known as a global leader in processing of personal data ethically, it is not desirable to exclude data processors whose improper use of foreigners' personal information can injure them. One would also anticipate that the European Union would demand a legally enforceable guarantee that such a clause would never be applied to those who reside in the EU, but how could such a guarantee be provided?

(ix) Duties and obligations missing from the Bill

Some of the rights provided under GDPR were not included in the Bill and the Srikrishna Report, and they are not included in the PDP Bill either. The Parliament could consider strengthening the Bill by adding them.

(x) Privacy by default

By mandating each Data Fiduciary to implement a privacy by design policy, the PDP Bill puts into practise Policy by Design.[48] "Privacy by Design" is a phrase that refers to "data protection through technology design." It demands that privacy be considered at every stage of the engineering and design process. However, PDP Bill do not go into details of the precise safety precautions that should be used while implementing privacy by design. Data Fiduciary may submit its privacy by design policy for certification to the Authority in accordance with Clause 22(2) of the PDP Bill, and the same will be displayed on the Data Fiduciary's website as well as the Data Protection Authority's website for confirmation.

The Srikrishna Bill required that data fiduciaries 'shall implement' privacy by design policies, including that 'obligations … are embedded in organizational and business practices.'[49] In other words, there was a legal obligation to implement privacy by design and to some extent the privacy by default.[50] It was a very modest implementation of GDPR's Article 25, but one worth including.[51]

All of these legislative requirements have been eliminated by the government Bill, leaving only the ambiguous necessity to "create a privacy by design policy",[52] which can then be "certified" by DPAI and published.[53] This seems to have no practical impact, and is useless window-dressing.

(xi) Protection from automated decisions

The right to be exempt from automated judgments that have major implications, at the very least without human participation, should be incorporated into the Bill, just like it is in GDPR Article 22. This is a crucial modern data privacy right because it is well known that algorithms, "big data," and AI will progressively dominate how personal data is processed in the future, necessitating a special legislative provision. Such safeguards are absent from the government Bill, which only has a right to data portability,[54] which has nothing to do with safeguarding against automated choices.

(xii) Right to object or block processing

The Bill does not include rights to object to or stop processing for legitimate reasons, same like GDPR's Article 21. The Srikrishna Report's notion that data principals could obtain interim orders from the DPAI based on their right to request corrections is unfeasible since requests to halt processing (temporarily or permanently) are not always the result of correctable errors.

(xiii) Restrictions upon direct marketing

The Srikrishna Report claimed that since PDP Bill's rigorous way to consent effectively demands an opt-in to the direct marketing, there is no need for a direct marketing opt-out. This Bill's broad latitude for the secondary use of personal data for "reasonable reasons" that are determined by regulations renders that argument less persuasive. This is such a significant use of personal data that the Bill should specify that marketing to existing customers always requires the provision of an easily used "opt out," where obtaining consent is not possible (as in

GDPR Article 21), or all marketing uses require 'opt-in' consent (as in Hong Kong).

V. Way Forward

Data protection rights must be able to fully benefit from representative actions by non-governmental organisations (NGOs) to enforce all facets of the law and seek remedies for data principals. Data protection rights affect every person in India and are frequently difficult to understand and enforce due to their often complex and technical nature. A fundamental tenet of the GDPR is to enable such NGO activities (Article 80), and enforcement efforts by such NGOs (such as NOYB in Austria, LQDN in France, and PI in the UK) are already among the most powerful forces behind GDPR enforcement. The current provision in Clause 64(3) of the Bill is too narrow, as it only deals with the compensation actions.

A number of suggestions are offered by the researchers as follows -

(i)   The phrase "or data principal affected by an order" should be added after "aggrieved by an order" in Clause 54(2) to clarify any misunderstanding.

(ii) Data principals should have the same liberty and right to approach an adjudicating officer to request the imposition of penalty on a data fiduciary in the same way they can approach under Clause 64 to get compensation.

(iii) Data principals should have this positive right as stated in the Act, to directly approach the Supreme Court to seek compliance orders, penalties or compensation, and not only the right of appeal provided under Clause 75 of the Act. This would provide a competitive substitute to the statute-based systems established by the DPAI, its AOs, and the Appellate Tribunal, if these systems are proved to be effective then they will make approaching to the courts unnecessary, but it will be a great advantage to data principals if they have the option of going to the courts directly instead, if the statute-based systems are not functioning.

(iv) While it is risky to guess about which components of the GDPR would be seen as essential for a successful "adequacy" finding, it is almost guaranteed that a stringent interpretation of the standards for the independence of supervisory bodies will be at the top of the list. If India wants such a result, it should consider this as well as the fact that having an independent supervisory body that can oversee both the public and private sectors is wise public policy. The entirety of Section 86 should be repealed, and the DPAI should be officially granted the same level of full independence as the majority of other DPAs worldwide.

(v) It must be made clear in the Bill that data principals have the right to hold data fiduciaries accountable for any obligations they have to them as well as for any violations of their express rights. Infringement of rights and obligations should be treated like each other.

(vi) Such rights of action must be completely extended to NGOs who represent data principals rather than being restricted to data principals acting alone. Public Interest Litigation (PIL) has a long history in India of being used effectively to uphold and alter the law.

Endnotes

1   *A Free and Fair Digital Economy Protecting Privacy, Empowering Indians*, Committee of the Experts under the Chairmanship of Hon'ble Justice B.N. Srikrishna, (19.02.2022), https://meity.gov.in/writereaddata/files/ Data_Protection_Committee_Report-comp.pdf.

2   Vinod Joseph & Protiti Basu, *The Personal Data Protection Bill 2019 - A Comparison with The 2018 Bill*, ARGUS PARTNERS, (02.06.2022),

https://www.mondaq.com/india/Privacy/876842/The Personal Data Protection Bill 2019 A Comparison With The 2018 Bill

3   IT Act, 2000, Section 43A.

4   IT Act, 2000, Section 72.

5   Information Technology (Reasonable Security Practices and Procedures and Sensitive Personal Data or Information) Rules, 2011.

6   Rohit Mahajan, Manish Sehgal and Shree Parthasarathy, *India Draft Personal Data Protection Bill, 2018 and EU General Data Protection Regulation A comparative view*, PDPB v. GDPR, Deloitte, (16.01.2023) https://www2.deloitte.com/content/dam/Deloitte/in/Documents/risk/in-ra-india-draft-personal-data-protection-Bill-noexp.pdf

7   *Comparative analysis: General Data Protection Regulation, 2016 and the Personal Data Protection Bill, 2018*, IKIGAI LAW, (19.01.2023), https://www.ikigailaw.com/comparative-analysis-general-dataprotection-regulation-2016-and-the-personal-data-protection-Bill-2018/#acceptLicense.

8   PDP Bill, 2019, Clause 15(2).

9   PDP Bill, 2019, Clause 53.

10   PDP Bill, 2019, Clause 54(1).

11   PDP Bill, 2019, Clause 54(2).

12   PDP Bill, 2019, Clause 72(1).

13   PDP Bill, 2019, Clause 63(1).

14   PDP Bill, 2019, Clause 63(3)

15   PDP Bill, 2019, Clause 63(5).

16   PDP Bill, 2019, Clause 63(1).

17   PDP Bill, 2019, Clause 64(1).

18   PDP Bill, 2019, Clause 64(2).

19   PDP Bill, 2019, Clause 64(7).

20   PDP Bill, 2019, Clause 3(20).

21   PDP Bill, 2019, Clause 75(1).

22   PDP Bill, 2019, Clause 83(2).

23   PDP Bill, 2019, Clause 63(1).

24   Dvara Research *Initial Comments*, 1.5, citing *Puttaswamy v Union of India* (2018), striking down Section 47 *Aadhaar Act.*

25   GDPR, Article 77.

26   GDPR, Article 78.

27 GDPR, Article 82(6).

28 GDPR, Article 52(1) & (2).

29 PDP Bill, 2019, Clause 86.

30 Dvara Research *Initial Comments*, 2.1.2.

31 *Ibid.*

32 PDP Bill, 2018, Clause 42.

33 PDP Bill, 2019, Clause 14.

34 PDP Bill, 2019, Clause 12 & 13.

35 PDP Bill, 2019, Clause 25(1).

36 GDPR, Article 33.

37 PDP Bill, 2019, Clause 25(5).

38 PDP Bill, 2019, Clause 11(6).

39 PDP Bill, 2019, Clause 11(2)(e).

40 GDPR, Article 7(3).

41 PDP Bill, 2019.

42 PDP Bill, 2019, Clause 17(1)(b).

43 GDPR, Article 15(3).

44 PDP Bill, 2019, Clause 3(20).

45 G. Greenleaf, "India's Personal Data Protection Bill, 2019 Needs Closer Adherence to Global Standards (Submission to Joint Committee, Parliament of India" (12.02.2022)

46 PDP Bill, 2019, Clause 37.

47 PDP Bill, 2019.

48 PDP Bill, 2019, Clause 22.

49 PDP Bill, 2018, Clause 23(1).

50 It was not a full 'privacy by default' requirement, because it does not state that most privacy-protective legal choice should be presented to the users as default. It is advantageous that users must not have to opt-in to get the maximum protection of privacy, but alternatively should have to opt-out to surrender their privacy.

51 *Supra* note 11 at 12

52 PDP Bill, 2019, Clause 22(1).

53 PDP Bill, 2019, Clause 22(4).

54 PDP Bill, 2019, Clause 19.

Chapter 19

Conceptual Framework of Right to Privacy with Special Reference to Violation in Matrimonial Privacy

Anita Verma
Assistant Professor
L.R. Group of Institutions
Solan, Himachal Pradesh, India

Abstract

Privacy is the right which the individual mainly possesses from their birth, available to the person even after marriage. The right to privacy plays an important role in marriage because everyone has the right to do what he wants to do even if he has the right to be alone for some time. The right to privacy is protected as an intrinsic part of the right to life and personal liberty under Article 21 and as a part of the freedoms guaranteed by Part III of the Indian Constitution. A husband cannot force his wife to tell him everything which she does not want to and at the same time, the wife did not force her husband to not do a certain act. Law cannot force any person to provide such information which consequently involves information related to them, every married woman, as well as every married man, has the right to privacy and another spouse cannot interfere or infringe the Fundamental right to privacy. The main aim of the present paper to know more about the concept and violation of Privacy in matrimonial relationship, how legislature and judiciary consider laws at the national and international levels to protect the Right to privacy.

Keywords: Privacy, Violation, Husband, Wife, Marriage.

I. Introduction

Privacy is an inherent and natural right and essential for the very essence of human being to blossom in its all possibilities. Every society which ignores the private spheres machinates human beings as an automaton which may be efficient in production of goods and values but essentially free will and human dignity are compromised. The concept of privacy cannot be circumscribed like a fixed mathematical formula; it is imagined in the cultural mind of every unique society like basic jural postulates. Privacy is a right that all human beings enjoy by virtue of their existence. It also extends to physical integrity, individual autonomy, free speech, and freedom to move, or think. This means that privacy is not only about the body, but extends to integrity, personal autonomy, data, speech, consent, objections, movements, thoughts, and reputation. Therefore, it is a neutral relationship between an individual, group, and an individual who is not subject to interference or unwanted invasion or invasion of personal freedom. All modern societies recognize that privacy is essential and recognize it not only for humanitarian reasons but also from a legal point of view.

Right to Privacy is an important right under the Right to Life and Personal. Liberty as also an integral part of Human Rights Law which is a matter of concern for everybody in the contemporary social scenario. Privacy does not only mean leading an isolated life, but specifically it denotes freedom from unauthorized and unwarranted interference into one's private life. The state cannot interfere in the boundaries provided by the constitution. In India, previously the right to privacy is not considered that much and various landmark judgments provide that in India there is no concept of the right to privacy exists. In 2017, India recognized the right to privacy as a fundamental right, but the world already recognized the right to privacy as an important right for a human being.

II. Meaning and Definition of Right to Privacy

First of all, we need to know what does the word Privacy mean. The right to privacy means one person has all the right to enjoy his personal life without the intervention of the state or any other private individual. Privacy is not singular concept but is multidimensional concept deserves more of enumeration than definition. According to etymological context, privacy has been taken from Latin term 'Privatus' which means separated from the rest' deprived of something, especially office, participation in the government' and from 'privo' which means 'to deprive'. It is the ability of an individual or group to seclude themselves or Information about themselves and there by reveal themselves selectively. It is the right of every person to enjoy his/her life to the full extent and no person can interfere in his life, without the permission of other person. The right to privacy is granted by our constitution itself and the branches from which is right came into existence is Article 14, 19 and 21.

According to Black's Law Dictionary "right to be let alone; the right of a person to be free from any unwarranted publicity; the right to live without any unwarranted interference by the public in matters with which the public is not necessarily concerned". *According to Article 21 of the Constitution of* "No person shall be deprived of his life or personal liberty except according to procedure established by law", it has been interpreted that the term 'life' includes all those aspects of life which go to make a man's life meaningful, complete and worth living.

In *Gobind vs. State of Madhya Pradesh*, Privacy, in its simplest sense, allows each human being to be left alone in a core which is inviolable, yet the autonomy of the individual is conditioned by her relationships with the rest of society. Those relationships may and do often pose questions to autonomy and free choice.

III. Evolution of Right to Privacy in India

India has essentially been gregarious society wherein cooperation and not competition, society and not solitude have been dominant themes of its culture and civilization. Hence, sometimes it is doubted whether privacy is a value of human relation in India. Certainly, it is wrong to suppose that the concept of privacy is alien to Indian culture. We can trace the concept of privacy in the Hindu Texts like Dharmashastras and Hitopadesha. The Evolution of Right to Privacy in India commentaries expounded the laws of privacy in Indian sub-continent. The kings were bound to uphold Dharma and to respect the privacy of the citizens. According to Hitopadesha, certain matters like worship, sex and family matters etc. should be protected from disclosure. There have been few studies that examines the various rules related to privacy in the ancient Indian society, but still if a close and cautious inspection of the overall way of life and the various obligations imparted on the individuals specially in their inter-personal relationship, shows that their exceptional rules that would respect one's personal privacy. Be it the Upanishads, or the Vedic culture, be it the Ramayana or Mahabharata or Manu Smriti, they have all considered privacy to be an important aspect of an individual's life. Kautilya in his Arthashastra prescribed a detailed procedure to ensure right to privacy while ministers were consulted. So, looking from the historical point of view, privacy can be considered to be civil liberty that is indispensable to the freedom and dignity of an individual.

From the ancient history of India, in the nineteenth and twentieth century, the so-called privacy was associated with that of inviolability of house of property. Like the Constitution of India Bill 1895 which can be regarded to be one of the earliest documents that stated that every citizen has in his house an inviolable asylum, then with the passage of time there was the Commonwealth of India Bill, 1925, which protected unwanted interference to one's dwelling without due process. A similar kind of rights was also projected in the Nehru Report of 1928. As we move towards the modern times especially the time when the time came for India to be independent and the time came for the framing of the Constitution for an independent India, it can be noted that the right to privacy was not specifically highlighted within the list of fundamental rights to be conferred to the citizens of India. Even though a debate and discussion did take place in the Constituent Assembly regarding right to privacy. The formal proceeding of the Constituent Assembly started with the drafting of the in December 1946 and the Constituent Assembly constituted various committees whose main work was to provide reports to the Drafting committee, which would in turn formulate a draft of the Constitution.

IV. Jurisprudential Growth of Laws Relating to Privacy in India

The proposed data protection framework is true to the ratio of the judgement of the Supreme Court of India in Puttaswamy's case. The Supreme Court held that the right to privacy is a fundamental right flowing from the right to life and personal liberty as well as other fundamental rights securing individual liberty in the constitution. Privacy itself was held to have negative aspect, (the right to be let alone), and a positive aspect, (the right to self-development). The sphere of privacy includes a right to protect one's identity. The right recognizes the fact that all information about a person is fundamentally her own, and she is free to communicate or retain it to herself. The core of informational privacy, thus, is a right to autonomy and self- determination in respect of one's personal data. The court observed the following-: "Formulation of a regime for data protection is a complex exercise which needs to be undertaken by the State after a careful balancing of the requirements of privacy coupled with other values which the protection of data sub-serves together with the legitimate concerns of the State." Privacy too can be restricted in well-defined circumstances. There is a legitimate state interest in restricting the right, the restriction is necessary and proportionate to achieve the interest and the restriction is by law.

(i) Phone Tapping and Right to Privacy

Phone tapping and right to privacy is affected by new technological developments relating to a person correspondence and hence has become a debating issue. In the Supreme Court observed that the Court will not tolerate safeguards for the protection of the citizen to be imperiled by permitting the police to proceed by unlawful or irregular methods. Telephone tapping being foray of right to privacy and freedom of expression and also government cannot impose restrictions on publishing defamatory materials against its officials that make it violative of Article 21 and Article 19(1) (a) of the Constitution.

(ii) Gender Priority on Privacy

Right to privacy includes gender priority that implies not merely the prevention the incorrect depiction of private life but the right to prevent it being described at all. Even a woman of easy virtue is entitled to privacy, and no one has the right to invade her privacy. Every female has the basic right to be treated with decency and proper dignity.

(iii) Health and Privacy

Health sector is an important matter of concern in privacy and also one of the major aspects of right to privacy. Health information not only includes information about the health or disability, but also the information related to health service one may receive. It's a human tendency that the information regarding health is considered highly sensitive by many people. The right to life is so important that it supersedes right to privacy. A doctor is under an oath or under medical ethics for not to disclose the secret information about the patient as the disclosure will adversely affect or put in danger the life of other people.

(iv)   Power to Search and Seizure

The Court held that any legislation prominent on the personal liberty of a citizen must in order to be constitutional, laid down the triple test by the Supreme Court in the case of *Maneka Gandhi v. Union of India*. This triple test requires any law interfering in the concept of Personal Liberty to meet certain standards: It must prescribe a procedure; The procedure must withstand the test of one or more of the fundamental rights conferred under Indian Constitution, which may be applicable in a given situation. In a recent case of *Navtej Singh Johar v. Union of India,* Supreme Court of India held that as it applied to consensual sexual conduct between adults in private is constitutional.

(v)   Right to Privacy and Security of State

India implemented a wide range of data sharing and surveillance schemes after the Mumbai attacks in 2008 so as to increase public safety by tackling crime and terrorism. So as to centralize the interception of communications, data and enable law enforcement agency to access. The Central Monitoring System is created. After the implementation of such system, it would be connected to the Telephone Call Interception System which will help monitor voice calls, SMS and MMS, fax communications on landlines, video calls etc.

V. Marital Life and Right to Privacy

Marriage is one of the most sacred and important relationships between a man and a woman. It also involves the families of both spouses. It is an accepted fact that the right to privacy in the present situation prevails everywhere. The bond of marriage is not such that needs to prove transparency before any third person. The conversations, the personal life of a married couple are the examples of privileged communication. Marital rights refer right of married couples against each other. These rights include right of cohabitation, maintenance, etc. After marriage the husband and wife are treated a family. There are various stages of married life. The first stage starts after marriage in which the husband and wife live together and perform their matrimonial obligations. If the husband is good and caring, then the life of wife would be very easy and happy. But if any of them are careless and they are not performing their matrimonial obligations in proper ways then there may be chances of their separation. As for as the marital privacy is concerned, it is recognized by most to the countries of the world. In *Griswold v. State of Connecticut,* a law which prohibited the use of contraceptives by a couple for preventing unwanted pregnancy was challenged on the ground of violation of privacy. The U.S. Supreme Court held that statute forbidding the use of contraceptive violates the right of marital privacy which is protected under Bill of Rights. Though after marriage spouses get some rights against each other but it is well accepted that marital rights are subject to right of privacy. In *State v. Perez ,* husband had recoded some moment of his wife when she was bathing in bathroom. When the wife got those recording, she filed a case before the court for violation of her right of privacy by the husband.

The issue before the court was raised whether a spouse has a reasonable expectation of privacy from being videotaped surreptitiously by the other spouse while alone in a shared, residential bathroom? The court by stating that the wife had a reasonable expectation of privacy from being surreptitiously videotaped by him while she was alone in their shared bathroom the court held that "A person is guilty of a gross misdemeanor who surreptitiously installs or uses any device for observing, photographing, recording, amplifying, or broadcasting sounds or events through the window or other aperture of a place where a reasonable person would have an expectation of privacy and has exposed or is likely to expose their intimate parts or the clothing covering the immediate area of the intimate parts; and does so with intent to intrude upon or interfere with the privacy of the occupant".

Sexual intercourse is common among husband and wife during the subsistence of marriage. It is also the obligation of spouses to cooperate with each other. In sexual act the consent of the partner is necessary, if it is without the consent of any of them then it may be a reason of mental and physical pain and suffering. Indian Penal Code criminalized the non-consensual sex. Recently Supreme Court has stated that husband is not the master of his wife, the relationship between husband and wife is based on mutual love and trust. Supreme Court in many cases dealt with the issues relating to matrimonial rights and right of wife to live the dignity. Courts are flooded with the cases relating to matrimonial rights and right to privacy of women after Justice K. S. Puttaswamy Case.

(i) Marital Communication Privilege

The Indian Evidence Act provides on marriage communication privilege, also known as spousal privilege. It states that a married person cannot be compelled to make public any communication that has happened between the person and their spouse, and also disallows the person from revealing such communication. It is a remnant of British colonial law, which itself was used in Britain. The concept of spousal privilege is rooted in two medieval concepts, as observed in *Trammel v. United States,* in which a person could not incriminate against himself as he has an interest in the proceedings, and that the married couple was one, given the absence of an independent legal existence of a woman. Because of this it was considered that a communication revealed by one spouse was anyway not admissible, then it followed that the other spouse could not do so too. Moreover, given the lack of rights for women at that time, they were considered to be 'property' of men, which gave way to the logic that a person's property will not be able to testify against the owner of that property.

There are two types of this privilege, Testimonial Privilege and Spousal Confidence privilege. Testimonial Privilege removes the duty on a witness to produce evidence against the accused, in this case, a spouse against a spouse. It exists to promote matrimonial harmony and prevent discord. This privilege is associated with older English laws.

(ii)   Violation of Right to Privacy in Matrimonial Cases

Individuals have a right to privacy during marriage. It is not uncommon for spouses to pry into each other's personal communications, records and things to uncover hidden information especially as the level of mistrust. Here some examples

a.   Opening mail addressed to the other spouse

This is a common situation when one spouse moves out of the house during marriage while his/her mail continues to be delivered there. It can be tempting to open mail addressed to the other spouse but it may amount to an offence.

b.   Accessing e-mail

Reading your spouse's e-mail without permission is another common temptation. It is generally okay if you routinely share a computer with open access, however, if the e-mail account is password protected and access is obtained without the account owner's knowledge or consent, whatever material obtained will violate the law and will be subject to evidentiary preclusion, an injunction or protective order.

c. Accessing text messages

Similar to e-mail, text messages will also be subject to evidentiary challenges on the same grounds. Incoming text messages can often be viewed on a cell phone screen when not in use, so there is no rule that would protect the privacy of a message that pops up on screen. However, if the cell phone text messages are viewed as a result of accessing a password protected device without permission, there could be legal ramifications and they will be challenged as ill-gotten information and precluded as evidence.

d. Recording Conversation

In Connecticut, it is illegal to record conversations without the consent of both parties to a telephone conversation. for example, a spouse may not secretly plant a recording device in the other spouse's car to record that spouse's conversations with a third party. The recorded conversation may be inadmissible in court and ruled unusable during the discovery process on the grounds that it would be unjust.

VI. International Covenants and Declarations on Right to Privacy

Various instruments collectively form a strong foundation for the protection of the right to privacy at the international level, though the specific interpretation and enforcement of this right may vary by region and country.

(i) Universal Declaration of Human Rights

"No one shall be subjected to arbitrary interference with his privacy, family, home or correspondence nor to attack upon his honour and reputation. Everyone has the right to protection of the law against such interference or attacks."

(ii)   International Covenant on Civil and Political Rights

"No one shall be subjected to arbitrary or unlawful interference with his privacy, family, home and correspondence, nor to unlawful attacks on his honor and reputation".

(iii)   European Convention on Human Rights

"Everyone has the right to respect for his private and family life, his home and his correspondence. There shall be no interference by a public authority except such as is in accordance with law and is necessary in a democratic society in the interests of national security, public safety or the economic well-being of the country, for the protection of health or morals or for the protection of the rights and freedoms of others."

VII. Emerging Legal Trends vis-a-vis Right to Privacy in India

Once after the recognition of right to privacy as a fundamental right, it will be enough to encroach into any sphere of activity. With the advancement of technology and the social networking sites the intrusion of such a right has become extremely difficult. The extent to which privacy matters in individuals is subjective and differs from person to person. The Information Technology Act, 2000 also includes Right to Privacy which makes unauthorized access into a computer resource as an offence.

The Indian Constitution includes right to press which sometimes come in conflict with right to privacy. Such question is well responded by bringing the concept of public interest and public morality and other provisions mentioned under the Constitution of India. The publication of personal information of an individual without his approval is justified, if such information forms part of public records including Court records. In several aspects, right to privacy may come in conflict with the investigation of police. Various tests such as Narco-Analysis, Polygraph test or Lie Detector test and Brain Mapping tests make unwarranted intervention into the Right to Privacy of a person.

(i)   The Privacy Bill, 2011

The Bill says, "every individual shall have a right to his privacy confidentiality of communication made to, or, by him it including his personal correspondence, telephone conversations, telegraph messages, postal, electronic mail and other modes of communication; confidentiality of his private or his family life; protection of his honor and good name; protection from search, detention or exposure of lawful communication between and among individuals; privacy from surveillance; confidentiality of his banking and financial transactions, medical and legal information and protection of data relating to individual."

(ii)   The Personal Data Protection Bill, 2018

A final report and a draft Bill were released by the committee in July 2018, which was called as the Personal Data Protection Bill, 2018. The Personal Data Protection Bill provided for the establishment of a Data Protection Authority to oversee activities that involve processing of data. It was felt to recognize the need to protect personal data under the fundamental right to privacy. There was also a need to create a collective culture that foster a free and fair digital economy was to be taken into consideration, respecting the informational privacy of individuals, progress and innovation.

(iii)   The Personal Data Protection Bill, 2019

The Bill provide for protection of the privacy of individuals relating to their personal data, specify the flow and usage of personal data, create a relationship of trust between persons and entities processing the personal data, protect the fundamental rights of individuals whose personal data are processed, to create a framework for organizational and technical measures in processing of data, laying down norms for social media intermediary, cross-border transfer, accountability of entities processing personal data, remedies for unauthorized and harmful processing, and to establish a Data Protection Authority of India for the said purposes and for matters connected there with or incidental thereto. It goes on to outline three key points:

The right to privacy is a fundamental right and it is necessary to protect personal data as an essential facet of informational privacy. The growth of the digital economy has expanded the use of data as a critical means of communications between persons. It is necessary to create a collective culture that fosters a free and fair digital economy, respecting the information privacy of individuals, and ensuring empowerment, progress and innovation through digital governance and inclusion and for matters connected therewith or incidental thereto.

(iv) **Digital Personal Data Protection Bill, 2022**

The Government has released the draft of the proposed Digital Personal Data Protection Bill, 2022 (DPDP Bill) for public comments. The Bill is expected to be introduced in the Parliament in the Budget Session 2023. The Bill has undergone multiple iterations. The first draft, the Personal Data Protection Bill, 2018 was proposed by the Justice Srikrishna Committee. The Committee was set up by the Ministry of Electronics and Information Technology (MeitY) with the mandate of setting out a data protection law for India. The Government revised this draft and introduced the Personal Data Protection Bill, 2019 in the Lok Sabha in 2019. The Bill was referred to a Joint Parliamentary Committee (JPC). The JPC Report was accompanied by a new draft Bill, namely, the Data Protection Bill, 2021 that incorporated the recommendations of the JPC. However, the Bill was withdrawn in

August 2022. Now, the Government has introduced the new Digital Personal Data Protection Bill, 2022.

VIII. Conclusions

No person who is or has been married, shall be compelled to disclose any communication made to him during marriage by any person to whom he is or has been married; nor shall he be permitted to disclose any such communication, unless the person who made it, or his representative in interest, consents, except in suits between married persons, or proceedings in which one married person is prosecuted for any crime committed against the other. The right to privacy in India is not explicitly mentioned in the Constitution, but it has been inferred from various fundamental rights, including the right to life and personal liberty under Article 21. Individuals have the right to autonomy and decision-making within their personal lives, including matters pertaining to marriage and intimate relationships. In the context of marital life, privacy is a fundamental right that encompasses the right to make personal choices within the confines of one's marital relationship. This includes decisions regarding intimate matters, family planning, contraception, and other aspects of personal life that may involve the spouse. Several landmark judgments by the Supreme Court of India have affirmed the right to privacy in the context of marital life. The right to privacy is not absolute and can be restricted under certain circumstances, such as for public safety or the protection of the rights of others. Balancing individual privacy with societal interests remains a crucial aspect of legal deliberations in India.

References

1 Justice K. S. Puttaswamy (Retd.) v. Union of India, (2017) 10 SCC 1.
2 Different aspects of Right to Privacy under Article 21 also available at: https://blog.ipleaders.in/different-aspects-of-right-to-privacy-under-article-21/. (Last Visited on 07.07.2023).

3   Kailash Rai, Constitutional Law of India, 290 (Central Law Publications, 11th Edn., 2013).

4   Govind vs State of Madhya Pradesh & Anr, AIR 1975 SC 1378

5   Durga Das Basu, Introduction to the Constitution of India, 90 (Lexis Nexis Co Pvt.Ltd, 21st Edn., 2013).

6   K.C. Joshi, The Constitutional Law of India, 197 (Central Law Publication,3rd Edn., 2016).

7   Jai. S. Singh, The Constitutional Law of India, 341-426 (Central Law Publications1st Edn., 2018).

8   Black's Law Dictionary 8th ed. South Asian Edition 2015, Brayan A. Garner, editor, ISBN 978-93-84746-31-5 Sai Printo Pack (P) Ltd. New Delhi.

9   1975 SCR (3) 946.

10 I.P Massey, Constitutionalizing of Right to Privacy in India, 310-311 (B. P. Sehgal).

11 Sanjeeb Panigrahi, The privacy paradigm, The Statesman (New Delhi, July 27, 2017).

12 http://www.vedpradip.com/articlecontent.php?aid=40. Last visited on 06.07.2023).

13 Kautilya The Arthsashtra. Also available at: https://ncjindalps.com/pdf/HUMANITIES/The%20Kautilya%20Arthashastra%20-%20Chanakya.pdf

14 Constitution of India Bill,1895.

15. Commonwealth of India Bill - Effort towards constitution of India. also available at: https://civilaspirant.in/commonwealth-of-india-Bill-effort-towards-constitution-of-india/. (Last visited on 07.07.2023).

16 Nehru Report (1928) - Modern India History Notes. Available at: https://prepp.in/news/e-492-nehru-report-1928-modern-india-history-notes. (Last visited on 07.-07.2023).

17 Justice K.S. Puttaswamy vs. Union Of India, ((2017) 10 SCC 1).

18 Kailash Rai, Constitutional Law of India, 301(Central Law Publications, 11th Edn., 2013).

19 Sabah S. Al-Fedaghi, The "Right To Be Let Alone" And Private Information, 98-107(Seventh International Conference on Enterprise Information Systems).

20 Jyoti Panday, India's Supreme Court Upholds Right to Privacy as a Fundamental Right—and It's About Time, (August 2017).

21 Gautam Bhatia, Indian Constitutional Law and Philosophy, September 2017.

22 R.M. Malkani v. State of Maharashtra, AIR 1973 SC 157

23 Arshia Jain, Right to Privacy and Phone Tapping, (Law Insider, 30 Jan. 2022).

24 Avtar Singh, The Constitution of India, 245(central Law Publication 1st Edn., 2019).

25 P.K Rana, "Right to Privacy in Indian Perspective" 2IJL 07 (2016)

26 1978 SCR (2) 621.

27 Constitution of India.

28 AIR 2018 SC 432.

29 Indian Penal Code, 1860.

30 Communication During Marriage. Also available at: https://old.amu.ac.in/emp/studym/99998369.pdf. (Last visited on 07.07.2023).

31 Aarvi, Right to Privacy within marriage in India, (2018).

32 381 U.S. 479 (1965).

33 22 U.S. 579 (1824).

34 United States vs. Perez.

35 Indian Evidence Act.

36 445 U.S. 40(19801).

37 Eashaan Agrawal, Privileged Communication: Discussing Marital Privilege. Also available at: https://www.nujssacj.com/post/privileged-communication-discussing-marital-privilege. (Last visited on 01.07.2023).

38 The Universal Declaration of Human Rights, 1948.

39 The International Covenant on Civil and Political Rights, 1966.

40 The Charter of Fundamental Rights of the European Union.

41 Information Technology Act, 2000.

42 Draft Personal Data Protection Bill, 2018

43 What Is the Personal Data Protection Bill 2019. Also available at: https://www.upguard.com/blog/personal-data-protection-Bill. (Last visited on 08.07.2023).

44 Digital Personal Data Protection Bill 2023 approved by the Cabinet, Also available at: https://currentaffairs.adda247.com/draft-digital-personal-data-protection-Bill-2023-approved-by-the-cabinet.

Chapter 20

A Study of Data Protection Laws in China

Riya Shukla

B.A.L.L.B. (Hons.) Student

SSOL, Sharda University

Greater Noida, UP, India

Abstract

The landscape of data protection in China is evolving rapidly, driven by new regulations and the increasing importance of safeguarding personal information in the digital age. Notably, the Personal Information Protection Law (PIPL) and the Cybersecurity Law have introduced stringent requirements for data handling, localization, and consent mechanisms. These regulations are reshaping the practices of businesses operating in or connected to China, compelling them to invest in robust cybersecurity measures, enhance transparency, and prioritize explicit consent from individuals. Privacy concerns are gaining prominence, particularly in the context of emerging technologies like artificial intelligence (AI) and the Internet of Things (IoT). Balancing innovation with individual privacy rights has become a pivotal challenge. The future outlook for data protection in China anticipates further refinements to existing regulations, possibly expanding the scope of personal data, clarifying data localization mandates, and addressing novel privacy challenges posed by advancing technologies. Stricter penalties for non-compliance and enhanced transparency are expected outcomes, providing individuals with greater control over their personal data. Moreover, as China seeks to engage in international data transfers, negotiations on data transfer agreements with other nations are on the horizon. The evolving data protection landscape in China reflects the intersection of technological advancements, privacy concerns, and global data governance standards. Staying informed and proactive in adapting to these changes is imperative for businesses and individuals navigating this dynamic environment.

Keywords: China, Data protection, Digital age, Privacy.

I. Introduction

Data protection has become a paramount issue in the modern digital age, where personal information is constantly being generated, collected, and shared across a multitude of platforms and services. China, as one of the world's most populous and technologically advanced countries, has recognized the importance of safeguarding individuals' privacy and data security. In recent years, it has implemented a comprehensive framework of data protection laws and regulations to address these concerns.

The People's Republic of China has witnessed exponential growth in its digital economy, driven by a thriving tech industry and a vast online population. With this growth, however, comes an increased risk of data breaches, unauthorized data access, and misuse of personal information. In response to these challenges, China has developed a legal and regulatory framework to protect the privacy and data rights of its citizens, as well as to govern the actions of organizations and businesses that handle personal data.

This introduction provides an overview of data protection in China, highlighting the key legislation, regulatory bodies, and evolving landscape of data privacy. It explores the country's commitment to balancing the need for data-driven innovation with the imperative to safeguard individual privacy rights. As we delve deeper into the topic, we will examine the principles, rights, and obligations outlined in China's data protection laws and regulations. Additionally, we will explore the implications of these laws for businesses, individuals, and the global data economy.

China's approach to data protection is of international significance, especially in a world where data flows transcend borders. Understanding the intricacies of China's data protection landscape is crucial for businesses operating within its jurisdiction, as well as for those engaged in cross-border data transfers and collaborations. Furthermore, it sheds light on the broader global discourse on data privacy and the challenges and opportunities that come with the digital age.

In this research paper, the researcher has delved into the intricacies of data protection laws in China, examining the legislative framework, data protection principles, rights of data subjects, cross-border data transfer mechanisms, and the enforcement landscape. Through a comprehensive analysis, the aim is to provide insights into how China's data protection laws impact various stakeholders and contribute to the evolving landscape of global data privacy.

II. Overview of Data Protection Laws

China's data protection landscape has evolved significantly in recent years, reflecting the country's growing recognition of the importance of safeguarding personal information in the digital age.

(i) Historical Development

China's journey towards formulating comprehensive data protection laws has undergone significant development over the years. In the pre-2000s era, China did not have specific data protection laws. Instead, the protection of personal data primarily relied on general provisions in civil laws and regulations. However, as the digital age emerged and technology became more pervasive, the need for robust data protection measures became increasingly evident.

In 2000s, China began to recognize the importance of data protection, especially in the context of e-commerce and online services. During this period, various regulations addressing different aspects of data protection were introduced. However, there was still no comprehensive framework to address the evolving challenges of the digital era.

2010s witnessed significant strides in China's data protection landscape. The country introduced a series of sector-specific laws and regulations, including the Cybersecurity Law (2017) and the Personal Information Security Specification (2018). These regulations laid the foundation for data protection in the digital age, emphasizing the importance of safeguarding personal information and enhancing cybersecurity.

In 2020s, China took a decisive step forward by passing the Personal Information Protection Law (PIPL) in August 2021. This law represents a landmark in China's data protection journey, providing a unified and comprehensive framework for personal data protection. It addresses critical issues such as the processing of personal information, cross-border data transfers, and the rights of data subjects, marking a significant milestone in China's data protection landscape.

(ii) Key Legislation and Regulations

China's data protection framework is supported by a set of key legislation and regulations that provide the legal foundation for protecting personal data and ensuring data privacy. The most notable among these is the Personal Information Protection Law (PIPL), which was enacted in 2021. The PIPL is the cornerstone of China's data protection efforts, covering various aspects of personal data processing, including consent, purpose limitation, data minimization, security measures, data subject rights, and cross-border data transfers.

Additionally, the Cybersecurity Law (CSL), introduced in 2017, plays a crucial role in ensuring the security of network and data systems. It includes provisions related to data localization and protection, emphasizing the importance of safeguarding data in cyberspace. Moreover, the Personal Information Security Specification (PISS), although not legally binding, provides valuable guidelines for organizations on how to protect personal information effectively.

In the realm of e-commerce, the E-commerce Law, enacted in 2019, contains provisions related to the protection of personal data in e-commerce transactions. Collectively, these laws and regulations form a comprehensive legal framework that governs data protection in China.

(iii) Regulatory Authorities

Effective enforcement and oversight of data protection in China are entrusted to several regulatory authorities. The Cyberspace Administration of China (CAC) plays a central role in regulating internet and cyberspace activities, including matters related to data protection. The Ministry of Public Security (MPS) is responsible for law enforcement in the realm of cybersecurity and conducts investigations into cybercrimes and data breaches. The State Administration for Market Regulation (SAMR) also contributes by regulating data protection in the context of consumer rights and e-commerce. Additionally, the National Information Security Standardization Technical Committee (NISSTC) develops and publishes national standards, including those related to data security and personal information protection.

These regulatory bodies collectively work to ensure compliance with data protection laws and regulations, investigate violations, and promote best practices in data security and privacy within China's digital landscape.

III. Data Protection Principles

China's data protection framework, as enshrined in the Personal Information Protection Law (PIPL) and related regulations, embodies several fundamental principles and safeguards aimed at preserving the privacy and security of personal data. Among these, two pivotal principles are consent and purpose limitation.

Regarding consent, the PIPL underscores the necessity of securing explicit and informed consent from individuals before the collection and processing of their personal information. This stipulation ensures that individuals have a clear understanding of how their data will be used and empowers them to make informed decisions about their personal information. Consent, in this context, must be freely given, specific, and revocable, allowing individuals to exercise control over their data and its usage.

Complementing the principle of consent is that of purpose limitation. Under this mandate, organizations are obligated to use personal data solely for the explicit purposes that have been disclosed to the data subjects. Any utilization of data that deviates from these specified purposes is prohibited. Purpose limitation acts as a safeguard against the misuse of personal data, ensuring that it is not processed for unrelated activities, thus preserving individuals' privacy rights, and preventing unwarranted intrusions into their personal lives.

Another fundamental aspect of China's data protection framework is the principle of data minimization and accuracy. The PIPL mandates that organizations collect and retain only the minimal amount of personal data necessary to achieve the intended purpose. This approach encourages responsible data handling practices that reduce privacy risks and limit the exposure of sensitive information. Additionally, the law underscores the importance of maintaining the accuracy of personal data. Organizations must take proactive measures to ensure that the information they collect, and process is up-to-date and correct. Data subjects are granted the right to request corrections to inaccuracies, promoting transparency and reliability in data processing.

Security and accountability are core pillars of China's data protection regime. Organizations are tasked with implementing robust security measures to safeguard personal data against breaches, unauthorized access, and other security threats. This entails the adoption of encryption, access controls, and routine security assessments to fortify data protection efforts. Simultaneously, accountability mechanisms require organizations to establish comprehensive data protection policies and compliance procedures. The appointment of data protection officers and the maintenance of detailed records of data processing activities promote transparency, compliance, and accountability in the handling of personal data.

In essence, China's data protection framework is built upon a foundation of consent, purpose limitation, data minimization, accuracy, security, and accountability. These principles collectively contribute to the country's commitment to ensuring the privacy and security of personal information in the digital age, striking a balance between the need for data-driven innovation and the imperative to safeguard individual privacy rights.

(i) Data Subjects' Rights

China's data protection framework, as defined by the Personal Information Protection Law (PIPL), grants data subjects a set of essential rights designed to empower individuals and protect their personal data. These rights reflect the global trend of enhancing data privacy and control.

Firstly, individuals in China have the right to access their personal data held by organizations. This right enables data subjects to request and obtain information about what personal data is being collected and processed, the purposes of such processing, and who has access to it. This transparency empowers individuals to monitor and understand how their data is being used.

Additionally, the right to rectification allows data subjects to correct inaccuracies in their personal data. It ensures that individuals have the means to update their information when it becomes outdated or erroneous, promoting data accuracy and reliability.

Furthermore, the right to erasure, often referred to as the "right to be forgotten," is a significant component of data protection in China. It grants individuals the ability to request the deletion of their personal data under specific circumstances. This right is vital in cases where data is no longer necessary for its original purpose or when consent is withdrawn, giving individuals greater control over their digital footprint.

China's data protection framework also recognizes the right to data portability, allowing individuals to request the transfer of their personal data from one service provider to another. This facilitates competition and choice by enabling individuals to switch services without losing their personal information, promoting a dynamic and user-centric digital ecosystem.

Lastly, individuals have the right to object to automated decision-making processes, which may significantly affect them. This right safeguards individual against solely algorithmic decisions that could lead to adverse consequences without human intervention or review.

It ensures that individuals have the opportunity to challenge and seek human intervention in such cases.

These rights collectively empower individuals by providing them with greater control over their personal data and the means to enforce their privacy preferences. They form a cornerstone of China's commitment to data protection and reflect the evolving landscape of privacy in the digital age.

IV. Data Processing and Cross-Border Transfers

Data processing and cross-border data transfers are critical aspects of China's data protection landscape, governed by the Personal Information Protection Law (PIPL) and related regulations. These regulations reflect China's commitment to balancing the free flow of data with the imperative to protect individuals' personal information.

In China, organizations engaging in cross-border data transfers must adhere to stringent requirements designed to safeguard the privacy and security of personal data. Firstly, data transfers must be based on a legitimate and specific purpose, with the informed consent of the data subjects. This means that organizations must clearly communicate the purpose of the data transfer to individuals and obtain their consent, ensuring that data is not transferred for unrelated or unauthorized activities.

Furthermore, data controllers and processors must conduct a security assessment before cross-border data transfers. This assessment evaluates the destination country's data protection laws and the recipient's data security capabilities. If the assessment determines that the transfer may pose risks to data security or the rights and interests of data subjects, additional measures, such as encryption or other security safeguards, may be required to mitigate these risks.

China's approach to cross-border data transfers also involves a system of mechanisms to facilitate such transfers while ensuring data protection. These mechanisms include:

(i) **Consent-Based Transfers:** Data subjects' explicit consent allows organizations to transfer their data across borders. However, this consent must be informed and specific to the purpose, and individuals can revoke it at any time.

(ii) **Standard Contractual Clauses:** Organizations can use standard contractual clauses approved by data protection authorities to ensure that data transfers comply with data protection regulations. These clauses establish contractual obligations between data exporters and importers to protect personal data.

(iii) **Certification Mechanisms:** China may establish certification mechanisms for countries, regions, and organizations that meet certain data protection standards. Data transfers to certified entities may be considered compliant.

(iv) **Legally Binding and Necessary Transfers**: Transfers deemed necessary for reasons such as the performance of a contract, protection of life or property, or public interest may be exempt from the requirement of consent.

(v) **Other Approvals**: In certain cases, organizations may obtain approval from relevant authorities for cross-border data transfers that do not fit within the other mechanisms.

Additionally, China's data protection framework includes data localization requirements, which mandate that personal information of Chinese citizens collected and stored by critical information infrastructure operators be stored within China's territory. These operators must conduct a security assessment when transferring data abroad, further emphasizing the importance of data security and protection. The data localization requirements aim to enhance data sovereignty, promote data security, and protect the interests of Chinese data subjects. However, it's important to note that these requirements primarily apply to a subset of organizations deemed critical information infrastructure operators, and not to all entities processing personal data.

In conclusion, China's approach to data processing and cross-border data transfers reflects a commitment to striking a balance between data-driven innovation and the protection of individual privacy rights. The requirements for data transfers, the availability of mechanisms to facilitate such transfers, and data localization provisions collectively contribute to the evolving landscape of data protection in China. By imposing strict safeguards and fostering transparency, China aims to ensure that personal data remains secure and that the rights and interests of data subjects are respected, even in the context of cross-border data flows.

V. Compliance and Enforcement

China's data protection landscape presents both opportunities and challenges for organizations seeking to comply with the evolving regulatory framework. Compliance challenges, regulatory enforcement mechanisms, and case studies of enforcement actions provide a comprehensive overview of the complex and dynamic nature of data protection in China.

(i) Compliance Challenges for Organizations

Navigating China's data protection regulations can be a daunting task for organizations, both domestic and international. One of the primary challenges lies in the complexity and rapid evolution of the regulatory environment. China has introduced a series of data protection laws and regulations, including the Personal Information Protection Law (PIPL), the Cybersecurity Law (CSL), and the Personal Information Security

Specification (PISS), which often overlap and require meticulous interpretation.

Data localization requirements present another compliance hurdle. Critical information infrastructure operators, as defined by CSL, must store personal data of Chinese citizens within China's borders. Complying with these requirements entails significant technical and operational adjustments for multinational companies, which may have previously stored data outside China. Furthermore, understanding and obtaining informed consent from data subjects can be challenging due to language barriers and cultural differences. China's diverse population and unique regional characteristics necessitate comprehensive consent processes that accommodate various demographics and regions.

(ii) Regulatory Enforcement Mechanisms

China employs a multifaceted approach to regulatory enforcement, empowering various governmental bodies to oversee data protection compliance. The Cyberspace Administration of China (CAC) is at the forefront of enforcing data protection laws and regulations, monitoring data processing activities and conducting investigations into potential violations. CAC plays a pivotal role in issuing guidelines, coordinating cross-agency efforts, and ensuring compliance within the digital realm. The Ministry of Public Security (MPS) focuses on cybersecurity and law enforcement, addressing cybercrimes and data breaches. In cases where data breaches are criminal offenses, MPS conducts investigations and imposes penalties, further emphasizing the need for robust data security measures.

The State Administration for Market Regulation (SAMR) oversees data protection in the context of consumer rights, e-commerce, and online marketplaces. SAMR plays a critical role in regulating data privacy in the business-to-consumer sphere. Furthermore, China's regulatory enforcement mechanisms include the imposition of penalties and fines for non-compliance. These penalties can range from warnings to substantial fines, suspension of business operations, confiscation of illegal gains, and even revocation of relevant licenses or permits. The severity of penalties underscores the government's commitment to enforcing data protection regulations and promoting compliance.

(iii)   Case Studies of Enforcement Actions

Several notable enforcement actions in China provide insights into how regulatory authorities address data protection violations. For example, in 2020, the CAC initiated investigations into multiple major tech companies for suspected monopolistic practices and potential violations of data protection regulations. These investigations resulted in fines and regulatory actions against several prominent technology firms, signaling the government's commitment to holding industry giants accountable for data protection and antitrust issues.

Additionally, in response to the 2018 Cambridge Analytica scandal, CAC launched investigations into Chinese tech companies to ensure they were not involved in similar data misuse practices. These investigations led to increased scrutiny and tighter regulations surrounding the handling of personal data in the tech industry. In another case, a Chinese online video streaming platform was fined for infringing users' privacy by collecting excessive personal information without obtaining proper consent. This case served as a reminder of the importance of transparent data collection practices and the consequences of non-compliance.

These enforcement actions illustrate China's proactive approach to addressing data protection violations, irrespective of the size or prominence of the organizations involved. They underscore the government's commitment to protecting individuals' privacy and promoting compliance with data protection laws and regulations.

VI. Comparison with International Data Protection Standards

China's data protection laws, including the Personal Information Protection Law (PIPL) and the Cybersecurity Law, have introduced several unique features and requirements that differ from international data protection standards, such as the European Union's General Data Protection Regulation (GDPR).

(i) Contrasts with GDPR (General Data Protection Regulation)

China's data protection landscape, as exemplified by the Personal Information Protection Law (PIPL) and other related regulations, presents significant contrasts with the General Data Protection Regulation (GDPR) of the European Union (EU). These contrasts have notable implications for global businesses operating in both regions, as they must navigate distinct regulatory frameworks to ensure compliance and protect individuals' privacy rights.

(ii) Extraterritorial Reach and Scope

GDPR's extraterritorial reach extends to organizations worldwide that process the personal data of EU residents, regardless of the organization's location. In contrast, while China's PIPL has certain provisions for extraterritorial application, it primarily focuses on protecting the personal data of Chinese citizens. Organizations outside China processing Chinese citizens' data are subject to its provisions.

(iii) Data Localization vs. Data Transfers

GDPR does not mandate data localization but imposes stringent rules on cross-border data transfers. Organizations must adhere to mechanisms like Standard Contractual Clauses (SCCs) or Binding Corporate Rules (BCRs) for international data transfers. In contrast, China's PIPL introduces data localization requirements, particularly for critical information infrastructure operators, who must store personal data of Chinese citizens within China's borders.

(iv) Consent Requirements

GDPR emphasizes obtaining explicit and informed consent from data subjects for data processing, with strict conditions for valid consent. China's PIPL also requires consent but does not stipulate GDPR's explicit consent standard, potentially allowing for broader interpretations.

(v) Data Subject Rights

While both GDPR and PIPL grant data subjects rights, there are differences in their implementation. For instance, GDPR includes the "right to be forgotten" explicitly, whereas PIPL refers to the right to erasure but lacks the same level of detail and clarity.

VII. Implications for Global Businesses

China's data protection regulations have profound implications for global businesses, necessitating extensive adjustments and considerations. For instance, international tech giants like Google and Facebook, which collect and process vast amounts of user data, must now reevaluate their data collection practices in China to comply with stringent consent requirements. Moreover, businesses with supply chain operations intertwined with China, such as Apple's reliance on Chinese manufacturing, must ensure that their partners adhere to data localization and security mandates, leading to complex contractual negotiations and increased compliance costs.

The impact extends to financial institutions operating in China, where customer data security is paramount. Citibank, for instance, faces challenges in managing cross-border data transfers while adhering to China's strict regulations, potentially affecting its digital banking strategies. Lastly, multinational retail companies like Walmart, while expanding their e-commerce presence in China, must navigate data transfer mechanisms and consumer consent requirements to maintain their competitive edge, as data-driven marketing strategies become increasingly regulated. In essence, China's data protection framework reshapes the landscape for global businesses, necessitating a comprehensive revaluation of data practices and strategies to thrive in this vast and regulated market.

China's data protection regulations are causing global businesses to rethink their operations in various sectors. In the healthcare industry, pharmaceutical companies like Pfizer and AstraZeneca are grappling with the need to store clinical trial data locally in China to comply with data localization requirements. This raises concerns about data integrity, security, and the ability to seamlessly collaborate across borders, potentially affecting drug development timelines and patient outcomes.

In the automotive sector, companies like Tesla, which collect extensive vehicle data for research and development purposes, must navigate the complexities of data transfer mechanisms to send telemetry data back to their headquarters outside of China. This can lead to delays in analyzing crucial vehicle performance data and implementing necessary improvements.

For financial technology (fintech) companies like PayPal and Square, managing customer financial data securely in China is crucial. Compliance with China's data protection laws adds layers of complexity to their payment processing systems, potentially requiring significant investments in data security infrastructure and legal resources to ensure adherence to Chinese regulations.

Global social media platforms like Twitter and LinkedIn face challenges in complying with China's data protection laws, particularly regarding the transparency and consent requirements for user data processing. These companies must adapt their advertising and user engagement strategies to align with these regulations, impacting revenue streams and user experiences.

Furthermore, even traditional manufacturing companies like General Electric (GE) are affected by China's data protection laws when it comes to operating smart factories and industrial IoT systems. GE must ensure that data generated in Chinese factories complies with local regulations while integrating it with their global data analytics infrastructure.

In summary, China's data protection regulations have far-reaching implications across industries, impacting data management, operations, compliance, and competitive strategies for global businesses, from pharmaceuticals and automobiles to fintech and social media, and even traditional manufacturing. Adapting to these regulations is essential for maintaining market presence and competitiveness in China's rapidly evolving business landscape.

VIII. Data Protection in Specific Sectors

In China, data protection regulations have a profound impact on various sectors, each facing its own unique challenges. In the healthcare and medical data sector, stringent data privacy laws require pharmaceutical companies, hospitals, and research institutions to carefully manage patient data. This involves data localization, ensuring that sensitive medical records stay within Chinese borders, and obtaining explicit consent for data processing. These regulations aim to safeguard patient confidentiality and data integrity but can pose hurdles for international healthcare companies conducting clinical trials or managing global health databases.

The financial services industry in China also faces significant data protection requirements. Banks, payment processors, and fintech companies must implement robust cybersecurity measures to protect customer financial data. Cross-border data transfers must adhere to specific regulations, and the extraterritorial reach of Chinese laws can affect international financial institutions operating in the country. Compliance costs are high, as companies invest in data security infrastructure and legal resources to navigate this complex landscape.

E-commerce and technology companies, which are the backbone of China's digital economy, must also adapt to data protection rules. Online marketplaces, such as Alibaba and JD.com, must prioritize consumer consent for data processing, particularly in targeted advertising. Tech giants like Tencent and Baidu face challenges with cross-border data transfers for their global operations, necessitating the use of data transfer mechanisms like Standard Contractual Clauses. Compliance is paramount to maintain trust among users and avoid regulatory penalties.

Data protection regulations in China touch upon critical sectors like healthcare, finance, and tech-driven e-commerce, impacting data handling practices, security measures, and international operations. These regulations reflect China's commitment to safeguarding personal data while requiring businesses to invest in compliance efforts, reshape data strategies, and navigate the intricate regulatory landscape to succeed in this dynamic market.

IX. Challenges and Future Developments

In China, data protection faces various challenges and is poised for significant future developments. One key challenge revolves around privacy concerns in emerging technologies, such as artificial intelligence (AI) and the Internet of Things (IoT).

(i) Privacy concerns in emerging technologies

Privacy concerns are prominent in China's rapid adoption of emerging technologies, particularly in the realms of artificial intelligence (AI) and the Internet of Things (IoT). AI-powered facial recognition is a noteworthy example. While it has been extensively used for various applications, from unlocking smartphones to tracking criminals, it has raised significant privacy questions. For instance, the widespread deployment of facial recognition in public spaces has led to concerns about constant surveillance and potential misuse of biometric data. This has prompted discussions about the need for comprehensive regulations to ensure responsible use and protect individuals' privacy rights.

In the context of IoT, the proliferation of connected devices and smart cities initiatives in China has created a vast web of data collection points. Smart devices, from home appliances to city infrastructure, constantly gather data on individuals' behaviors and activities. Privacy concerns arise regarding the security and control of this data, as well as the potential for unauthorized access or misuse. For instance, smart home devices may inadvertently record sensitive conversations or activities, raising questions about who has access to this data and how it is protected.

As China continues to lead in the development and deployment of these technologies, striking a balance between innovation and privacy protection becomes increasingly critical. It highlights the need for robust cybersecurity measures, transparent data usage policies, and regulatory frameworks that ensure individuals' personal data is handled responsibly, ultimately safeguarding their privacy in the age of AI and IoT.

(ii)   Potential Amendments to Data Protection Laws

China's data protection landscape is likely to witness potential amendments to its existing data protection laws, such as the Personal Information Protection Law (PIPL) and the Cybersecurity Law, as technology evolves, and privacy concerns persist. These amendments could encompass various aspects of data protection, including the definition of personal information, consent requirements, and regulations governing cross-border data transfers. For example, as China's digital economy continues to grow, there may be a need to fine-tune regulations to address emerging challenges like data security breaches, algorithmic decision-making transparency, and the protection of sensitive health and biometric data. Such amendments would reflect the evolving nature of data privacy concerns and ensure that China's legal framework remains relevant and effective in safeguarding individuals' personal data in an increasingly data-driven society. Businesses operating in China should closely monitor these potential amendments to adapt their data handling practices and compliance strategies accordingly.

Certainly, here are some potential examples of amendments to China's data protection laws:

a. Expanded Definition of Personal Information

As technology advances, the definition of personal information may be broadened to include new data types. For instance, laws could explicitly cover information generated by emerging technologies like wearable devices, biometric data, and even genetic information.

b. Enhanced Consent Requirements

Amendments might impose stricter requirements for obtaining user consent. For example, businesses could be required to obtain explicit consent for each specific data processing purpose, making it more challenging to rely on broad, blanket consent.

c. Data Localization Clarifications: Amendments could provide more clarity on data localization requirements. This might include specifying which types of data must be stored within China and which can be transferred abroad, helping businesses better navigate cross-border data management.

d. Stricter Penalties: To deter non-compliance, potential amendments may introduce even harsher penalties for data breaches and privacy violations. This could include larger fines and legal consequences for responsible individuals within organizations.

e. Algorithmic Transparency: As AI systems become more prevalent, amendments may require greater transparency in the use of algorithms that process personal data. Regulations could mandate explanations for automated decisions that impact individuals.

f. Protection of Health Data: With the rise of digital health technologies, amendments may introduce specific provisions to protect the privacy of health-related data, ensuring that it is handled with the utmost care and security.

g. Cross-Border Data Transfer Framework: To facilitate international data flows, future amendments could establish a framework for data transfer agreements with other countries, aligning China's data protection laws with global standards.

h. Data Breach Reporting: Amendments might stipulate stricter requirements for reporting data breaches promptly to authorities and affected individuals, ensuring a more transparent response to data security incidents.

i. These examples illustrate potential areas of focus for amendments to China's data protection laws, reflecting the evolving landscape of data privacy and technology. It's important for businesses operating in China to stay informed about these potential changes and proactively adapt their data handling practices and compliance strategies accordingly to remain compliant and protect individuals' personal data.

X. Conclusions

China's data protection landscape is undergoing significant changes with the introduction of new regulations like the Personal Information Protection Law (PIPL) and the Cybersecurity Law. These laws introduce stricter requirements for businesses handling personal data, including data localization, enhanced consent mechanisms, and heightened penalties for non-compliance. Moreover, privacy concerns are growing, particularly regarding emerging technologies like AI and IoT, necessitating a delicate balance between innovation and individual privacy rights.

For businesses operating in or with connections to China, compliance with data protection laws is paramount. Failure to do so can result in significant fines and reputational damage. Companies must invest in data security infrastructure, adapt data handling practices, and ensure that they are transparent in their data processing operations. Individuals in China can expect better protection of their personal data, enhanced control over how it is used, and greater transparency from businesses regarding data collection and processing. However, they must also be prepared for more stringent consent requirements, which may impact the convenience of using digital services.

The future of data protection in China is likely to involve further refinements and amendments to existing regulations. This could include expanding the definition of personal information, clarifying data localization requirements, and addressing emerging privacy challenges posed by AI and IoT. Stricter penalties for data breaches and privacy violations may be introduced to deter non-compliance. Additionally, as China seeks to facilitate international data flows, negotiations on data transfer agreements with other countries are expected. Overall, data protection in China is expected to evolve in tandem with technological advancements, privacy concerns, and global data governance standards. Businesses and individuals alike should stay informed and proactive in their approach to data privacy in this rapidly changing landscape.

References

1   Feng, Yang. "The future of China's personal data protection law: challenges and prospects." Asia Pacific Law Review 27.1 (2019): 62-82.

2   Calzada, Igor. "Citizens' data privacy in China: The state of the art of the Personal Information Protection Law (PIPL)." Smart Cities 5.3 (2022): 1129-1150.

3   Pernot-Leplay, Emmanuel. "China's approach on data privacy law: a third way between the US and the EU?." Penn St. JL and Int'l Aff. 8 (2020): 49.

4   Geller, Anja. "How Comprehensive Is Chinese Data Protection Law? A Systematisation of Chinese Data Protection Law from a European Perspective." GRUR International 69.12 (2020): 1191-1203.

5   Creemers, Rogier. "China's emerging data protection framework." Journal of Cybersecurity 8.1 (2022): tyac011.

6   Weber, Philip Andreas, Nan Zhang, and Haiming Wu. "A comparative analysis of personal data protection regulations between the EU and China." Electronic Commerce Research 20 (2020): 565-587.

7   De Hert, Paul, and Vagelis Papakonstantinou. "The data protection regime in China." Depth Analysis//Brussels Privacy Hub Working Paper 1.4 (2015).

8   Sun, Liyuan, Hongyun Zhang, and Chao Fang. "Data security governance in the era of big data: status, challenges, and prospects." Data Science and Management 2 (2021): 41-44.

9   Hsu, Piao-Hao. "Emerging China data protection law: soft power from EU GDPR?." Tamkang Journal of International Affairs 25.1 (2021).

10   Roberts, Huw, et al. "Governing artificial intelligence in China and the European Union: Comparing aims and promoting ethical outcomes." The Information Society 39.2 (2023): 79-97.

11   Cheng, Jing, and Jinghan Zeng. "Shaping AI's future? China in global AI governance." Journal of Contemporary China 32.143 (2023): 794-810.

12   Gong, Jiankun, et al. "Do privacy stress and brand trust still matter? Implications on continuous online purchasing intention in China." Current Psychology 42.18 (2023): 15515-15527.

Chapter 21

Facial Recognition Versus Privacy: Striking a Balance

Fiza Ahsan

B.A.L.L.B. (Hons.) Student
SSOL, Sharda University
Greater Noida, UP, India

Rishita Tripathi

B.A.L.L.B. (Hons.) Student
SSOL, Sharda University
Greater Noida, UP, India

Abstract

We live in the world full of cameras from surveillance cameras to the very own mobile devices camera. Face Recognition Technology is a technology which have a rapid growth in the world over the years. All the devices nowadays have face recognition technology system, Face recognition technology (FRT) is a biometric system used originally for the fight against the crimes by identifying and verifying an individual identity from their facial feature, unique features and poses etc. Face recognition technology emerges as one of the most used technologies in the world. Like every technology have pros and cons this one has too. It is beneficial for security, safety. In banking and devices access become much easier. Now, one can access it by their faces and no need of the passwords which is hackable. As a result, an individual privacy is in threat and individual liberty is in stake. Individual identity can easily be known by any third party. Humans can use it inappropriate way. Without knowing, they are always under the face recognition technology system without their consent. The future of facial recognition technology will be influenced by a complex interplay of technological advancements, regulatory frameworks, ethical considerations, public attitudes, and societal needs. Ongoing dialogue and collaboration among various stakeholders will be crucial in shaping a future that balances innovation with responsibility.

Keywords: - Face Recognition, Technology, Privacy, Liberty, Threat.

I. Introduction

Face Recognition Technology introduced by the pioneers of face recognition were Woody Bledsoe, Helen Chan Wolf and Charles Bisson. In 1964 and 1965, the above three began to work using computers to recognize human face. Face recognition technology commonly known as FRT. This technology widespread in the world very fast. In nowadays FRT Plays a dominant role in the people's life because in devices we use I.e., in mobile phones and apps etc. They have the feature of FRT through which a person can unlock their device and apps. It uses gradually spreads in the world. FRT is the cutting-edge technology has the potential to revolutionize various industries to security to law enforcement to customer experience to personal privacy. FRT Also known as Fr is an advance method to identify or verify a person identity by using facial features such as mouth, eyebrow, nose and contour of jawline. These features Then captured and converted to the digital data points creating a facial template specific to individual.

Originally, this technology used to fight crimes. Facial recognition now mundane or material uses such as accessing bank account and social media platform but there are worries that could also present a threat to privacy and individual liberty the technology also raises questions about the use of database. Facial recognition technology has a wider ray of application and benefits it has the positive impact of our daily life. In the field of security and law enforcement it helps in enhance surveillance system, border control and even locating missing persons. Airports and other high security areas utilize this technology to identify the individual on watch list and enhanced security measures. Moreover, it helps in reducing identity theft or online shopping fraud in various sectors be it banking, insurance or online shopping. Another industry greatly influenced by FRT is the retail sector. buy recognizing recurring customers' business can provide personalized experiences and targeted advertisement based on individual preferences. this not only improves customer satisfaction but also increases sales revenue. Additionally, FRT application in smart homes and smart devices where it enables a more intuitive and secure user experience. S any emerging technology also brings.

II. Main Features of Face Recognition Technology and History

The process of FRT contains three main steps- This system captures image of video of a person face through a camera or surveillance device. It analyzes the image and extracts key facial features creating multidimensional mathematical representation face print unique to that person. This face print compared to the database to known individual to find the match. If there is a match the person identity is confirmed and if not, then the person's identity remained unrecognized.

FRT is a biometric technology used to authenticate or identify an individual form of picture or a facial image. This technique makes it possible to authenticate identity to check an individual who he or she claims to be. The images are compared to an existing Image in a data base. an individual can also be identified in a group comma place image or database from stool or from a video software generate image based on the facial unique traits. The base of the years the distance between the two pupils, shape of the nose eyebrows and mouth and even skin grains, here and clothes are not taken into account. The next step to compare face print in a database. To reduce the margin of the error the quality of image must be good the face must be well let. Full face should be at the correct distant from the lenses.

It is a quick and efficient verification system. It is a technology potentially capable of matching a human face from a digital image or a video frame against a database of faces. Facial recognition systems have seen wider uses in recent times on smartphones and in other forms of technology such as robotics. In recent years FRT has developed a lot and use of FRT has grown exponentially. Facial recognition systems are employed throughout the world today by governments and private companies. Use of FRT is rapidly increasing day-by-day and the market for FRT is growing as organizations employ the technology for many reasons like verifying and/or identifying individuals to Grant them access to online accounts, authorizing payments, monitoring employee attendance and much more. The global facial recognition market size is forecasted to each $12-67 Billion by 2028. Face recognition has the potential to benefit people and society however significant drawbacks with this technology are concerned surrounding privacy while there is no doubt that these concerns are valid. We need to be careful of sweeping generalizations.

The use of facial recognition systems has also raised controversy with claims that the system violates citizens' privacy. These claims have led to the ban of facial recognition systems in several cities in the United States. Growing suicidal concerns led social networking company meta platforms to shut down its Facebook facial recognition systems in 2021, deleting the face scan data of more than one Billion users. The change represented one of the largest shifts in facial recognition usage in the technology's history. The European Parliament passed a resolution in October 2021 prohibiting police from implementing large scale facial recognition in public places or border checks.

III. Face Recognition Privacy – The Major Concern

Recently the face recognition privacy has become one of the major concerns for many people. It has become a source of anxiety for many people. These anxieties are reflected by many of the institutions all around and worldwide that have implemented same strict regulations for the protection of people's biometric data.

Europe's General Data Protection Regulation (GDPR) is one of the examples of such anxiety and concern. According to Europe's GDPR, people are entitled to their privacy and if there's any breach of such privacy then it will be met with consequences. The worries which people have about privacy are not unreasonable. It is very obvious that no one would feel good that they are being watched or their personal information could be leaked.

One of the major concerns about security is hacking. For example, if someone's camera with face recognition was compromised somehow, a hacker might be able to get their hands on identifying information about anyone in the system of that camera. One more concern is that the face recognition can be helpful to law enforcement agencies but on the other side the police can access face recognition footage only with permission from the owner of the camera. Another possibility being, imagine if these were sent to the cloud where they could be matched up with some other technologies like, for example- Amazon's recognition software that can match the faces with massive database. This is a very worrying and one of the major concerns. Despite of the disturbing things that have been mentioned here the specific face recognition technology used, how it is deployed and what all are the computing methods which is it relies on, all these play a role in determining the privacy risks.

Here, a question may arise that- Is there really a privacy concern with face recognition? So, the short answer to this question Maybe/or is, yes. As discussed, and mentioned earlier privacy concerns surrounding the face recognition technology are absolutely valid. From hacking concerns to having keen observation to issues like how data is stored, face recognition can put personal identifiable information at risk in same or the other instances.

There are different types of FRT constraint facial recognition technology and unconstrained facial recognition technology. constrained facial recognition technology, the software that compares the first presented data point of access to the database of an authorized person. To determine whether there is a match access is granted and if there is no match excess is granted then there is a security alert can be triggered. Unconstrained facial recognition technology, the software compares faces in the preset watch lists to face in the video surveillance footage. In real conditions such as pose variation low resolution, facial expression, different hair styles, illumination changes, blurring and occlusion between people. Typically, people in vehicles will appear in different angles, poses and quality levels. This is much harder. it is usually seen in service it is usually seen in surveillance camera.

IV. Facial Recognition Technologies Privacy Principles

Following are the major privacy principles with respect to FRT -

(i) Consent

Obtain express or affirmative consent from an individual when enrolling in a program that uses FRT for identification or verification purposes. identifying an individual to third party who would not otherwise have known that individual identity. new unique or different biometric identifier should be created and maintained over a period of time without the consent. Exemption where there is no requirement of express or affirmative consent. when notice is provided but no consent is required. Collection of data including face recognition templates for security, fraud and assets protection. it does not require a consent.

When sharing the identification of an individual occurs within a vendor management framework where third party is contracted service partner to provide goods and services requested by an individual. Where notice is provided but opt out consent is sufficient.

Limited use of FRT that create facial template data tied unique Persistence identifier are exempted.

(ii) Use - respect for context

Use of FRT in such a way that it is fair and not unfair to consumers including weighing the privacy risks against clear and articulate benefits to consumers and provides consumer to make choices.

(iii) Transparency

Companies implementing FRT system or software should develop and publish their privacy policies explaining their use of FRT in clear and detailed way of the data collected. company which designs and develops FRT software and hardware system should take reasonable steps by making recommendation and give guidance, facilitate transparency and privacy compliance of not used by third party their FRT.

(iv) Data Security

Company will come with comprehensive data security Programme that is only designed to protect the security, privacy, confidential information by an unauthorized access or use or unintended or inappropriate disclosure. Companies will also update their policies from time to time to improve the privacy policies so that there are no loopholes left.

(v) Privacy by design

Companies' privacy by design should be incorporated in the company from very beginning who manufactures, resell, install FRT system. Company should implement privacy by design into all organizational institutions' practices. It should promote the consumer privacy and data security through all the organization.

(vi) Integrity and access

Implement reasonable measures or steps to maintain the accuracy of FRT. Offer individual reasonable access to review or request a correction or delete of inaccurate FRT data. Law enforcement access, FRT template data may be disclosed, or commercial data base searched in response to law enforcement entities with a court order or court order warrant, a writ to appear in the court or summon issued by a judicial officer to comply with other state or local laws for law enforcement action.

(vii) Accountability

Establish or implement the transparent policies procedures and practices to ensure that these principles underlie the use of FRT by the company in all consumers facing contexts to the fullest extent application. Take reasonable steps to ensure the use of FRT and data by the organization and in partnership with all the third-party service providers or business partners comply to these principles.

(viii) Benefits of Facial Recognition Technology

Companies and organizations have developed consumer-facing applications of facial recognition technology that have improved the customer experience. FRT applications include enhancements in safety and security, access and authentication, tremendous leaps in diagnosing diseases, and targeted marketing and customer service.

Of the listed applications of FRT, those related to safety and security are often the most hyped benefits associated with the technology. Law enforcement agencies use FRT for finding missing persons, identifying wanted criminal and for early threat detection. In medicine, FRT can be used to diagnose diseases that cause detectable changes in appearance and is expected to become an invaluable diagnostic tool for all types of conditions.

On a consumer level, FRT can be used as a tool for locking personal devices and accessing accounts or services, like those in banking. Face recognition requires no hackable passwords, and the technology can be used at physical branches and ATMs. FRT also has valuable applications in retail. Using your face to pay can drastically cut down checkout lines. Customer experiences can also be improved through customized shopping experiences.

V. Loopholes in Face Recognition Technology

The use of facial recognition technology tends to be harmful due to the following practices:

(i) Race and Gender

Many at times it has been reported that its accuracy rates fall based on race and gender and due to this a person is misidentified as someone else, which can result in false positive and also if a person is not verified as themselves it tends to be a false negative.

False positive can lead to bias against the person who has been misidentified. Cases relating to false negative can lead to debarring of the person from essential schemes which uses FRT for access. For example- failure of the biometric based authentication under Aadhaar which led to failure from receiving services to the people who were in need and in turn it led to starvation deaths.

There is a biasness in facial recognition system. No doubt our society itself is biased and discriminatory and this is embedded into data sets as well.

Research has proved that gender and racial biasness arises from the fact that under-represented groups just do not produce enough data that systems can train on.

(ii) Violation to Privacy

Though through legal frameworks like data privacy regimes, government plans to talk about the questions of privacy and try to sort out all the related issues but it aims to achieve with the use and help of technology somewhere or the other it comes into conflict with one another.

(iii) Reliability

The reliability of data collected is taken into deliberation as the data collected may be used by court of law during criminal trial.

(iv) Absence of data protection law

FRT system in the absence of data protection law is also one of the major concerns.

VI. Conclusions

The face recognition is the cutting-edge technology used to authenticate the and verify an individual identity. It is an advance method, which is used to identify individual by their facial features. It is originally used to fight against crimes but now widely used in the banking and security sectors. It has pros and cons attached to it. It is beneficial for easy access of the devices and in banking sectors also. It also emerges as a threat to individual privacy and liberty as their privacy be infringed by leaking of some confidential information to the third party which should be remain private. This information used by third party in an inappropriate manner by which individuals' privacy be violated. Face recognition technology includes privacy policies, so every institution and individual should comply to it.

References

1   https://xailient.com/blog/face-recognition-and-privacy-is-there-cause-for-concern/.

2   https://www.sciencedirect.com/science/article/abs/pii/S0167404820302273.

3   https://itbrief.co.nz/story/benefits-vs-risks-of-facial-recognition-technology.

4   https://www.verdict.co.uk/is-facial-recognition-a-threat-to-privacy/?cf-view.

5   .https://www.isaca.org/resources/news-and-trends/newsletters/atisaca/2022/volume-51/facial-recognition-technology-and-privacy-concerns#:~:text=Data%20breaches%20involving%20facial%20recognition,faces%20cannot%20easily%20be%20changed.

6   https://www.liberties.eu/en/stories/facial-recognition-privacy-concerns/44518.

7   https://www.asisonline.org/security-management-magazine/monthly-issues/security-technology/archive/2021/december/facial-recognition-in-the-us-privacy-concerns-and-legal-developments/.

8   https://senstar.com/senstarpedia/facial-recognition-problems/.

9   https://sitn.hms.harvard.edu/flash/2020/racial-discrimination-in-face-recognition-technology/.

10   https://www.csis.org/analysis/questions-about-facial-recognition.

11   https://blogs.microsoft.com/on-the-issues/2018/12/06/facial-recognition-its-time-for-action/.

12   https://www.axios.com/2023/01/07/facial-recognition-issues-problems.

Chapter 22

A Review on Patient's Legal Rights vis-a-vis Privacy in the U.S.A. and India

Ananya Sharma
B.B.A.L.L.B. (Hons.) Student
SSOL Sharda University
Greater Noida, UP, India

Jitesh Sharma
Advocate
Punjab and Haryana High Court
Chandigarh, India

Abstract

Electronic medical record (EMR) software is a digital version of paper-based patient health records. Paper has historically been used to create medical records. Medical records can now be converted from paper to electronic formats thanks to modern advancements in information and communication technology. So, it is crucial in healthcare settings to protect user data privacy. The key difficulties are figuring out how to increase the ability of EMR systems to protect patient data privacy while maintaining these systems' functionality and interoperability. If EMRs are embraced by healthcare organizations, they can offer numerous advantages to doctors, patients, and healthcare services. But worries about patient information security and privacy may prevent certain healthcare organizations from adopting EMRs at high rates. One of the major issues of EMR is the security of a large amount of sensitive health data stored in many places and formats. The development of technologies intended to safeguard privacy and enable the wider use of EHR has also resulted in significant changes over time, particularly in this area. But many of the same privacy ethics issues still exist today, despite technological advancements.

Keywords: Confidentiality, Electronic health records, Privacy, Security, international.

I. Introduction

An e-health record is described as a digital representation of a patient's medical history that a healthcare professional has maintained for a period of time. It includes all the crucial administrative clinical data necessary to the care that a person receives from a specific provider, including demographic data, updates on treatment, medications, significant signs, health history, vaccination reports, laboratory results, and radiology reports. Many medical centers and organizations use paper to keep track of information pertaining to health, which has resulted in a significant paper trail. As a result, most organizations have developed an interest in switching from traditional medical records made of paper to e-health records. The conversion of paper medical data to electronic data gives rise to fresh privacy worries. Only authorized employees of the treatment provider organization, particularly physicians, clinical support staff, and administrative staff, have access to paper records for specific patients. Individual patients' electronic medical files may also be accessed by the same staff members. Electronic documents, nevertheless, can be released in very large quantities, as opposed to traditional documents, for instance through the unintentional loss or theft of electronic storage devices. The danger of breach rises as more protected health information (PHI) is made accessible to networks of providers, their business partners, and for ancillary purposes like investigation and advertising. Patients who use records on paper can be reasonably sure that they're aware of where their documents are kept and that just a handful of others will have access to this data. Patients cannot be assured of who has their data, how it is being utilized, or how well it remains secure from unintentional release or breach of security as PHI is increasingly stored, shared, and used electronically.

In the medical field, confidentiality offers two goals. The initial objective is to protect the privacy of patients, so they do not feel vulnerable or ashamed. The following objective is to foster open dialogue between the physician and the person being treated. It is challenging to uphold confidentiality in today's technologically advanced healthcare delivered by interdisciplinary teams. The right to privacy, security, and anonymity are major problems that need to be dealt with in computerized medical records.

Any breach in the security of the patients' data could result in legal issues, which is why doctors are typically highly concerned that a person who is not authorized could access the information of patients that is contained in the digital health record database and exploit it. Most physicians who use computerized records for medical purposes think that paper records are considerably more secure and confidential than electronic records, hence they favor paper records over electronic ones. This demonstrates how seriously the problem of security and privacy on EMR is taken. Patients may elect not to disclose material to avoid inappropriate usage if they are not guaranteed privacy.

As a result, many nations are working to change their system of healthcare by using technological advances. The usage of IT has assisted people receive better care, enhanced overall wellness, and lowered the expense of medical care. The current advancements in technological innovation have led to the digitization of health records, which has led to the creation of new or enhanced methods for the efficient collecting, analyzing, conserving, consulting, and dissemination of health data. Health department administrators undertaking assessments of health and formulating policy have considerably greater access to digital medical data since it is more versatile and can be disseminated throughout healthcare institutions. Patients can also access this information. The majority of the work to date has pointed to advantages of a digitalized framework for medical outcomes. However, these digitalized health records expose medical records to information technology-related vulnerabilities. Potential users of health IT are highly concerned about IT-related security and privacy, which has a detrimental impact on the trust in electronic medical records. The usage of electronic health records may not be completely supported by both patients and doctors as a result of this decline in trust, endangering the significance of technological advances. This could then end up in a decline in overall health surveillance or health study, as well as less efficient medical provision.

Therefore, it has been stated that prior to its introduction, cyber-safety procedures used in conjunction with digital medical records ought to be thoroughly understood. The Health Information Technology for Economic and Clinical Health Act and the Health Insurance Portability and Accountability (HIPAA) Act mandated many safety precautions because the data kept in the EHR is extremely critical. HIPAA defines the three foundations that it employs to apply administrative protections, structural safeguards, and technical safeguards to ensure that the confidential medical data remains protected. The three pillars, commonly known as the healthcare security safeguard themes, cover anything from methods for securing computer locations to using antivirus software to preserve patient data.

II. Privacy of Electronic Health Records

The importance of privacy as an underlying concept and prerequisite for the unrestricted overseas transfer of identifiable information has long been acknowledged by the Organization for Economic Co-operation and Development (OECD). The fundamental goal was to offer a set of rules that would both safeguard privacy and encourage the unrestricted exchange of information. These recommendations offer clear principles for member nations to adopt as the foundation for national laws globally in order to accomplish these goals. Consequently, the recommendations of the OECD play a crucial role in the creation of laws like GDPR.

Privacy is fundamental to the Convention of Human Rights, according to the United Nations General Assembly (UNGA). However, in modern times, privacy has taken on a variety of meanings that are determined by every territory or nation. As a result, numerous studies have been conducted on the privacy of clinical data.

Determining how much of the data belongs to the patient and how much can belong to healthcare organizations, as well as whether the data owner's assent is necessary if the data are used for research, has not been simple. To guarantee the confidentiality of the information and, at the same time, to ensure its preservation, the data acquired must be shielded from unlawful access. Only those uses that the patient has authorized should be allowed to utilize this medical data. The regulations and processes for access to information must be followed in order to guarantee that only authorized individuals or apps are granted access to the patient's health information.

The term "health information technology" refers to all information technology platforms that are used to assist the delivery of medical treatments and the administration of the medical industry. As a result, the data that medical information technology stores are extremely vulnerable. The data contains details about the health status of the individual as well as details about the ailments, diagnoses, tests, and therapies that have been administered. As a result, this information must be safeguarded to prevent manipulation, enabling patients to continue disclosing details about their employment and personal wellness while taking their ethical and legal responsibilities into account. The ever-changing nature of healthcare data has an adverse effect on maintaining the privacy of medical documents.

Privacy, security, and anonymity are the typical challenges that must be addressed in the electronic health record EHR system. Despite their close connections, security and privacy are actually very distinct. Accessibility to an individual's confidential data is limited and permitted for only those who are authorized, which is how security is defined. Contrarily, privacy implies the privilege that someone has to choose for oneself when, how, and to what extent others have access to their confidential data.

III. Legal Framework in the U.S.A.

HIPAA was initially presented in 1996. Employees are permitted by law to maintain their medical insurance policies even when they change jobs. It is also known as Public Law 104-191. It created guidelines for safeguarding medical data (Office for Civil Rights, 2003). In order to guarantee the caliber of medical care and the safety of the public, guidelines were created to protect individual medical records while enabling dissemination of data. DHHS enacted the Privacy Rule, which codified the HIPAA obligations. Medical care clearinghouses, specific business partners of covered businesses, and all health insurance policies that offer or reimburse for medical services are included under the Privacy Rule. It also pertains to medical professionals who share patient data for business activities, medical claims, or similar specific goals. The

HIPAA privacy provisions cover PHI, or protected health information, that contains a person's either direct or potential personal identification. The rule authorizes exposure for a number of public interest purposes, such as certain administration, law enforcement, and health-related initiatives; military operations; and science. It also authorizes release for treatment, reimbursement, and medical care operations. Consent from patients is not necessary for some uses of publication, although it is necessary for some other uses of exposure. Patients have the right to ask for information to, rectification of, and a review of breaches of their PHI. They may also ask for constraints on disclosure for the HIPAA-allowed uses, and if an entity accepts the limitations, it is accountable for carrying them out. Health information technology was still in its infancy when HIPAA was passed. The HIPAA security provisions were improved in various ways by HITECH, which Senate legislated shortly after realizing the possibility of health information technology to increase the affordability of treatment. Firstly, it broadened the scope of business associates to encompass institutions that regularly access and transmit PHI, such as RHIOs and HIOs, and it subjected business associates to the same safety standards and sanctions as health care plans and medical facilities. Secondly, it laid up guidelines for informing folks whose PHI has been compromised. Thirdly, it instructed DHHS to inform covered companies and people of their legal obligations with relation to the confidentiality and safety of PHI. Furthermore, it allowed patients more rights about their data in digital form and instructed DHSS to issue guidelines on the determination of the minimal required data. Additionally, with a few exceptions, it outlawed the selling of PHI or EHRs. Last but not least, HITECH required stricter compliance with existing PHI confidentiality laws and increased the severity of sanctions for infractions.

Additional confidentiality constraints are imposed at the national level for records of care given in narcotics and alcohol rehabilitation centers, as well as for therapies for dependency on drugs or alcohol, a drinking disorder, and, in the case of veterans, contracting HIV or sickle cell syndrome. However, the transmission of PHI among the DoD and the VA is exempt from these rules. When it is "considered essential by fitting military leadership officials to guarantee that there is adequate execution of the military mission" or upon separation from service in order to support the identification of perks qualification, the Department of Defense oversight permits dissemination of shielded PHI for service members by any covered entity.

The regulations concerning the confidentiality of PHI vary between states. Other jurisdictions have imposed limitations on the sharing of PHI regarding particularly delicate subjects like mental disorders or AIDS, while certain states do not have any regulations or have embraced the HIPAA regulations. Although the care delivered in MHS or VA hospitals is exempt from the state's requirements, civilian clinicians may still feel constrained by them, except when they are caring for patients who are on active military service.

Scholars that are interested in security and HIE have noticed that as data transfer capacities advance, some loopholes in federal privacy laws exist, which undermine public confidence in HIE. To guarantee that PHI communicated over a geographically dispersed system is adequately safeguarded after it leaves the authority of the initial holder of the data, for instance, network layout attributes and the supervision and oversight procedures have yet to be defined and verified. McGraw et al also point out the necessity of modifications to policies that will reinforce HIPAA confidentiality safeguards and foster confidence in HIE. For instance, by outlining health-related operations, increasing limitations on further utilization of PHI for advertising, and reliving confidentiality norms to make sure there is little chance of re-identification of unidentified information.

IV. Laws governing the Confidentiality and Privacy of a patient in India

The Indian Medical Council (Professional Conduct, Etiquette and Ethics) Regulations, 2002 in chapter 7- (7.14) state that a licensed physician is not allowed to share any patient information that was learned during the course of providing care or while using their professional judgement.

The implications of the infraction are discussed in Chapter 8.2 of the text. The document demonstrates that if an allegation arises about the professional misconduct of a licensed physician and that grievance is put before the Medical Council of Disciplinary Action, the rightful medical council will initiate an inquiry and additionally provide the licensed physician a chance to be heard in person or by a pleader after receiving the complaint. And if the licensed doctor is discovered to have engaged in unlawful conduct during the duration of the investigation or proceeding, the Medical Council may order that his medical practice be shut down permanently or just temporarily, depending on the circumstances.

Additionally, in accordance with chapter 8-(8.5), the rightful Council may forbid the doctor from engaging in the procedure or practice that is the subject of investigation or inspection if the decision about the allegation made against him is still unresolved.

There are no particular rules in India that guarantee the safety and anonymity of information pertaining to patients aside from the "code of ethics," although the ministry of health presented the Digital Information Security in Healthcare Act (DISHA) in 2018; it has not yet been completed.

Given the prevalence of electronic medical files that go beyond paper, it is anticipated that it will provide an extensive legal structure to protect patient privacy.

The public will have entire control of their medical information once it is finalized. For instance, the DISHA Act will entirely safeguard data if it is entered into the medical system by a doctor when a patient visits the doctor, and the doctor enters the test results into a digital medical record.

Absolute secrecy and anonymity are impossible in the medical sector as, even though doctors are aware that failing to share this information with the public could lead to the transmission of hazardous illnesses like HIV/AIDS and tuberculosis, they would begin maintaining all patient medical records concealed or confidential. Therefore, it is occasionally necessary to share information about patients in the greater good of the general welfare.

V. Recent Judgments in India

This part shall discuss the significant judgments relating to privacy of medical information –

(i) Mr. Surupsingh Hrya Naik v. State of Maharashtra

In this particular case, there existed a contradiction between the Right to Information Act of 2005 and the Medical Council Code of Ethics. In the present case, it was debated whether releasing the health records to the public would violate the right to privacy in accordance with the Right to Information Act. Therefore, the Bombay High Court determined that in this case, the right to information will take precedence over the right to anonymity and safety. As in the current case, the petitioner endured the punishment that the Honorable Supreme Court had placed on him, during which time he was hospitalized and underwent surgery because he had been having troubles with his heart, hypertension, and blood sugar levels. In this case, it was decided that the medical files of an individual who has been handed down a sentence found guilty or is in detention of the police or a court should be made accessible to the person requesting the details if the person was admitted to a nursing facility or hospital during that time. However, the hospital or nursing home should be maintained by the State, a government agency, or another organization. The information that is documented in writing can only be withheld in extraordinary circumstances with good reason. The Right to Information Act can therefore take precedence over the Medical Council Code of Ethics in this instance.

(ii)   Radiological and Imaging Association v. Union of India

In this case, the petitioner contested both the Collector and District Magistrate of Kolhapur's circular, which demanded that all sonography machines be equipped with SIOBs (silent observers) as part of the "save the baby" campaign to improve the region's sex ratio and that all radiologists and sonologists submit the online Form F required by the Pre-conception and Pre-natal Diagnostic Techniques Rules (PNDT). On the basis that it breaches their patients' privacy, the petitioner objected to this. The Bombay High Court ruled that the ultrasound machine's silent spectator, which stores the images and prevents them from being transmitted online to any servers, must only be widened at the Collector's or Civil Surgeon's request and in the presence of the relevant radiologist, sonologist, or doctor in charge of the Ultrasound Clinic. After all relevant facts were taken into account, it was determined that neither the patient's right to privacy nor the doctor's obligation of secrecy had been violated. It was further noted that the public interest must prohibit or restrict the right to privacy and that it must evolve in tandem with every single PNDT rule. Additionally, because the country's diminishing sex ratio was deemed an urgent public concern that might supersede the right to privacy, the deployment of a silent observer system on a sonograph provides the essential protections and safeguards and does not breach any privacy rights.

(iii)   Mr. X v. Hospital Z

In this specific case, the respondent took a specimen of the appellant's blood because it was going to be transfused into another individual. The appellant's specimen's blood examination revealed that the appellant is infected with HIV. Therefore, the applicant's wedding was annulled as hospital officials disclosed information to his family without his explicit authorization, and somehow the information ended up reaching the girl's family. The appellant was severely chastised and outcast by the local population as a result, forcing him to quit his job and move to another town.

In order to seek compensation for the respondent's actions, the appellant petitioned the National Consumer Dispute Redressal Commission. The appellant claimed that the respondent had violated his responsibility to keep the health data of the patient discreet by illegally disclosing the patient's vital information without permission from the patient. However, the National Consumer disagreement Redressal Commission rejected their argument and said that a civil court would be the appropriate venue for resolving such a disagreement.

Following that, the appellant entered the plea in the Supreme Court and asserted that the "obligation of diligence" is applicable in the medical field and also includes the "duty of confidentiality." It was further argued that because the duty was broken or infringed the respondent in question was responsible for the harm that ensued.

Therefore, the Court determined that Ms. A's basic right to information will prevail over the appellants' fundamental right to privacy in the event of a dispute between those two rights when it comes to the fatal illness that was endangering Ms. A's existence. The Court also decided that the responsibility for upholding confidentiality in the doctor-patient relationship is not absolute and can be compromised for a greater cause or concern, leading to the respondents being found not guilty. Therefore, the respondent was found not guilty for the same reason.

But the Court found the appellant guilty under Sections 269 and 270 of the Indian Penal Code since he knew he had a communicable illness but still chose to get married.

VI. Conclusions

The clinician has an obligation to uphold the privacy and confidentiality of their patients as their top concern. The doctor-patient relationship is built on trust and includes a lot of information or data that needs to be kept private to avoid abuse. Every industry is transitioning quickly from paper to digital records, and this shift from paper to the internet necessitates safeguarding information and constant monitoring. Only the Indian Medical Council (Professional Conduct, Etiquette and Ethics) Regulations, 2002 are currently in effect in India; several other Acts have not yet been created or brought into action.

However, there are a lot of loopholes in the regulations that have been formed or are being developed since, on one end of the spectrum, they guarantee the right to privacy while, on the other, they state that anybody may request access to documents under the Right to Information Act of 2005.

In order to create an effective structure, the State must actively participate in the creation of legislation governing the security and privacy of patients as well as facilitate interaction between various organizations operating in the private as well as public sectors. Together, the efforts of these industries or other interested parties can guarantee the establishment of a solid and potent fundamental structure in the nation on which the Right to Privacy and Confidentiality in Healthcare can be effectively built.

References

1 https://cis-india.org/internet-governance/blog/privacy-in-healthcare-policy-guide.

2 http://samch.doh.gov.ph/index.php/patients-and-visitors-corner/patients-rights.

3 http://contacttracing.ashm.org.au/why-are-privacy-and-confidentiality-important.

4 http://archive.nmji.in/archives/Volume-27/Issue-1/27-1-SFM-III.pdf.

5 https://www.sciencedirect.com/science/article/pii/S1110866520301365.

6 https://www.researchgate.net/publication/15651989_Confidentiality_of_Medical_Records_The_Patient's_Perspective.

7 https://link.springer.com/article/10.1007/s10389-022-01795-z.

8 https://www.ncbi.nlm.nih.gov/pmc/articles/PMC9327609/.

9 https://www.jstor.org/stable/10.7249/j.ctt3fgzjz.10.

10 Indian Medical Council (Professional Conduct, Etiquette and Ethics) Regulations, 2002.

11 Indian Penal Code, 1860.

12 Surupsingh Hrya Naik v. State of Maharashtra AIR 2007 Bom 121.

13 Radiological and Imaging Association v. Union of India, AIR 2011 BOM 171.

14 Mr. X v. Hospital Z, (1998) 8 SCC 296.

Chapter 23

Protection of Digital Personal Data in India: An Analysis

Arvind Gupta

Advocate On- Record

Supreme Court of India

New Delhi, India

Vanshika Sah

B.A.L.L.B. (Hons.) Student

SSOL, Sharda University

Greater Noida, U.P., India

Abstract

'Data' the word comprising of four alphabets is more valuable than any precious metal and equal to any luxury in the world in the postmodern era. It can be compared to almost every precious thing and commodity and can be termed renewable oil also because the oil can, for once not be reused but data can be reused without losing its relevance. Thanks to its sustainable nature, (copy paste and saved), internet connection, and the biggest platform in the world, the cloud, the data or the directory of the identity becomes so vulnerable to threats not only originating from outside the borders but also inside them. Not long ago, it was reported that illegitimate tapping of individuals was being done using the Aadhaar details of individuals. This research paper highlights that why the government institutions should be also included as a part of the data collection bodies under the scanner as a part of this data protection Bill, its key points from a different analytical perspective and loopholes. It is pertinent to mention that several amendments were suggested by a joint parliamentary committee on the same lines because of which, the Bill has been withdrawn. The data will be collected through Doctrinal method.

Key Words- Internet, Government, Privacy, India, Bill, deliberations.

I. Introduction

India with over 820 million internet users, soon to touch 1.2 billion, has become the world's largest connected democracy, with the leading presence on the global internet. Unlike China›s walled off and heavily censored internet (or intranet), the internet in India is open and accessible and interconnected to the global digital network, just like western democracies.[1] The personal data protection Bill was initially in the year 2019 by the Ministry of Electronics and Information Technology. The Bill was further analyzed by a Joint Parliamentary Committee in consultation with experts and stakeholders. Later the same was subjected to a report by Justice Krishna Committee. As of 2022, the Parliament has tabled up another draft of the same Bill with corrections manifold. The Bill proposes the setting up of a Data Protection Authority of India for the same.

As of the 2020, almost 43.5% of the population was actively using the internet and definitely these many people generate data. All of them willingly or unwillingly generate some amount of data and this data is unprotected as of the second decade of this digital millennia. Where strong territorial jurisdictions like the European Union have a stringent data protection law in place such as the General Data Protection

Regulation or the GDPR, the largest democracy lacks one. The absence invites unwanted dangers and perils as well as several precariousness and uncertainty in the digital arena.

The Cambridge Dictionary defines data as '*Data refers to the facts or information, especially facts or numbers, collected to be examined and considered and used to help in decision- making or information in an that when put together can be used for analysis and predictions to some extent, and used to help decision- making, or information in an electronic form that can be stored and used by a computer.*'[2] Thus data is the new gold. There are instances of data being sold online for money openly in the market.

It is no hidden fact that the government agencies have a legitimate authority to tap on the data of the citizens but what makes it wrong is the possibility of this data being misused. The recent findings that say that the *Aadhar data of over 11 crore farmers has been leaked due to a fault in the government website PM KISAN Samman Nidhi.*[3]

The National Identity card scheme, which was flagged by the Honorable Prime Minister, Shri Narendra Until 2017, A part of the initiative's website, according to the security researcher, was returning farmers' Aadhaar numbers. *"A dashboard feature on the PM Kisan website allows you to view various charts and data. In a blog post, he claimed, "An endpoint in the dashboard was leaking Aadhaar numbers of all the farmers based on area (state, district, village).* These examples highlight one thing that, the body that had been set up to ensure protection of our data is itself under the apprehension that at the end of the day, it is also a political party, and it needs to be in power.

German law only allowed such searches to be conducted in criminal cases, but an amendment now allows searches even in instances when authorities have no basis for suspicion. Bijan Moini, a lawyer representing GFF, now fears that a "national database of biometric features is being compiled."[4]

II. National Database of Biometrics

One might ask what is a national database of biometrics? Biometrics is the measurement and statistical analysis of people's unique physical and behavioral characteristics. The technology can be used to identify and access control or for identifying individuals who are under surveillance. The focus needs to be shifted on the fact that this data contains very sensitive information about individuals. The General Data Protection Regulation of the European Union, in fact defines 'biometric data' as such personal data that results from specific technical processing relating to the physical, physiological or behavioral characteristics of a natural person, which allow or confirm the unique identification of that natural person[5]. *More than a Billion Indians have uploaded their biometric details to a national database in exchange for a unique 12-digit ID that authorities say will transform how citizens interact with government and propel the world's second-most populous country into the digital age.*[6]

In 2018, according to an internet security threat report issued by security software company Symantec, India was amongst the top five nations in the world that are facing cyber threats and targeted attacks[7]. There needs to exist a fine balance in between the individual liberty and authority of the government. It can now be asked that if the government that the citizens have themselves elected and bought into power cannot be trusted, then who can be?

An important aspect to the same can be understood by the fact that the latest addition to the policy of Data Protection in India in the form of Digital Personal Data Protection Bill, 2023 does not regulate harms arising out of processing of personal data. It is pertinent to note that

The Shrikrishna Committee which served as the paramount analyst in the first ever draft of the Data Protection Bill highlighted that if the data is not protected then, it may lead to unreasonable surveillance and profiling as well as identity theft. The Personal Data Protection Bill, 2019 had defined harm to include: mental injury, identity theft, financial loss, reputational loss, discriminatory treatment, and observation or surveillance not reasonably expected by the data principal.[8] The 2019 Bill required data fiduciaries to take measures to prevent, minimize, and mitigate risks of harm. These included undertaking evaluation of these risks in impact assessments and audits.

III. The Way Ahead

In India, the collection, storage and handling of biometric data is governed by the information technology law contained under the Information Technology Act, 2000 (IT Act), primarily through the rules framed under it. The Information Technology (Reasonable Security Practices and Procedures and Sensitive Personal Data or Information) Rules, 2011 (Privacy Rules) lays out the specific conditions that regulate personal information and sensitive personal data or information, including biometric data.[9]

The Digital Personal Data Protection Bill, 2023 has several answers to this recurring problem of data handling. It proposes the following measures to safeguard the Personal Sensitive Data of individuals.

(i) Extra and Intra Territorial Application of Law

This Bill will apply to the personal data which is being collected either in India and is digitized but it will also apply to such processing outside India if a foreign Intermediary is carrying on trade in India. This is a remarkable step forward in the direction of data protection not only within the digital boundaries of India but also outside the nation as the globalization is hitting its peak.

(ii) Consent as a paramount necessity in processing of Personal Data

The Bill provides for lawful processing of personal data and only upon the consent of the individuals whose data is being processed. Clause 6(1) elucidates that,

"The consent given by the Data Principal shall be free, specific, informed, unconditional and unambiguous with a clear affirmative action, and shall signify an agreement to the processing of her personal data for the specified purpose and be limited to such personal data as is necessary for such specified purpose"[10] Consent refers to the will of individual users who choose whether their data can be used or not. This does not include voluntary sharing of data to the State for purposes such as permits, licenses and services.

Table 1: Comparison of various drafts of the Data Protection Law

The Draft Personal Data Protection Bill, 2018	The Personal Data Protection Bill, 2019	Recommendations of the Joint Parliamentary Committee	The Digital Personal Data Protection Bill, 2023
Scope and Applicability			
• Processing of personal data: (i) within India, (ii) outside India if it is for business carried on, offering of goods and services, or profiling individuals, in India	• Expands the scope under the 2018 Bill to cover certain anonymised personal data	• Expands the scope under the 2018 Bill to include processing of non-personal data and anonymised personal data	• does not cover offline personal data and non-automated processing
Reporting of data breaches			
• Fiduciary to notify the Data Protection Authority about a breach which is likely to cause harm, the Authority will decide whether to notify the data principals or not	• Same as 2018 Bill	• All breaches, regardless of potential harm, must be reported to the Authority, within 72 hours	• Every personal data breach must be reported to the Data Protection Board of India and each affected data principal, in prescribed manner
Exemptions from provisions of the Bill for the security of the state, public order, prevention of offences etc.			
• Processing must be authorised pursuant to a law, and in accordance with the procedure established by law, and must be necessary and proportionate	• The central government, by order, may exempt agencies where processing is necessary or expedient, subject to certain procedure, safeguards, and oversight	• Adds that order should specify a procedure, which is fair, just, and reasonable	• The central government may exempt by notification, does not require any procedure or safeguards to be specified
Right to Data Portability and Right to be Forgotten			
• Data principal will have the right to data portability (to obtain data in interoperable format), and right to be forgotten (to restrict disclosure of personal data over internet)	• Provided for both rights	• Provided for both rights	• Not provided
Harm from processing of personal data			
• Harm includes monetary loss, identity theft, loss of reputation, and unreasonable surveillance • Data fiduciaries to take measures to minimise and mitigate risks of harm • Data principal has a right to seek compensation in the event of harm	• Same as 2018 Bill	• The central government should have powers to prescribe additional harms	• Not provided

(iii) Right to Withdrawal of Consent

Clause 6(4) of the Bill provides that a Data Principal at all points of time shall have the right to withdraw their consent with the ease of being doing so in the same manner in which such consent was given. There lies a lacuna in this provision that the consequences of the withdrawal shall not affect the legality of processing of the personal data based on consent before its withdrawal. An illustration to understand the same can be, if an individual is the user of an online shopping website owned by an E- Commerce service provider, and consents to the processing of their personal data for purposes of fulfilling their supply order and makes payment for the same but withdraws the consent soon after, the individual may be stopped for using the website in such case but this shall not affect the supply of goods already ordered and paid for by the individual before such withdrawal.

(iv) Obligations of data fiduciaries

The entity determining the purpose and means of processing, (data fiduciary), must: (i) make reasonable efforts to ensure the accuracy and completeness of data, (ii) build reasonable security safeguards to prevent a data breach, (iii) inform the Data Protection Board of India and affected persons in the event of a breach, and (iv) erase personal data as soon as the purpose has been met and retention is not necessary for legal purposes (storage limitation). In case of government entities, storage limitation and the right of the data principal to erasure will not apply.

(v) Data Protection Board of India

The central government will establish the Data Protection Board of India. Key functions of the Board include: monitoring compliance and imposing penalties, directing data fiduciaries to take necessary measures in the event of a data breach, and hearing grievances made by affected persons. Board members will be appointed for two years and will be eligible for re-appointment. The central government will prescribe details such as the number of members of the Board and the selection process. Appeals against the decisions of the Board will lie with TDSAT.[11]

(vi) Parliamentary Oversight

It will be arbitrary to say that there is no mechanism of checks and balances on processing of data by the State. The Supreme Court in 2017[12] held that any infringement to the right of privacy should be proportionate to the need for such enforcement. It cannot be said that all data that is processed by the State is for illegitimate purposes or creating a surveillance map as the purpose at large is maintaining public order and convenience. Although this Bill does not provide for government agencies to delete personal data after the purpose of processing has been met, it provides for parliamentary oversight of such processing from time to time to keep in check the actual purpose of its usage.

IV. Conclusions

It is a settled principle that state rule and public welfare go hand in hand. While on one hand it cannot be denied that India with a population of over 1.5 billion is not unprotected in terms of Digital private data, it can also be said that the government needs more stringent and independent bodies to regulate data management and processing. It is high time we realize that this aspect of modern India is mammoth enough to have a special and exclusive chunk of Indian Administrative Paradigm. It is now becoming necessary that Indian Political contenders include data protection in their manifestos as a part of their power plan. It is high time that individuals and government go hand in hand against the masked world of data invasion and eliminate potential data privacy threats because datum is the new human.

Endnotes

1   'Finger on the Future' – an epiphany of Ideas – The Times of India, Editorial November 30, 2022

2   Cambridge Dictionary, American dictionary

3   'New Aadhar data leak exposes 11 crore Indian Farmers' sensitive Info' – Zee News, June 14, 2022

4   https://www.dw.com/en/in-germany-controversy-still-surrounds-video-surveillance/a-50976630 German controversy over video surveillance, DW, October 24, 2019

5   Biometric Data: Regime In India, Mondaq, October 28, 2019 https://www.mondaq.com/india/privacy-protection/857992/biometric-data-regime-in-india

6   India Billion Strong Biometric Database – Phys Organisation

7   "India ranks 3rd among nations facing most cyber threats: Symantec," Economic Times, April 4, 2018.

8   Clause 2 (20), Clause 2 (38), Clause 15, The Personal Data Protection Bill, 2019, as introduced in Lok Sabha

9   Biometric Data: Regime In India, Mondaq, October 28, 2019 https://www.mondaq.com/india/privacy-protection/857992/biometric-data-regime-in-india

10 Digital Personal Data Protection Act 2023

11 https://prsindia.org/files/Bills_acts/Bills_parliament/2023/Legislative_Brief_Digital_Personal_Data_Protection_Bill_2023.pdf PRS India for Analysis of Digital Personal Data Protection Act, 2023

12 Justice K.S. Puttaswamy (Retd) vs. Union of India, W.P. (Civil) No 494 of 2012, Supreme Court of India, August 24, 2017.

Chapter 24

Financial Privacy in the Fintech Era: Navigating New Horizons

Muskan Agnihotri
B.A.L.L.B. (Hons.) Student
SSOL, Sharda University
Greater Noida, UP, India

Aishna Kumar
L.L.M. (Corporate Law) Scholar
University of Edinburgh
Scotland, U.K.

Abstract

For business purposes financial institutions need to collect specific details of people for lending purposes. However, they need to ensure that the data is protected and not shared with any 3rd party without user consent. Maintenance of confidentiality of such sensitive information is paramount in this context. Financial privacy is all about keeping your financial information and transactions private, it also covers banks, credit unions, savings institutions, insurance agencies, etc. It also handles huge data of its consumers to protect them from various scams financial privacy is important as it gives personal freedom to control your money and financial choices without someone snooping on your business. Another way is economic growth when people know their financial information is safe, they are more likely to invest, create businesses, and help the economy to grow. The inclusion of financial privacy helps include everyone in the financial system. The requirement for financial institutions to maintain the privacy of data grows along with the frequency of processing personally identifiable information as well as the threats of cybercrimes.

Key Words – Bank, Government, Law enforcement, Privacy.

I. Introduction

Progressive financial institutions are actively exploring innovative technologies to enhance operational efficiency, speed up service delivery, and elevate the overall customer experience. The rapid advancement of information technology has spurred the adoption of digital banking solutions, reshaping the financial services sector with a strong focus on managing customer experiences. In response to competition from consumer giants like Amazon, Facebook, and Google, the financial services industry is keen on enhancing its online customer support capabilities. Notably, most financial services executives consider improving the customer experience as the primary motivator for embracing digital transformation in banking. The emergence of smart analytics empowers financial companies to tap into the vast troves of consumer data, enabling them to gain deeper insights into customer behavior and offer more tailored services. Technology has also facilitated the creation of innovative financial products and services. A significant challenge lies in developing more efficient payment systems. Additionally, there's growing anticipation that robo-advisory services will play a substantial role in the industry's future, alongside the increasing popularity of blockchain-based offerings. Data assumes a pivotal role for data-centric firms like fintech companies, which heavily rely on personal data for activities such as risk assessment and pricing. However, the landscape of privacy regulations could potentially pose obstacles to fintech growth and competition within the financial sector. To rebuild trust and foster progress, financial service providers and regulators must collaborate to construct an infrastructure that prioritizes transparency, data sharing, and privacy safeguards. Importantly, this collaborative effort should incorporate the perspectives of individuals whose data is utilized, ensuring that their values are respected in policies and that they receive tangible benefits from the use of their data. Across the globe, regulatory bodies are taking steps to enact data privacy laws.

II. Dimensions of Financial Privacy

The various dimension of financial policies are as follows: -

(i) **Banking Secrecy:** A bank is a trusty vault that does not share your information with others unless it's required by the law. The bank keeps secrets, like a trustworthy friend.

(ii) **Data Privacy:** Keeping your personal and financial data safe from prying eyes. It is like putting a lock on your diary to make sure no one reads your secrets.

(iii) **Cryptocurrency**: Cryptocurrencies are like digital money. Some, like Bitcoin, offer more privacy than traditional money. Think of it as a secret handshake in the digital world.

(iv) **Cash Transactions**: When you pay with cash, there is no digital trail.

(v) **Financial Surveillance**: Governments and banks watch financial transactions to catch the frauds.

III. Benefits Of Technology in Financial Services

The benefits of technology in financial services are extensive and have revolutionized the industry in numerous ways, enhancing efficiency, accessibility, and overall customer experiences.

First and foremost, technology has greatly improved convenience and accessibility for consumers. Online and mobile banking platforms enable customers to access their accounts, conduct transactions, and manage investments from anywhere at any time, reducing the need for physical visits to brick-and-mortar branches. This increased accessibility has been especially important during events like the COVID-19 pandemic, where digital channels became essential for banking services. Technology has also significantly streamlined operational processes within financial institutions. Automation, artificial intelligence (AI), and machine learning are being used for tasks such as account verification, fraud detection, and risk assessment. These technologies not only reduce the potential for human error but also enhance the speed and accuracy of these processes, ultimately benefiting both financial institutions and their customers. In the realm of customer service technology has introduced chatbots and virtual assistants that provide quick and efficient responses to customer inquiries, improving response times and reducing operational costs. Additionally, predictive analytics can help financial institutions anticipate customer needs and offer personalized recommendations and products, enhancing the overall customer experience. Moreover, cost efficiency as a notable advantage of technology in financial services. Digital transactions are often more cost-effective than traditional in-person or paper-based processes, allowing financial institutions to reduce overhead costs. This cost savings can be passed on to customers through lower fees and better interest rates.

The advent of mobile payments and digital wallets has also transformed the way people make transactions. These technologies offer a convenient and secure way to make payments, reducing the reliance on physical cash and checks, which are often less secure and more cumbersome.

For investors, technology has democratized investing by enabling access to a wide range of financial products and investment opportunities. Online trading platforms and robo-advisors provide individuals with the tools and information they need to make informed investment decisions, often at lower costs than traditional investment management services. Furthermore, technology has facilitated the development of innovative financial products and services such as peer-to-peer lending, crowdfunding, and blockchain-based cryptocurrencies. These innovations have created new avenues for funding and investment, challenging traditional financial models and fostering financial inclusion. In terms of security, advancements in encryption, biometrics, and authentication technologies have made online financial transactions more secure than ever before. These measures help protect customer data and prevent unauthorized access, instilling trust in digital financial services.

IV. Digitalization in the Financial Service Industry

Nowadays, more people have access to the Internet, and they are using it more frequently, creating a larger potential customer base for online financial services.

(i) **Mobile Technology**: The expansion of mobile technologies has been a driving force behind digitalization. With the rise of smartphones and mobile apps, consumers can access financial services anytime and anywhere, making transactions and financial management more convenient.

(ii) **Data Processing and Analytics**: Enhanced data processing and analytical capabilities are enabling financial institutions to gather and analyze vast amounts of data. This data-driven approach allows for better risk assessment, personalized financial products, and improved decision-making.

(iii) **Connectivity and Smart Devices**: The increasing connectivity between smart devices and financial services platforms has led to innovations such as mobile payments, wearables for banking, and IoT (Internet of Things) integration with financial services. These innovations enhance user experience and provide new opportunities for financial institutions to offer services.

(iv) **Mobile Money**: The mention of mobile money accounts exceeding one Billion globally in 2019 illustrates the growing popularity of mobile-based financial transactions. This trend is gradually replacing traditional cash-based transactions and is indicative of the shift toward digital financial services.

(v) **Innovation Across the Industry**: Digitalization has impacted every aspect of the financial services industry. This includes payments with instant payment methods and digital wallets, credit services like crowdfunding and online lending, insurance with the rise of insurtech, investment services through robo-advisors, and core banking services offered by online digital banks. These innovations cater to changing consumer preferences and expectations.

(vi) **Biometric Identification:** Biometric authentication methods, such as fingerprint and facial recognition, have improved security and convenience in financial transactions. They are often used for identity verification and access control.

(vii) **Back-End Technologies**: Cloud computing and big data have become essential for the financial sector. Cloud services provide scalability and cost-efficiency, while big data analytics offer insights into customer behavior and market trends.

Digitalization is reshaping the financial services industry by making services more accessible, convenient, and efficient for consumers. It is also driving innovation in various areas of finance, leveraging technology to create new opportunities and improve the overall financial experience.

V. Adoption Of Financial Technology

The adoption of financial technology, often referred to as fintech, has been a transformative force in the financial services industry, reshaping the way businesses and consumers interact with financial products and services. This adoption can be elaborated upon in detail through several key aspects:

(i) **Enhanced Accessibility and Convenience:** Fintech has significantly improved access to financial services. With the proliferation of smartphones and the internet, individuals can now access their bank accounts, make payments, and invest in financial markets from the comfort of their homes or on the go. This level of accessibility has broadened financial inclusion, reaching underserved populations who previously had limited access to traditional banking services.

(ii) **Digital Payments:** The adoption of fintech has led to the rise of digital payment solutions, such as mobile wallets, contactless payments, and peer-to-peer payment apps. These technologies have made payments faster, more convenient, and often more secure than traditional cash or check-based transactions.

(iii) **Robo-Advisors**: Fintech-driven robo-advisors have democratized investing by providing automated, algorithm-based investment advice and portfolio management. These platforms offer cost-effective solutions for individuals looking to invest their money and have opened up the world of wealth management to a broader audience.

(iv) **Peer-to-Peer Lending and Crowdfunding**: Fintech platforms have facilitated peer-to-peer lending and crowdfunding, enabling individuals and businesses to access funding from a diverse pool of investors. This alternative financing model has disrupted traditional lending institutions, offering more flexible borrowing options.

(v) **Blockchain and Cryptocurrency**: Blockchain technology, which underpins cryptocurrencies like Bitcoin and Ethereum, has introduced decentralized and secure methods of transferring and storing value. While cryptocurrencies have generated significant attention, blockchain has broader applications, including supply chain management and transparent record-keeping.

(vi) **Risk Assessment and Fraud Prevention**: Fintech leverages data analytics, artificial intelligence, and machine learning to assess credit risk, detect fraudulent activities, and enhance security. These technologies allow financial institutions to make more informed lending decisions and protect customers from fraudulent transactions.

(vii) **Personalized Financial Services**: Fintech companies harness data analytics to offer personalized financial recommendations and services. By analysing a user's financial behavior and preferences, these platforms can suggest suitable financial products, such as credit cards, loans, or investment options, tailored to individual needs.

(viii) **Cost Efficiency**: Fintech has driven cost efficiencies across the financial industry. Online and mobile banking, automated processes, and reduced overhead costs have translated into lower fees and better interest rates for consumers. Additionally, fintech's ability to automate tasks has increased operational efficiency for financial institutions.

(ix) **Regulatory Challenges**: The adoption of fintech has brought regulatory challenges as governments and financial authorities work to establish guidelines and safeguards for these rapidly evolving technologies. Ensuring consumer protection, data privacy, and financial stability are among the key concerns in fintech regulation.

(x) **Global Expansion**: Fintech is not confined to specific regions or markets. It has seen widespread global adoption, with fintech startups and established financial institutions alike seeking to tap into new markets and demographics.

(xi) **Digital Experience Platforms for Banks**

A Digital Experience Platform (DXP) for banks is a comprehensive solution designed to enhance the online and digital interactions between banks and their customers. It is a crucial component of a bank's digital transformation strategy, aiming to deliver seamless, personalized, and efficient digital experiences across various channels, including websites, mobile apps, and online banking portals. Here, we'll elaborate on the key elements and benefits of a DXP for banks.

(xii) **Omni Channel Engagement**: A DXP enables banks to provide a consistent user experience across multiple digital touchpoints, ensuring that customers receive the same level of service and information whether they're using a website, mobile app, or any other digital platform. This helps in building trust and loyalty among customers.

(xiii) **Personalization**: DXPs leverage customer data and analytics to offer highly personalized experiences. They can tailor content, offers, and recommendations based on an individual's financial history, preferences, and behavior. This personalization enhances engagement and customer satisfaction.

(xiv) **Content Management**: Banks can easily manage and update content across their digital channels using a DXP. This feature is essential for ensuring that customers have access to accurate and up-to-date information, including product details, interest rates, and regulatory updates.

(xv) User-Friendly Interfaces: A DXP simplifies the design and management of user interfaces. It allows banks to create intuitive and user-friendly interfaces that improve the overall customer experience, making it easier for customers to navigate and complete tasks online.

(xvi) Security and Compliance: Security is paramount in banking, and a DXP includes robust security features to protect customer data and transactions. It also assists banks in adhering to regulatory compliance requirements, ensuring that all digital interactions meet legal and industry standards.

(xvii) Data Integration: DXPs integrate with various data sources within the bank, including core banking systems, CRM databases, and third-party financial tools. This data integration enables real-time access to customer account information and transaction history, facilitating more informed decision-making and enabling customers to track their financial activities.

(xviii) Analytics and Insights: A DXP provides powerful analytics tools that help banks understand customer behavior, preferences, and engagement patterns. Banks can use these insights to optimize their digital services continually, drive marketing strategies, and identify opportunities for upselling and cross-selling.

(xix) Customer Self-Service: Customers can access a range of self-service features through a DXP, such as applying for loans, opening accounts, transferring funds, and managing investments. This reduces the need for customers to visit physical bank branches, enhancing their convenience and satisfaction.

(xx) Scalability: DXPs are scalable solutions, allowing banks to adapt to changing customer needs and market dynamics. Whether the bank is serving a few thousand customers or millions, the platform can accommodate growth without significant disruptions.

(xxi) Competitive Advantage: In a highly competitive banking landscape, a well-implemented DXP can set a bank apart by delivering exceptional digital experiences. This can attract new customers, retain existing ones, and drive revenue growth.

VI. Basic Tenets for Safe Framework in Organization

Establishing a safe framework for financial privacy within an organization is essential to protect sensitive financial data and ensure compliance with data privacy regulations. This framework is built upon several basic tenets that serve as foundational principles for safeguarding financial privacy. Let's explain these tenets in detail:

(i) **Data Minimization**: The principle of data minimization involves collecting and retaining only the minimum amount of financial data necessary for legitimate business purposes. Unnecessary data should not be collected, reducing the risk of data breaches and privacy violations. This tenet emphasizes the importance of limiting data collection to what is essential for providing financial services.

(ii) **Informed Consent:** Obtaining clear and informed consent from customers is fundamental. Individuals should be aware of what data is being collected, how it will be used, and with whom it may be shared. Consent should be voluntary, explicit, and revocable, allowing individuals to control their data.

(iii) **Data Security**: A robust data security infrastructure is critical. Financial institutions must implement encryption, access controls, intrusion detection systems, and regular security audits to protect financial data from unauthorized access, breaches, or cyberattacks. Adhering to industry best practices and staying updated on security threats is essential.

(iv) **Transparency**: Transparency in data practices builds trust. Financial organizations should be transparent about their data collection, usage, and sharing policies. Privacy policies and terms of service should be easily accessible and written in plain language to ensure customers understand how their data is handled.

(v) **Data Retention and Deletion**: Financial institutions should establish clear policies for data retention and deletion. Data should not be stored longer than necessary for the intended purpose, and customers should have the ability to request the deletion of their data when they no longer want it to be retained.

(vi) **Access Control and Authentication**: Limiting access to financial data to authorized personnel is crucial. Multi-factor authentication and strict access controls should be implemented to ensure that only individuals with a legitimate need can access sensitive financial information. Regular access reviews should be conducted to prevent unauthorized access.

(vii) **Government, Law enforcement**: Financial organizations must maintain the accuracy of customer data. Inaccurate data can lead to errors in financial transactions and affect customers' financial well-being. Processes for updating and correcting data should be established, and customers should have the ability to review and correct their information.

Incorporating these basic tenets into an organization's framework for financial privacy helps create a strong foundation for protecting customer data, maintaining compliance, and building trust with customers. Financial institutions should continuously review and update their privacy practices to adapt to changing regulations and emerging security threats in the ever-evolving landscape of data privacy.

VII. Protium's model of data privacy

When it comes to the protection and preservation of personal information, Protium's secure in-house tech capabilities built on public cloud, prove to be robust. This keeps data security and access to data as primary building blocks of our engineering finance philosophy, which is a cohesive collaboration of tech, risk, and data and analytics to ascertain specific customer requirements and create innovative solutions.

(i)   Encryption of KYC data

Protium takes personal information security seriously and data is protected through encryption, which scrambles all the classified information into an unreadable format using public cloud infrastructure. Not only personal information, but all information from the point of capturing to storing in databases is encrypted. The entire KYC data is encrypted and only applications can access this information via secure APIs using subscription model. The encryption keys are access controlled, that means the data once scrambled cannot be read by anyone. These services are hosted in private networks in public cloud and are protected by strong firewalls and network configurations.

(ii)   Role-based access control (RBAC)

RBAC is a technique for controlling application access based on the roles of specific users inside an organization. Its foundation is the idea of positions and privileges. Access is determined by the role and responsibility of the employee. Based on the roles, the use of certain programs, as well as network access, could be restricted — sensitive information, thus, has lesser chances of being leaked or falling into the wrong hands. The access is reviewed at regular intervals to keep it updated and safe.

(iii) Secure Microservices

Since microservices make the backbone of protium's lending stack for any operation performed and all the information travels through these services. All APIs are authenticated, authorized and encryption enabled with subscription-based access model. HTTPS with TLS 1.2

is a requisite for the use and construction of APIs, and as such, it would act as safeguard against leakages. Microservices secrets/credentials are never a part of application code instead are derived at run time based on application roles and environment.

(iv) Secure Computing devices

Protium ensures secure computing devices for its employees by encrypting and protecting the computing infrastructure with the latest antivirus solutions. The devices are segmented at the network layer to limit the impact of any affected devices, and constant monitoring is in place to oversee the computing, branch, and firewall infrastructure.

(v) Data Loss Prevention (DLP) Practices

Data Loss Prevention is the action of identifying and stopping breaches of confidential credentials. Protium's stringent password encryption and cloud backup strategies provide the greatest degree of protection, visibility, and control while allowing their employees to collaborate freely and with confidence.

VIII. Conclusions

Financial privacy is a paramount concern in today's digital age, where the financial services industry is increasingly reliant on data-driven technologies. Protecting the privacy of individuals' financial information is not only a legal and ethical imperative but also essential for maintaining trust and security within the financial sector. Financial institutions and organizations must prioritize data minimization, informed consent, robust data security, transparency, and regulatory compliance as core principles of their approach to financial privacy. These principles form the foundation of a safe and responsible framework for handling sensitive financial data, ensuring that individuals have control over their information and are well-informed about how it is used. The impact of fintech and digital transformation in the financial industry has brought both opportunities and challenges for financial privacy. While these technological advancements have improved accessibility, convenience, and efficiency, they have also raised concerns about data collection, profiling, and security. Striking the right balance between innovation and data protection is crucial., As the regulatory landscape continues to evolve, financial institutions must adapt to new requirements and best practices, focusing on accountability and governance to ensure ongoing compliance. Ultimately, fostering a culture of respect for financial privacy and a commitment to safeguarding customer data will be key to building and maintaining trust in the financial services sector. By upholding these principles, organizations can navigate the complexities of financial privacy in the digital era and provide customers with the assurance that their financial information is handled with care and integrity.

References

1   https://jsis.washington.edu/news/data-privacy-security-and-regulation-in-financial-technology/.

2   https://www.hindustantimes.com/ht-insight/economy/financial-privacy-must-be-recognised-in-india-101627104522659.html.

3   https://privacyinternational.org/learn/financial-privacy.

4   https://www.winston.com/en/legal-glossary/financial-privacy.

5   https://privacyinternational.org/learn/financial-privacy.

6   https://epic.org/the-right-to-financial-privacy-act/.

7   https://protium.co.in/2023/04/14/data-privacy-in-financial-services/.

8   https://www.privacycompliancehub.com/gdpr-resources/five-reasons-why-fintechs-need-to-prioritise-privacy/.

9   https://www.pwc.in/assets/pdfs/consulting/cyber-security/banking/security-challenges-in-the-evolving-fintech-landscape.pdf.

10 https://www.idx.us/knowledge-center/data-privacy-concerns-in-booming-fintech-industry.

11 https://jsis.washington.edu/news/data-privacy-security-and-regulation-in-financial-technology/.

12 https://www.fortanix.com/blog/data-security-and-a-fintech-state-of-mind.

13 https://economictimes.indiatimes.com/industry/banking/finance/banking/data-privacy-should-not-be-compromised-in-using-fintech-fm/articleshow/86583586.cms?from=mdr.

14 https://clevertap.com/blog/fintech-data-privacy/.

15 https://www.fintechnews.org/6-data-security-issues-fintech-firms-are-facing/.

Chapter 25

Demystifying Data Mining: Comprehensive Insights

Akash Kumar

B.A.L.L.B. (Hons.) Student
SSOL, Sharda University
Greater Noida, UP, India

Sumit Kumar

Project Manager
HCL Technologies
BVBA Brussels, Belgium

Abstract

Data mining, a pivotal discipline in the realm of big data, entails the exploration and analysis of large datasets to uncover hidden patterns, extract valuable insights, and support informed decision-making. This Chapter provides a concise overview of the multifaceted world of data mining, encompassing techniques such as association rule mining, clustering, classification, and anomaly detection. The impact and applications of data mining span diverse domains, including retail, healthcare, finance, manufacturing, e-commerce, social media, agriculture, and energy. However, data mining is not without its challenges and ethical considerations. Ensuring data quality, safeguarding privacy, and addressing bias are among the persistent concerns in this field. Ethical practices, transparency, and compliance with regulations like GDPR are integral to responsible data mining. As data continues to proliferate, data mining remains an indispensable tool for transforming raw data into actionable knowledge. This abstract encapsulates the essence of data mining, its power, challenges, and ethical imperatives in the context of a data-driven world.

Keywords- Data, Extraction, Mining, Processing, digital, global.

I. Introduction to Data Mining

Data mining is a multidisciplinary field that blends techniques from statistics, machine learning, and database management to discover hidden patterns, relationships, and valuable insights within large datasets. It involves the automated or semi-automated exploration of data to uncover trends, anomalies, and knowledge that can be leveraged for decision-making and prediction.

In today's data-driven world, organizations collect and store vast amounts of data, ranging from customer preferences and purchase history to sensor readings and social media interactions. Data mining plays a pivotal role in transforming this raw data into actionable information. It helps businesses enhance their operations, optimize marketing strategies, improve customer experiences, and make informed decisions.

II. Components of Data Mining

Key components of data mining include:

(i) **Data Collection:** Gathering and pre-processing data from various sources, ensuring data quality and consistency.
(ii) **Exploratory Data Analysis (EDA):** Understanding the data's structure, distribution, and potential relationships through visualization and statistical analysis.

(iii) **Data Cleaning and Transformation:** Removing noise and handling missing values or outliers to prepare data for analysis.

(iv) **Pattern Discovery:** Employing algorithms and techniques to uncover patterns, such as associations, clusters, and trends, within the data.

(v) **Modelling and Prediction:** Building predictive models that can make informed forecasts based on historical data.

(vi) **Evaluation:** Assessing the accuracy and performance of data mining models using appropriate metrics.

(vii) **Deployment:** Integrating data mining results into practical applications, such as recommendation systems, fraud detection, and medical diagnosis.

Data mining finds applications in various domains, including business and marketing, healthcare, finance, and scientific research. It enables organizations to gain a competitive edge by making data-driven decisions, identifying opportunities, and mitigating risks.

As the volume and complexity of data continue to grow, data mining remains a critical tool for extracting knowledge and valuable insights from these vast datasets, facilitating innovation, and driving progress in diverse fields.

III. Understanding Data Mining

Data mining is a process that involves the exploration and analysis of large datasets to discover patterns, relationships, and valuable information that might not be immediately apparent. Its primary goal is to transform raw data into actionable knowledge, allowing organizations and individuals to make informed decisions, predict future outcomes, and uncover hidden insights.

(i) Key Concepts of Data Mining

a. **Data:** Data mining starts with data, which can come from various sources, including databases, spreadsheets, sensors, websites, and more. This data can be structured (e.g., relational databases) or unstructured (e.g., text or images).

b. **Patterns and Relationships:** Data mining algorithms are used to identify patterns and relationships within the data. These patterns can take the form of associations (items that tend to occur together), clusters (groups of similar data points), classifications (predicting categories for new data), and more.

c. **Predictive Modelling:** One of the primary uses of data mining is predictive modelling. By analyzing historical data, data mining can create models that make predictions about future events or trends. For example, it can be used to predict customer churn, stock prices, or disease outbreaks.

d. **Machine Learning:** Many data mining techniques are closely related to machine learning, which involves training algorithms to improve their performance over time. Supervised learning, unsupervised learning, and reinforcement learning are common machine learning approaches used in data mining.

e. **Data Pre-processing:** Data often requires cleaning and pre-processing before it can be effectively mined. This involves handling missing values, dealing with outliers, and transforming data into a suitable format.

IV. Application of Data Mining

Data mining has a wide range of applications across various industries:

(i) **Business and Marketing:** Businesses use data mining to improve customer targeting, optimize marketing campaigns, and make strategic decisions based on market trends.

(ii) **Healthcare:** In healthcare, data mining can assist in disease diagnosis, drug discovery, and patient outcome prediction.

(iii) **Finance:** In finance, it's used for fraud detection, risk assessment, and stock market forecasting.

(iv) **Science and Research:** Data mining is employed in scientific research to discover patterns in experimental data and make scientific predictions.

V. Challenges of Data Mining

While data mining is a powerful tool, it also comes with challenges, including:

(i) **Data Quality:** Ensuring that the data used for mining is accurate, complete, and reliable is crucial.

(ii) **Privacy Concerns:** The mining of personal data has raised privacy concerns, leading to the development of privacy-preserving data mining techniques.

(iii) **Scalability:** Handling and mining large datasets can be computationally intensive, requiring specialized hardware and algorithms.

In summary, data mining is a process of discovering hidden knowledge and patterns within data, making it a valuable tool for businesses, researchers, and decision-makers across various domains. Its applications continue to grow as data becomes increasingly important in our modern world.

VI. Evolution of Data Mining and Its Historical Context

Data mining has evolved over the decades as -

(i) **Early Beginnings (1960s - 1980s)**

The concept of data mining can be traced back to the 1960s when statisticians and researchers started developing techniques to analyze large datasets. Early efforts focused on methods like cluster analysis and regression analysis to extract insights from data. Data mining was primarily used in fields like economics, biology, and social sciences.

(ii) Emergence of Databases (1980s - 1990s)

In the 1980s, the development of relational databases and the Structured Query Language (SQL) made it easier to manage and access large datasets. Researchers started applying data mining techniques to relational databases, leading to the emergence of the field of knowledge discovery in databases (KDD). KDD encompassed data cleaning, data preprocessing, pattern discovery, and evaluation.

(iii) Introduction of Data Warehouses (1990s)

Data warehousing gained popularity in the 1990s, allowing organizations to store and consolidate data from various sources in a central repository. Data mining techniques were increasingly applied to data warehouses to uncover business insights. The term "data mining" itself gained recognition during this period.

(iv) Rapid Technological Advancements (Late 1990s - Early 2000s)

The late 1990s and early 2000s saw significant advances in computing power and storage capacity, enabling the processing of massive datasets. Machine learning algorithms, particularly decision trees and neural networks, became integral to data mining. Commercial data mining software and tools started to emerge, making data mining more accessible to businesses.

(v) Data Mining in Business (2000s - Present)

Data mining found widespread application in business and marketing, with companies using it for customer segmentation, recommendation systems, and market basket analysis. Social media platforms and e-commerce sites leveraged data mining to personalize user experiences. The rise of big data further expanded the scope of data mining, as organizations sought to extract insights from large volumes of structured and unstructured data.

(vi) Ethical and Privacy Considerations (2010s - Present)

In recent years, ethical concerns related to data mining, including privacy, bias, and transparency, have come to the forefront. Regulations like GDPR and increased awareness of data privacy have led to more responsible data mining practices. Researchers and practitioners are developing privacy-preserving data mining techniques.

(vii) Integration with Artificial Intelligence (AI) (Present)

Data mining is closely integrated with AI and machine learning, where advanced algorithms can automatically learn patterns and make predictions from data. AI-driven data mining is applied in fields such as healthcare for disease prediction and autonomous vehicles for data-driven decision-making.

The evolution of data mining reflects the growing importance of data in various industries and the ongoing development of techniques to extract meaningful insights from this data. Today, data mining continues to evolve, driven by advancements in AI, big data technologies, and a growing emphasis on responsible data usage.

VII. Techniques of data mining

Some popular data mining techniques are as:

(i) Association Rule Mining

Definition: Association rule mining is used to discover interesting relationships or associations between items in large datasets. It identifies patterns like "if X, then Y," where X and Y are items or events that often occur together.

Applications: Market basket analysis in retail (e.g., identifying products frequently bought together), recommendation systems, and identifying co-occurring medical conditions in healthcare data.

(ii) Clustering

Definition: Clustering involves grouping similar data points together based on their characteristics or attributes. The goal is to find natural groupings or clusters within the data.

Applications: Customer segmentation in marketing, image segmentation in computer vision, and species classification in biology.

(iii) Classification

Definition: Classification is a supervised learning technique that assigns predefined labels or classes to data points based on their attributes. It's used for predictive modeling and decision-making.

Applications: Email spam detection, sentiment analysis in natural language processing, and disease diagnosis in healthcare.

(iv) Anomaly Detection

Definition: Anomaly detection (also known as outlier detection) identifies data points that deviate significantly from the norm or expected behavior. It focuses on identifying rare and unusual events.

Applications: Fraud detection in finance (e.g., detecting unusual credit card transactions), network intrusion detection in cybersecurity, and equipment failure prediction in manufacturing.

These data mining techniques can be implemented using a variety of algorithms and methods. For example:

Association Rule Mining: Algorithms like Apriori and FP-growth are commonly used for association rule mining.

Clustering: K-Means, DBSCAN, and hierarchical clustering are popular clustering algorithms.

Classification: Decision Trees, Random Forest, Support Vector Machines, and neural networks are widely used for classification tasks.

Anomaly Detection: Techniques include statistical methods, clustering-based approaches, and machine learning algorithms like Isolation Forest and One-Class SVM.

Each technique has its strengths and weaknesses, making them suitable for different types of data and applications. Data mining practitioners select the most appropriate technique based on the specific problem they aim to solve and the characteristics of their dataset.

Case studies and examples of data mining applications in specific domains

(v) Retail: Market Basket Analysis

Case Study: A large retail chain wanted to optimize its product placement and increase sales. They employed market basket analysis, a data mining technique, to uncover associations between products frequently purchased together.

Application: By identifying common product combinations, the retailer could strategically place these items close to each other in stores and promote them as bundles or discounts. This led to a significant increase in cross-selling and overall revenue.

(vi) Healthcare: Disease Diagnosis

Case Study: A hospital network aimed to improve early diagnosis of heart disease. They collected patient data, including medical history, test results, and lifestyle information.

Application: Data mining and classification techniques were used to develop predictive models. These models helped identify patients at high risk of heart disease, allowing for timely intervention and preventive measures.

(vii) Finance: Fraud Detection

Case Study: A credit card company sought to reduce fraudulent transactions. They collected transaction data, including customer behavior and transaction details.

Application: Data mining techniques, including anomaly detection and pattern recognition, were applied to identify unusual or suspicious transaction patterns. This proactive approach significantly decreased the incidence of credit card fraud.

(viii) Manufacturing: Predictive Maintenance

Case Study: A manufacturing plant wanted to minimize equipment downtime and maintenance costs. They collected sensor data from machines and historical maintenance records.

Application: Using predictive maintenance models based on data mining, the plant could predict when equipment would require maintenance or replacement, allowing for scheduled maintenance and reducing unexpected breakdowns.

(ix) E-commerce: Recommendation Systems

Case Study: A major e-commerce platform aimed to enhance user engagement and sales by improving its product recommendation system.

Application: Data mining and collaborative filtering techniques were applied to analyze user behavior and preferences. As a result, the platform started providing personalized product recommendations, resulting in increased sales and customer satisfaction.

(x) Social Media: Sentiment Analysis

Case Study: A social media company wanted to understand user sentiment toward its platform and identify trends in user feedback.

Application: Text mining and sentiment analysis techniques were used to analyze user-generated content. This allowed the company to gauge public sentiment, detect emerging issues, and improve user experiences.

(xi) Agriculture: Crop Yield Prediction

Case Study: A farming cooperative aimed to optimize crop yields. They collected data on soil quality, weather conditions, and crop types.

Application: Data mining and regression analysis were used to create predictive models for crop yields. This helped farmers make informed decisions about planting and resource allocation, resulting in increased productivity.

(xii) Energy: Load Forecasting

Case Study: A utility company needed accurate load forecasting to optimize energy production and distribution.

Application: Time series analysis and data mining techniques were employed to predict energy demand. This allowed the utility to adjust energy generation and distribution plans efficiently, reducing costs and improving reliability.

These case studies demonstrate the diverse range of applications for data mining across various domains. By leveraging data mining techniques, organizations can make data-driven decisions, optimize processes, and gain valuable insights to achieve their objectives.

VIII. Impact and benefits in the specific domains mentioned

The impact and benefits of data mining may be discussed under the following heads:

(i) Retail (Market Basket Analysis)

Impact: Market basket analysis has revolutionized retail by optimizing product placement and promotions.

Benefits: Increased sales due to targeted cross-selling and product bundling, Improved customer satisfaction through personalized recommendations, Reduced inventory costs by optimizing stock levels.

(ii) Healthcare (Disease Diagnosis)

Impact: Data mining in healthcare has enhanced early disease diagnosis and patient care.

Benefits: Early detection of diseases, leading to better treatment outcomes, Cost reduction through preventive measures, avoiding costly treatments, More efficient resource allocation in healthcare facilities.

(iii) Finance (Fraud Detection)

Impact: Data mining has transformed fraud detection in the financial industry.

Benefits: Reduced financial losses due to fraudulent transactions, Enhanced customer trust and confidence in financial institutions, Improved regulatory compliance and risk management.

(iv) Manufacturing (Predictive Maintenance)

Impact: Predictive maintenance has improved manufacturing efficiency and equipment reliability.

Benefits: Reduced downtime and production disruptions, Lower maintenance costs through scheduled, proactive maintenance, Extended equipment lifespan and optimized resource utilization.

(v)   E-commerce (Recommendation Systems)

Impact: Recommendation systems have revolutionized online shopping experiences.

Benefits: Increased sales and revenue through personalized product recommendations, Enhanced user engagement and retention, Improved customer loyalty and brand reputation.

(vi)   Social Media (Sentiment Analysis)

Impact: Sentiment analysis has enabled social media platforms to understand user sentiment and trends.

Benefits: Real-time monitoring of user feedback and sentiments, Timely response to emerging issues and user concerns, Targeted content and advertising, improving user experiences.

(vii) Agriculture (Crop Yield Prediction)

Impact: Crop yield prediction has revolutionized farming practices.

Benefits: Increased agricultural productivity and food security, Optimized resource utilization, including water and fertilizers, Better risk management for farmers.

(viii) Energy (Load Forecasting)

Impact: Load forecasting has improved energy production and distribution.

Benefits: Efficient energy resource allocation and production planning, Reduced energy wastage and costs, Enhanced grid stability and reliability.

In each of these applications, data mining has led to improved decision-making, cost savings, increased revenue, and enhanced user experiences. It has empowered organizations to extract valuable insights from data and leverage them for competitive advantages, ultimately contributing to efficiency, effectiveness, and innovation in various domains.

IX. Challenges and Ethical Considerations of Data Mining

Data mining, like any data-driven technology, comes with its own set of challenges and ethical considerations. It's important to be aware of these issues to ensure responsible and ethical use of data mining techniques. Here are some of the key challenges and ethical considerations associated with data mining:

(i) Key Challenges

Data Quality: Data used for mining can often be noisy, incomplete, or inconsistent. Poor data quality can lead to inaccurate or biased results.

Data Privacy: Maintaining the privacy of individuals and organizations is a significant concern. The mining of personal data can lead to privacy breaches and potential harm.

Data Volume and Complexity: As datasets grow in size and complexity, data mining algorithms may struggle to process them efficiently, requiring scalable solutions.

Overfitting: Complex models can overfit the training data, resulting in models that perform well on training data but poorly on new, unseen data.

Bias: Biased data or biased algorithms can result in unfair or discriminatory outcomes, particularly in classification tasks.

Interpretability: Some data mining models, especially deep learning models, can be challenging to interpret, which may hinder decision-making and regulatory compliance.

Ethical Use: Ensuring that data mining is used for ethical purposes and does not lead to unintended consequences is an ongoing challenge.

(ii)   Ethical Considerations

Privacy: Protecting the privacy of individuals is paramount. It's essential to obtain informed consent when collecting personal data and to employ privacy-preserving techniques when necessary.

Transparency: Data mining processes should be transparent and explainable. Users should understand how data is used and decisions are made.

Fairness: Efforts should be made to mitigate bias and ensure that data mining models do not discriminate against certain groups or individuals.

Data Ownership: Clarify data ownership and usage rights to avoid disputes and ensure responsible data sharing.

Security: Protect data from unauthorized access and breaches. Ensure that sensitive data is encrypted and stored securely.

Regulatory Compliance: Abide by relevant data protection and privacy laws, such as GDPR in Europe or HIPAA in healthcare.

Accountability: Individuals and organizations involved in data mining should be accountable for their actions and decisions.

Harm Mitigation: Consider the potential harm that could result from data mining activities and take steps to mitigate these risks.

Data Anonymization: Anonymize data whenever possible to reduce the risk of identifying individuals.

Data Governance: Implement strong data governance practices to ensure that data is used ethically and responsibly throughout its lifecycle.

Continuous Monitoring: Regularly monitor data mining processes to detect and address ethical issues that may arise over time.

It's crucial for organizations and individuals engaged in data mining to be aware of these challenges and ethical considerations. Ethical data mining practices not only protect individuals' rights and privacy but also contribute to trust, accountability, and the responsible use of data-driven technologies.

X. Conclusions

Data mining has evolved into an indispensable tool for extracting valuable insights from vast datasets across diverse domains. This research journey has showcased the power and versatility of data mining techniques, including association rule mining, clustering, classification, and anomaly detection. We have seen how these techniques are applied to uncover hidden patterns, support decision-making, and drive innovation in areas such as retail, healthcare, finance, manufacturing, e-commerce, social media, agriculture, and energy.

As data continues to grow in volume and complexity, data mining plays a pivotal role in transforming raw data into actionable knowledge. However, this transformation is not without its challenges and ethical considerations. Addressing issues of data quality, privacy, bias, and transparency is imperative to ensure responsible data mining practices.

Our study has provided valuable insights into the impact and benefits of data mining across various applications. It has demonstrated how data mining empowers organizations to make informed decisions, optimize processes, and enhance user experiences. Moreover, it underscores the importance of ethical data mining practices to protect individuals' rights and maintain trust in data-driven technologies.

As we look ahead, the field of data mining continues to evolve, with new techniques and applications emerging regularly. This research serves as a foundation for future exploration and innovation in data mining, inviting researchers and practitioners to delve deeper into this dynamic field. By fostering responsible data mining practices, we can harness the full potential of data-driven insights while upholding ethical standards and societal well-being. In the era of big data, data mining remains a vital ally, helping us navigate the data landscape, discover knowledge, and shape a data-driven future.

References

1   https://blog.box.com/how-does-data-mining-affect-privacy.
2   https://www.mbaknol.com/information-systems-management/data-mining-and-privacy-concerns/.
3   https://hitachi-systems-security.com/what-is-data-mining-and-how-does-it-impact-privacy/.
4   https://www.consumernotice.org/data-protection/mining/.
5   https://www.ibm.com/topics/data-mining.
6   https://www.sas.com/en_in/insights/analytics/data-mining.html.
7   https://www.wgu.edu/blog/data-mining-business-analytics2005.html.
8   https://www.tibco.com/reference-center/what-is-data-mining.
9   https://blog.box.com/how-does-data-mining-affect-privacy.
10 https://www.mbaknol.com/information-systems-management/data-mining-and-privacy-concerns/.

Chapter 26

Online Shopping Applications: Privacy Issues and Suggestions

Aisha Hassan Ali

B.B.A.L.L.B (Hons.) Student
SSOL, Sharda University
Greater Noida, UP, India

Dr. Sanskriti Mishra

Assistant Professor
SSOL, Sharda University
Greater Noida, UP, India

Abstract

Shopping online is one of the most convenient things that modern technology has afforded us. For shoppers who do not like the hassle of walking around crowded places or lining up in long queues, online shopping is the best alternative to shopping malls and other public venues. Additionally, in recent years, shopping online has become much more convenient via mobile payment solutions. But while you may enjoy scouring the Web for cheap deals in the comfort of your home, your shopping accounts and financial transactions could be compromised by countless prying eyes. Due to the nature of e-commerce and the thousands of options for online shops, it can sometimes be hard to tell if you're dealing with a legitimate merchant or a bogus one. Just as shoppers need to take security measures when shopping in brick-and-mortar stores, online users should also be aware of the risks involved when it comes to online transactions. Ideally, we all think of securing our credit card information, and that's good. But that's not the only privacy concern we should think of.

Keywords- Buyers, Online, Safety, Shopping, behaviour.

I. Introduction

We underline the importance of protecting your privacy when it comes to shopping online. Web threats are no longer limited to malware and scams. Attackers know that the more you do any online activities, you also increase the risk of revealing more information about yourself, especially when you're looking to make a purchase. Hence, searching for items alone could lead you from one website to another, and this increases the chance of stumbling upon a malicious one.

Now that data breaches and incidents of hacking and identity theft are becoming more common, online shoppers should protect themselves against likely attacks that could threaten their privacy. There are a number of different methods that can be used to invade a user's privacy, and sooner or later, an unaware user is bound to run into threats such as phishing, online scams, spam, Internet fraud, and malicious URLs.

(i)   The Negative Aspects of Shopping Apps

As consumers get more adept at spotting scams, fraudsters are forced to devise new methods to take our money. Effective PC security programs, such as Panda Safe Web, can detect and block phony websites before scammers can take advantage of us.

However, we are increasingly purchasing on smartphones and tablets rather than desktop computers. Recognizing this, fraudsters have begun to construct a variety of mobile-focused attacks aimed at stealing personal information and money.

(ii) Transactions on mobile apps are declining during online sales

Because consumers are frequently unable to view the entire picture on a mobile app, they may miss out on special offers or overlook hidden fees. "The smaller screen size and uncertainty about missing important purchase details make you much more ambivalent about completing the transaction than when you are looking at a big screen," said Nikolaos Korfiatis, professor at the University of East Anglia in Norwich, UK.

II. How to ensure Privacy and Security When Shopping Online?

One of the best conveniences that contemporary technology has given us is the ability to shop online. Online shopping is the perfect substitute for going to shopping malls and other public locations if you do not like the trouble of doing so or standing in long lines. Additionally, in recent years, mobile payment options have made internet buying much more convenient. Even though you may like buying things from the comfort of your home, your bank information and shopping accounts may be compromised by several nosy eyes. Due to the nature of e-commerce and the countless alternatives for online stores, it can occasionally be challenging to determine whether you're dealing with a trustworthy or fraudulent seller.

Online users should be aware of the hazards associated with online purchases, just as customers need to take security precautions while purchasing in physical establishments. We should all ideally consider protecting our credit card information, and that's a good thing. But there are other privacy issues we should consider as well.

We emphasize the significance of protecting your privacy when purchasing online in honor of Data Privacy Day. Malware and phishing scams are no longer the only online hazards. Attackers are aware that as you engage in more online activities, you run the danger of disclosing more personal information, particularly if you're considering making a purchase. Thus, looking for products on your own could take you to multiple websites and the likelihood, of coming across one that is harmful rises as a result. Online buyers should defend themselves against potential attacks that could compromise their privacy in light of the increasing frequency of data breaches, hacking events, and identity theft. A user's privacy can be violated using a variety of techniques, and sooner or later, an unwary use is certain to come across risks including phishing, online fraud, spam, and dangerous URLs.

(i)   Pointers for Safe Online Purchasing

Here are some general pointers for protecting your online purchasing privacy and security:

a.   **Double-check URLs** - If you haven't already saved your favorite shopping site's payment page and are still putting in names, double-check the URL because fraudsters can quickly replace payment sites and apps with fake ones. Checking the security lock indicator (HTTPS instead of HTTP) is one approach to determine if a site is safe. The former denotes a safe one.

b.   If you are a frequent mobile shopper, make sure to utilize the official online shopping app and avoid third-party apps for secure transactions.

c.   Always use strong and secure passwords as attackers can simply break into internet accounts such as banking and social media accounts. Because these accounts include sensitive and personal information, it is critical that you use unique, difficult-to-crack passwords across all of them.

d. **Use a secure network** - If you're paying using a mobile device, be sure you're using the official payment app and connecting to a secure and private network.

e. **Think before you click** - getting scammed online may result in an invasion of your privacy. Think twice before clicking on unverified posts, messages, or advertisements, and avoid suspicious-looking offerings. They are most likely used as bait to direct you to phishing sites and other malicious websites. Rather of depending on social media posts, consult authoritative websites.

(ii)   Ensuring use of safe shopping apps on Smart Phone

Here are some pointers:

a. **Only get apps from official sources**
   To protect its customers, both the Apple App Store and the Google Play Store employ a stringent approval process. When you submit an app to one of these official stores, it is examined to guarantee that it is not contaminated with malware and does not collect personal information without your permission. This is fantastic news for iPhone users because every app available for download has been thoroughly tested to assure its safety. Especially since Apple devices can only install apps from the App Store.

b. Android users, on the other hand, are not restricted to the Google Play store; they may download and install apps from anywhere. Despite the fact that this is significantly more versatile, not all app shops or websites apply the same security checks. Cybercriminals exploit this weakness by tricking Android users into downloading infected apps from email attachments or fake app stores. Once installed, this malware allows scammers to steal credit card details, or to encrypt your files so you cannot access them without paying a ransom.

c. When it comes to downloading shopping apps you must ensure they come from the official app store – otherwise you could be inviting cybercriminals onto your phone.

d. Install a mobile Security Tool

You wouldn't dream of leaving your PC unprotected against malware – so why ignore your mobile phone? Just this week mobile hacking hit the headlines again as government officials tried to highlight the risks.

e. To protect themselves from fraudulent shopping apps, Android users must install a mobile security program. Panda Mobile Security detects installed apps for malware and notifies you of potential issues before your data is stolen.

f. You can also restrict what each app does with Panda Mobile Security, prohibiting them from accessing your data or activating your camera or microphone. You may also restrict apps, good or malicious, from publishing your data to the cloud, adding an extra layer of security.

(iii) Maintain vigilance.

In addition to installing security software on your phone, you should exercise vigilance when using apps, site downloads, and email attachments. You shouldn't download unfamiliar programs from untrusted websites in the same way that you wouldn't open attachments from individuals you don't know on your PC. Criminals will invest more time and effort to assaulting our phones as they become a significant component of our buying habits. So, protect yourself now, before they strike you.

III. Privacy issues of Buyers vis-à-vis Online Shopping

According to new research, consumers would rather shun brands than risk their personal data. Brands that handle customer data in a transparent and ethical manner, on the other hand, will be best positioned to navigate and engage with this wary consumer base.

Shopper dissatisfaction with online retail sites is concentrating on two primary issues. Spoiler alert: client experience does not figure into this rating. The need for caution is what causes shoppers to abandon their shopping carts.

Due to consumers' reliance on online buying, there is a heightened emphasis on prudence while supplying brands with personal information. The findings of a recent study of over 1,100 respondents performed by marketing technology firm Wyng and released on Tuesday corroborated this reality. According to the "Wyng 2021 Report: State of Consumer Data Privacy," consumers would rather abandon purchases or shun firms that do not provide promises of preserving buyers' personal information.

(i)   Increased Bounce Rates

First, according to Qubit's research, shops may be underprepared to handle the predicted surge in online purchasing and shifting customer mindsets. According to the report, disappointed customers frequently switch to competitors. Wyng's research now demonstrates strong shopper linkages to privacy issues, which lead to cart abandonment. It was revealed that 70% of buyers skip transactions owing to privacy concerns.

It is well known that consumers are becoming increasingly sensitive about their privacy. Wendell Lansford, co-founder of Wyng, said he was not shocked that 80 percent of consumers felt that data security and privacy are important. "What surprised me was that 70% of those polled said they had abandoned purchases because they didn't trust how their data would be handled." That should serve as a wake-up call to any marketers or merchants who aren't already taking privacy seriously," he told the E-Commerce Times. The remaining 20% said it was still "somewhat important."

(ii)   Factors enhancing Consumer Awareness

According to the survey, consumer awareness of data privacy has increased due to a variety of factors, including high-profile breaches, greater laws, and tech titans drawing lines in the sand. Wyng's research, however, indicates more than just awareness. Based on how brands handle data, there has been a discernible shift in purchasing habits.

According to the Wyng study respondents, privacy is unique to each individual consumer. Certain identifying information is deemed more sensitive by them. Individual identification must be kept private, according to survey respondents. They placed a high emphasis on everyday touchpoints. Perhaps as a result of the recent increase in robocalls, 83 percent of respondents felt more protective of their phone number than any other form of identification. Identifiers identifying location, email addresses, online queries undertaken, and biometric information such as speech patterns and facial traits received the same or higher percentages.

More than 60% of respondents stated they would disclose more personal data with an e-commerce site if they could easily access the data they shared. Those polled also stated that they wanted sites to allow them to edit or delete their personal information at any time.

Even first-party behavioral data has flaws since it is collected in ways that consumers are unaware of and used in ways that consumers are unaware of. The actual gold standard is zero-party data. Since customers have full transparency and control over zero-party data, it gives brands a future-proof foundation for their data strategy.

IV. Concerns about privacy and security

Furthermore, concerns about privacy and security risks on mobile phones lead people to put products in their shopping baskets but then leave without paying. This makes it difficult for online retailers who invest extensively in mobile. While mobile shopping is supposed to make the process easier.

(i)   Online shopping data

For the study, published in the Journal of Business Research, the researchers studied online shopping data from random consumers. Customers are becoming more and more demanding and, with mobile shopping in particular, they don't forgive failures so offering a streamlined, integrated service is really important. Customers are becoming increasingly demanding, and they don't forgive failures in mobile shopping in particular, so offering a streamlined, integrated service is critical," the researcher stated about making the right choice or whether the site is secure enough leads to 'emotional ambivalence' about the transaction - and this means customers are much more likely to simply abandon their shopping carts without completing a purchase.

V. Conclusions and Suggestions

Many users share a lot of information on shopping apps on their phones, although this is easily avoided. Over half of all customers now use their cell phones to compare products and prices, and shopping applications are growing increasingly popular. Consumers may quickly adjust their privacy permissions by going to "Settings" on their smartphone and viewing the "Privacy" page, according to the researchers. They can alter their smartphone permissions for specific apps from here, such as disabling access to their camera, microphone, contacts, location, web browser history, and other online activity. Consumers who are wary of popular shopping apps can bypass them by visiting the retailer's website in a browser instead.

Marketers and merchants are currently living in the privacy-first era. The path forward is to collect zero-party data in order to develop relationships and provide consumers with transparency, choice, and power. In an increasingly cautious consumer market, firms that provide consumers a clear value exchange for their personal data will be best positioned to prosper in the long run. Brands must Consumers should be cautious about their privacy when using shopping applications

Suggestions

(i)   Brands must abandon their reliance on third-party data and instead invest in zero-party and first-party data. Use of zero-party data to personalize experiences without jeopardizing consumer confidence can help them with their consumer data initiatives while also protecting consumer privacy.

(ii)   When providing personal information to brands, customers must be transparent. That transparency has the potential to have a long-term influence.

(iii)   Consumer trust must be rebuilt through transparency and control over their personal data by brands.

(iv)   Brands must provide assurance that consumer data will be protected and not shared or sold. This will assist brands in avoiding cart abandonment and other short-term issues, as well as building brand confidence over time.

References

1   https://www.ecommercetimes.com/story/privacy-concerns-key-reason-buyers-flee-online-retailers-87301.html

2   https://www.trendmicro.com/vinfo/pl/security/news/online-privacy/be-aware-of-your-privacy-and-security-when-shopping-online

3   https://www.pandasecurity.com/en/mediacenter/mobile-security/dark-side-shopping-apps/

4   https://www.ecommercetimes.com/story/privacy-concerns-key-reason-buyers-flee-online-retailers-87301.html

5   https://consumerandsociety.com/2022/12/05/experts-warn-consumers-about-their-privacy-on-shopping-apps/

6   https://economictimes.indiatimes.com/tech/internet/why-online-shopping-sales-are-less-on-mobile-apps/privacy-and-security-issues/slideshow/62400433.cms?from=mdr

7   https://www.wyng.com/blog/wyng-2021-report-state-of-consumer-data-privacy-attitudes/.

8   https://www.wyng.com/blog/wyng-2021-report-state-of-consumer-data-privacy-attitudes/

9   https://www.elsevier.es/es-revista-european-research-on-management-business-489-articulo-privacy-concerns-online-purchasing-behaviour-S2444883416300134

Guarding Personal Privacy in India: A Journey through Evolution, Challenges and the Imperative for a Rights-Centric Approach

Bishnanand Dubey
Assistant Professor
College of Law and Legal Studies
Teerthanker Mahaveer University
Moradabad, UP, India

Abstract

Protecting individual privacy has become a crucial concern in a society that is becoming more connected and data-driven. The evolution of personal privacy protection in India is examined in this essay, from historical models to current difficulties. It draws attention to the necessity of a rights-centric strategy in the Indian environment. The paper starts out by looking at the historical foundations of privacy in India, using prehistoric writings and legal customs as sources. It explains how the idea of privacy has changed over time, interacting with sociological, technological, and legal developments. Particular focus is placed on how important rulings like those in the Puttaswamy and Kharak Singh cases have shaped the nation's privacy landscape. In conclusion, this study argues that a thorough rights-centric strategy is required given how personal privacy has evolved in India. It advocates for a comprehensive incorporation of privacy principles into the legal system, as well as coordinated efforts to address new issues and give people the tools they need to safeguard their privacy in the digital era.

Keywords – India, Privacy, Report, digital, Supreme Court.

I. Introduction

The idea of privacy as a fundamental right in India was recognized by the hon'ble Supreme Court in 2017 in the landmark Puttaswamy case. [1] Right to privacy is an important part of Article 21 which provides right to life and personal liberty as per the decision of court in Puttaswamy case. This decision reversed earlier rulings that had claimed that in India, the right to privacy was not a basic one.

The implications for individual autonomy and dignity of the Right to Privacy being recognised as a Fundamental Right in India.[2] Privacy and data protection are essential for individuals to control their personal information, be free from unreasonable surveillance and intrusion, and exercise their rights to freedom of expression and association. In this digital age its now important to cautious for collecting and processing data as it can have bad impact on individuals and Societies freedom.

India has recognized the right to privacy as a fundamental right in conformity with international human rights standards including the UDHR and the ICCPR. The recognition of the Right to Privacy has led to the development of new privacy laws and regulations, such as the Personal Data Protection Bill in India. This legislation seeks to regulate the collection, use, storage, and sharing of personal data by both the government and private entities and establish a Data Protection Authority to oversee and enforce these regulations.

The expansion of the definition of privacy violations and the severity of criminal punishments are results of the Privacy's Right as a fundamental right. India faces numerous difficulties in implementing and upholding the right to privacy, including a lack of knowledge and comprehension, insufficient legal systems, and ineffective enforcement practises. Additionally, marginalised groups need to have their privacy rights addressed and protected since they suffer particular privacy risks and vulnerabilities, including those involving women, children, and LGBTQ+ people.[3]

II. Importance of Privacy in the context of India

Privacy is important in the context of India for several reasons. Firstly, it is an indispensable human right that is essential for the protection of individual autonomy, poise, and personal freedom. It enables individuals to have control over their personal information and to live their lives free from undue interference and surveillance.[4] It can have a chilling effect on fundamental rights like freedom of expression, association, and access to information, privacy is crucial for the enjoyment of these rights.

Privacy is increasingly important in the digital age due to the use of biometric identification systems, CCTV cameras, and social media platforms. Privacy is essential for protecting marginalized communities, such as women, children, and LGBTQ+ individuals, from online harassment and exploitation, as well as discrimination and persecution depending on their gender identification or sexual orientation.[5]

India's the world's largest democracy with a diverse population and history of social and political unrest, making privacy essential for individuals to exercise their rights and freedoms without fear of persecution. The India's Apex Court's held that the Privacy's right is a fundamental right has led to the development of new privacy legislation and regulations in India. These regulations and guidelines are intended to control how both public and private institutions collect, utilize, store, and share personal data.[6]

The Personal Data Protection Bill outlines penalties for non-compliance, creates a Data Protection Authority, and regulates how individuals' personal data is processed in order to protect people's privacy. Since the use of surveillance technologies by the government and law enforcement agencies generates serious privacy concerns, it is evident how important privacy is to surveillance and national security. Privacy protection is essential if surveillance is to be conducted in a manner that upholds democratic principles and fundamental rights.[7] India is a developing nation with high rates of poverty, inequality, and social exclusion, making privacy crucial for defending the most defenseless and marginalized facets of society.

III. Historical context of the Right to Privacy in India

India's history of the right to privacy starts with Article 21, which defined the rights to life and personal freedom, in the 1950 Indian Constitution. The right to privacy is a crucial component of Article 21, according to the Supreme Court.

(i) Initial Case-Laws

The Supreme Court acknowledged privacy as a fundamental right for the first time in the case of Kharak Singh v. State of Uttar Pradesh.[8] In the case, the Uttar Pradesh Police Regulations which gave the police the power to spy on people—were challenged as being unconstitutional. The right to privacy was implied in the guarantee of personal liberty under Article 21 of the Constitution, according to the highest court's ruling, even though it was not specified specifically in the Constitution.

In numerous cases over the years, such as Govind v. State of Madhya Pradesh (1975)[9] and R. Rajagopal v. State of Tamil Nadu (1994)[10], The S.C. upheld the Right to Privacy as a Fundamental Right, emphasizing that the preservation of the Right to Privacy was necessary for the exercise of other Fundamental Rights, like the Freedom of Speech and Expression.

In 2017, the Supreme Court made a historic decision in the case of Justice K.S. Puttaswamy (Retd.) v. UOI[11], signifying the most important turning point in the history of the right to privacy in the country. The Aadhaar scheme, which attempted to establish a national biometric identity system, was challenged in the lawsuit as to its constitutional legitimacy. The Supreme Court ruled that the Aadhaar program must respect these rights and that the right to privacy is a basic right guaranteed by the Indian Constitution.

The 2017 Supreme Court ruling recognizing the right to privacy as a fundamental freedom has had a big impact on India's legal and regulatory framework defending privacy. The Personal Data Protection Bill, which attempts to give India a comprehensive framework for data protection, is one of the privacy laws and regulations that emerged from this.

Indian law's development of the right to privacy has been a protracted and difficult process. The Right to Privacy has grown over time as a result of numerous significant judgments rendered by the Indian judicial system.

In the early years of India's independence, the concept of privacy was not specifically stated in the constitution. The right to privacy was nonetheless viewed by the Supreme Court of India as an essential element of the life and personal freedoms protected by Article 21 of the Constitution.[12]

The Kharak Singh v. State of Uttar Pradesh case from 1963 marked a crucial turning point in the development of India's legal right to privacy. Although the right to personal liberty, which includes the right to privacy, is protected by Article 21 of the Indian Constitution, the Supreme Court found that this was not the case in this particular case. By declaring that a person's right to privacy was essential to their right to life and personal liberty and could only be restricted in accordance with due process of law, the court's decision established a precedent.

The Supreme Court of India further reaffirmed the Right to Privacy's fundamental validity in Govind v. State of Madhya Pradesh (1975)[13]. The court found that the right to privacy must be maintained in order for other fundamental rights, such the right to free speech and expression, to be practiced. The court hi

ghlighted that infringing someone's right to privacy should only occur in dire circumstances in order to preserve public safety or national security.

The decision strengthened the law's acceptance of the right to privacy as a crucial facet of a person's fundamental rights.

In R. Rajagopal v. State of Tamil Nadu (1994), the Supreme Court of India upheld the Right to Privacy as a fundamental right. The case involved a dispute over the publication of a book that revealed private details about the life of a famous person. According to the court's decision, under Article 21 of the Constitution, a person's right to privacy is intrinsically tied to their right to life and personal freedom. The court determined that if confidential information is disclosed without permission, a person's right to privacy may have been violated. The ruling created a framework for defending privacy when it was threatened and stressed how important it is as a basic right.[14]

The Supreme Court's ruling in Justice K.S. Puttaswamy (Retd.) v. Union of India was a significant turning point in the evolution of the Right to Privacy in Indian law. The court determined that the Right to Privacy was a fundamental constitutional right that must be upheld in order to preserve a person's dignity and autonomy. This judgment was the most important development in the history of privacy rights in India.

India's recognition of the Right to Privacy as a Fundamental Right has made it feasible to enact the Personal Data Protection Bill and other new privacy-related laws and regulations.

Indian law has changed over time, and numerous court decisions have recognized the Right to Privacy as a Fundamental Right.

(ii) Supreme Court's landmark judgment in the Puttaswamy case

The Indian Supreme Court recognized that the Right to Privacy is a Fundamental Right guaranteed by the Indian Constitution after ruling that the government's use of biometric identification in the Aadhaar program infringed this right. The government's collection of citizen biometric data for a national identity scheme is the subject of the current lawsuit, Justice K.S. Puttaswamy (Retd.) v. Union of India (2017).[15]

The Puttaswamy case marked a turning point in the development of India's right to privacy because of the Supreme Court's groundbreaking ruling stating that the right to privacy is a basic right protected by the Indian Constitution. In its decision, the court emphasized how important privacy is to a person's autonomy, dignity, and capacity to exercise other rights, such as the right to free speech and expression.

The Supreme Court's judgement, which reversed the rulings in M.P. Sharma v. Satish Chandra (1954)[16] and Kharak Singh v. State of Uttar

Pradesh (1963)[17], was rendered by a nine-judge panel. The Puttaswamy case reversed earlier findings that claimed there was no right to privacy protected by the Indian Constitution. The Puttaswamy case verdict by the court had broad ramifications for India's legal and regulatory system controlling privacy. It cleared the path for the creation of new privacy rules and regulations and provided India with a strong legal basis for privacy protection.

To build a thorough data protection framework and strengthen individual privacy protections, the Indian government formed a committee to prepare the Personal Data Protection Bill, 2019.

The Supreme Court acknowledged the right to privacy as a basic right protected by the Indian Constitution in its historic Puttaswamy decision. This choice had a significant impact on India's legal and regulatory framework protecting privacy, resulting in the adoption of new privacy laws and regulations.[18]

(iii) Supreme Court's position on the use of Aadhaar and the Right to Privacy

Citizens can use Aadhaar, a distinctive identifying system, as a reliable form of identity to obtain government services and subsidies. The use of Aadhaar has been controversial due to potential misuse of personal data and the violation of individual privacy rights. In 2018, the Supreme Court of India held that Privacy's right is Fundamental right. The Supreme Court ruled that the requirement to link Aadhaar with bank accounts and mobile phone numbers was illegal since it violated the right to privacy.

The government cannot need Aadhaar to obtain government benefits and services, the court decided, as doing so would be against private rights. Although it permitted the optional use of Aadhaar for some government services, people cannot be barred from receiving those services if they do not own an Aadhaar card. A key turning point for the Right to Privacy in India was the Supreme Court's decision on the use of Aadhaar, which established that privacy is a fundamental right and any limitations on it must be just and necessary. The decision established a framework for using Aadhaar and other identity systems that strikes a balance between the need for identification and privacy rights.

A pivotal moment in shaping the Right to Privacy in India was the landmark Supreme Court ruling on the use of Aadhaar. This ruling not only recognized privacy as a fundamental right but also emphasized that any restrictions on it must be fair and essential. In essence, it established a key balance between the defense of people's privacy rights and the imperative requirement for identification through systems like Aadhaar. This ruling offered a framework that upholds both of these principles, protecting individual privacy while enabling appropriate identification procedures.

In the cases of R. Rajagopal v. State of Tamil Nadu and People's Union for Civil Liberties v. Union of India, the Supreme Court ruled that the right to privacy cannot be used as a means of restricting press freedom or free speech. The court additionally determined that a person's right to privacy cannot be used as an excuse to impede their freedom of movement.

The government introduced the Personal Data Protection Bill to address the issue of how to balance privacy rights with other fundamental rights in the era of contemporary technologies. The purpose of this law is to control how private data is gathered, used, and transferred. A nuanced approach that considers the particular facts and circumstances of each case is necessary to strike a balance between privacy rights and other fundamental rights. The Supreme Court has provided instructions on how to achieve this balance.

IV. Implications of acknowledging the Right to Privacy as a fundamental right?

The legal and regulatory frameworks governing privacy in India have undergone multiple important changes as a result of the recognition of the right to privacy as a basic right. Some of the main implications are listed below:

(i) **Protection of Personal Autonomy:** A essential right, the right to privacy allows people to decide how their personal data is collected, used, and disclosed.

(ii) **Strengthening Other Fundamental Rights:** By recognizing the Right to Privacy as a fundamental right, the protection of individuals from arbitrary interference by the state is strengthened, thus reinforcing other fundamental rights.[19]

(iii) **Limiting State Surveillance:** State monitoring is constrained by the recognition of the Privacy's Right as a Fundamental Right, and the state is required to give justification for any such actions. This protection ensures that people are protected from unwarranted surveillance and that surveillance is carried out in a way that is consistent with democratic norms.

(iv) **Establishment of Privacy Regulations**: The formation of privacy rules and legislation has a legal foundation thanks to the acknowledgment of the right to privacy as a basic right, which enables people to pursue legal redress when their privacy is violated.

(v) **Protection of Personal Data:** The cornerstone of personal data protection in the digital age is the acknowledgement of the right to privacy as a basic right. Personal data must be safeguarded against unauthorized access, use, and disclosure.[20]

V. Implementation and Enforcement of the Right to Privacy in India

The Right to Privacy's implementation and enforcement in India is a crucial problem because the right only has value if it is successfully implemented. Here are some important details about the application and enforcement of India's Right to Privacy:

(i) **Data Protection Framework:** A complete framework for data protection in India would be established if the Personal Data Protection Bill, 2019, which is presently being discussed in Parliament, is passed. In addition to articulating important data protection principles like the requirement for informed consent prior to data collection and the right to data portability, the Bill would create a Data Protection Authority tasked with enforcing compliance with rules.

(ii) **Judicial Oversight:** In India, the judiciary plays a significant role in protecting the right to privacy. Through the landmark Puttaswamy decision, the Supreme Court played a vital role in establishing it as a basic right and establishing rules for its preservation. The courts have since made decisions in a large number of issues involving privacy rights.[21]

(iii) **Public Awareness:** The public's awareness and comprehension of the right are necessary for its implementation and enforcement. Civil society organizations and the government both have a part to play in raising awareness and fostering understanding.

(iv) **Government Action:** The government is heavily involved in putting the Right to Privacy into practice and enforcing it, making sure that laws are followed and that there are effective remedies available. It must also ensure that its own actions are lawful, adequately controlled, and compliant with privacy laws.

(v) **Corporate Responsibility:** Businesses and organizations must collect, retain, and use personal data in a responsible manner, and privacy rules must be transparent and easy to understand. Additionally, they must offer sufficient security measures to guard against unauthorized access, use, and disclosure of personal data.

VI. Challenges in implementing and enforcing the Right to Privacy

Despite being accepted as a fundamental right and the actions taken to address it, India's Right to Privacy implementation and enforcement nevertheless face a number of significant challenges. A few of the main issues are listed here:

(i)   **Lack of awareness**: One of the major challenges to adopting and sustaining the right to privacy is a lack of public awareness. The right to privacy is not well known, and many people may not know how to use it. People may find it difficult to protect their privacy rights due to ignorance, while businesses may find it easier to do so.[22]

(ii) **Weak enforcement mechanisms:** Concerns about the efficiency of enforcement procedures have been raised in light of the creation of a data protection framework and a Data Protection Authority (DPA). It is yet unclear how effective the DPA will be in enforcing privacy laws since it is still in the process of being constituted. Concerns exist over the regulatory agencies' ability to look into and enforce privacy infractions.

(iii) **Government surveillance:** There are worries that excessive and insufficient monitoring accompany government surveillance in India. For their potential to infringe privacy rights, surveillance programs like the Central Monitoring System (CMS) and the National Intelligence Grid (NATGRID) have drawn criticism. The government is accountable for making sure that its

445

surveillance programs adhere to privacy rules and are overseen appropriately.[23]

(iv) **Lack of transparency:** Inadequate disclosure by businesses and organizations regarding the collection and use of personal data makes it difficult to enforce the right to privacy. Individuals find it challenging to make educated decisions about sharing their data as a result.[24]

(v) **Limited resources:** In order to investigate and enforce violations of the right to privacy, regulatory authorities may run into funding issues and a shortage of qualified employees.

VII. Ongoing initiatives to improve India's data protection and privacy legislation.

There have been ongoing efforts to strengthen privacy and data protection laws in India. Here are some key developments:

(i) **The Personal Data Protection Bill:** The Personal Data Protection Bill was introduced by the Indian government in 2019 in an effort to regulate how personal data is treated by both public and commercial organisations. Requests for amendments to ensure that privacy rights be better protected are being considered by a legislative committee.

(ii) **The Data Protection Authority:** As was already reported, the government has suggested creating a Data Protection Authority to supervise the Personal Data Protection Bill's implementation. The authority will have the authority to look into privacy violations, enforce them, and apply sanctions for non-compliance.

(iii) **Other Developments**

a. **Supreme Court judgments:** The SC has repeatedly defended the right to privacy, as seen by the 2020 decision to invalidate a section of the Information Technology Act that permitted the arrest of people for publishing objectionable content online. The clause, in the opinion of the court, violated both the right to privacy and the freedom of speech.[25]

b. **Industry initiatives:** The establishment of the Data Security Council of India (DSCI) and the implementation of privacy policies and processes to comply with relevant legislation are examples of actions the private sector has made to enhance data protection and privacy.[26]

c. **Public awareness campaigns:** Awareness programs have been developed to educate the public about their rights to privacy and the importance of protecting personal information. The Ministry of Electronics and Information technologies launched "Digital India" to promote the safe and secure use of digital technologies.

VIII. Intersectionality and Privacy in India

Different forms of discrimination overlap and compound one another through a process known as intersectionality, which gives members of marginalized groups a variety of ways to experience oppression. It is particularly crucial in the context of privacy since marginalized groups may be more at risk for privacy violations as a result of the various forms of discrimination they encounter.

Due to the interplay of gender and sexual orientation, women and members of the LGBTQ+ community in India face increased dangers when it comes to privacy issues. Because of their gender identity or sexual orientation, this may result in harassment or discrimination, which may violate their privacy. Due to discrimination and monitoring, which can result in rights violations, people who identify as members of religious or racial minorities may be more vulnerable to threats to their privacy.

The SC recognized the importance of recognizing the Right to Privacy as a fundamental right for all individuals, including those who belong to marginalized groups, in privacy-related cases in India.[27] With the Personal Data Protection Bill, which mandates that data fiduciaries obtain individuals' explicit consent before processing sensitive personal data pertaining to caste, religious beliefs, and sexual orientation, India has made efforts to address intersectionality in privacy and data protection laws.

IX. Conclusions

A democratic society that respects human dignity, autonomy, and liberty must uphold the right to privacy. People enjoy additional protection from invasive government actions and uninvited data gathering and use by commercial companies since the right to privacy was proclaimed a basic right in India. Finding the proper balance between privacy and other fundamental rights is one of the biggest obstacles, India must overcome in order to protect its citizens' right to privacy. Adopting a strategy that puts individual rights first is essential for effectively addressing these issues. To do this, any privacy limits must take into account aspects including proportionality, necessity, transparency, and accountability.

A positive start toward strengthening India's privacy and data protection legislation is the passage of the Personal Data Protection Bill. It is intended to protect people's privacy and rights. A fine balance between privacy and these other crucial rights can also be achieved thanks to the Supreme Court's investigations into how freedom of speech and expression interact with other fundamental rights. To protect individual privacy and promote democratic and humanistic ideals, privacy must be recognized as a fundamental right in India. We must continue to work to make privacy laws like the Personal Data Protection Bill stronger if we want to ensure that it is a basic right in the long run. Finding the ideal balance between privacy and other fundamental rights, such as the freedom of speech and expression, is equally important. This continued dedication to upholding private rights will play a crucial role in determining how privacy will develop in India.

To find a balance between privacy and other fundamental rights, a rights-based approach that takes into account the concepts of human dignity, autonomy, and liberty is required. The right to privacy must be assessed against other rights, such as equality, the freedom of speech and expression, and access to information.

Suggestions - Rights-based approach to privacy

A rights-based approach to privacy suggests that we must take other rights and interests into account when evaluating privacy rights. There is no fundamental, universal right to privacy that always takes precedence over all others. Instead, it's a right that occasionally needs to be balanced with other freedoms, such as the ability to obtain information or the freedom of speech. Personal data may occasionally need to be shared in order to safeguard or advance these other rights, or to improve accountability and openness. Therefore, depending on the exact circumstances, it's vital to strike the correct balance between privacy and these other crucial factors. It's not always easy to make a choice; instead, one must carefully evaluate how many rights can coexist and occasionally how they should be balanced against one another.

A rights-based approach to privacy requires a thorough and contextual analysis of each incident in order to maintain the harmony between privacy and other fundamental rights. This approach recognizes the value of privacy while also understanding the need to strike a balance between privacy and other fundamental rights in order to build a society that is just and equitable.

Endnotes

1   *K.S. Puttaswamy v. Union of India*, (2017) 10 SCC 1.

2   Krishnadas Rajagopal, "The Lowdown on the Right to Privacy", *The Hindu*, July 29, 2017, *available at*: https://www.thehindu.com/news/national/the-lowdown-on-the-right-to-privacy/article19386366.ece (last visited on Mar. 10, 2023).

3   *Navtej Singh Johar v. Union of India*, AIR 2018 SC 4321.

4   The Constitution of India (1950), s. Article 21.

5   *K.S. Puttaswamy v. Union of India*, (2017) 10 SCC 1.

6   *Ibid.*

7   The Constitution of India, 1950.

8   AIR (1963) SC 1295.

9   AIR1975 SC 1378.

10 AIR 1995 SC 264.

11 (2017) 10 SCC 1.

12 The Constitution of India, 1950.

13 AIR 1975 SC 1378.

14 AIR 1975 SC 1378.

15 (2017) 10 SCC 1.

16 AIR 1954 SC 300.

17 AIR (1963) SC 1295.

18 (2017) 10 SCC 1.

19 The Constitution of India, 1950.

20 (2017) 10 SCC 1.

21 *Ibid.*

22 The Constitution of India, 1950.

23 "No Blanket Permission to Any Agency for Surveillance Under NETRA, NATGRID: Centre to HC", *The Economic Times*, June 22, 2022, *available at*: https://economictimes.indiatimes.com/news/india/no-blanket-permission-to-any-agency-for-surveillance-under-netra-natgrid-centre-to-hc/articleshow/92393282.cms (last visited on Mar. 11, 2023).

24 *K.S. Puttaswamy v. Union of India*, (2017) 10 SCC 1.

25 AIR 2020 Delhi 156.

26 Shiv Shankar Singh, "Privacy and Data Protection in India: A Critical Assessment" 54 *Journal of the Indian Law Institute* 663-677 (2011).

27 *K.S. Puttaswamy v. Union of India*, (2017) 10 SCC 1.

Chapter 28

The Paradoxical Context of the Restriction of the Right to Privacy and Data Protection in India

Dr. Zheer Ahmed

Assistant Professor

School of Liberal Arts

MIT-World Peace University

Pune, Maharashtra, India

Abstract

India, a country known for its rich cultural heritage and robust democratic system, confronts a complex predicament pertaining to the curtailment of the right to privacy and safeguarding of personal data. This study examines the complex dynamics of a paradoxical situation, exploring the subtle relationship between privacy rights, national security imperatives, and technology breakthroughs in India. The right to privacy is acknowledged as an inherent aspect of the right to life and personal liberty by the Indian Constitution, specifically in Article 21. Nevertheless, this inherent entitlement is situated within a paradoxical framework wherein, on the one hand, it is vigorously protected by the judicial branch, as evidenced by the significant Puttaswamy ruling that acknowledges privacy as a basic right. However, it is important to note that there are limitations imposed on it, which are justified on the grounds of national security. This creates a complex equilibrium between the rights of individuals and the need for collective security. This study examines the historical and legal progression of the right to privacy in India, with a particular focus on its acknowledgment as a basic right and the subsequent legislative measures, such as the Personal Data Protection Bill, aimed at governing data processing and strengthening data safeguarding. The dichotomy becomes evident as these endeavours coexist with the comprehensive surveillance infrastructure and biometric databases, such as Aadhaar, utilised by the government. Moreover, the Chapter assesses the repercussions of this paradoxical situation on individuals, enterprises, and the civic community in India.

Key Words: Rights, Privacy, Data Protection, Judiciary, India.

I. Introduction

The discourse on the right to privacy and data protection is not new. The Indian judiciary has played a significant role to define article 21 in accordance with right to privacy. In this regard, In the case of *Justice K.S. Puttaswamy (Retd.) v. U.O.I.*[1], a significant ruling was made by a nine-judge panel of the Supreme Court on August 24, 2017. This ruling upheld the constitutional recognition of the right to privacy. The Constitution of India, specifically Part III, recognises privacy as a fundamental aspect of our rights. These rights encompass various aspects such as equality (Articles 14 to 18), freedom of speech and expression (Article 19(1)(a)), freedom of movement (Article 19(1)(d)), protection of life and personal liberty (Article 21), among others. The right to privacy is a multifaceted concept that is encompassed within the terms of numerous legislations. The recognition of the right to privacy is acknowledged both domestically and globally through numerous conventions. The right to privacy is acknowledged and protected by several legal frameworks, including tort law, criminal law, and property law. Despite the absence of an explicit constitutional provision regarding the right to privacy in India, it is acknowledged as a fundamental right derived from various other rights. Similar to other fundamental rights, the right to privacy is not absolute and can be subject to reasonable limitations imposed by the State. Fundamental rights are seldom absolute in nature[2]. The phrase 'reasonable restrictions' implies that limitations may be placed on the exercise of a right, but these limitations should not be arbitrary or disproportionate[3].

Under specific conditions and criteria, the privacy interests of an individual can be superseded by competing State and individual interests. It is unable to surpass the boundaries of the rights from which it originates. The exercise of individual liberty, as allowed via fundamental rights, must be conducted in a manner that does not infringe upon the liberty of others. The right to privacy is inherently subject to acceptable limitations. The determination of appropriate constraints on privacy rights is under the purview of the courts. The criterion of reasonableness varies based on the facts of each scenario[4]. In essence, this implies that the entitlement to privacy will be subject to limitations outlined in Part III of the Constitution. The complete prohibition of private rights for individuals is not permissible; instead, it must satisfy specific criteria established by the courts, such as the proportionality test and the balancing test. The State is additionally obligated to examine the technique employed by them to determine if it infringes upon Articles 14 and 21. In other words, a regulation that limits the right to privacy must not be arbitrary but rather must be deemed just, fair, and reasonable.

During the deliberations of the Constituent Assembly, Dr. B.R. Ambedkar emphasised the necessity of acceptable restraints, asserting that the enjoyment of rights to their greatest extent is contingent upon the presence of such limitations[5]. The speaker emphasised the importance of striking a balance between individual liberty and the implementation of social control in order to achieve social justice. According to Charles A. Beard, freedom is a concept that is relative rather than absolute[6]. Frederic A. Ogg had a similar perspective, whereas P. Orman Ray asserted that liberty should not be regarded as a licence, and emphasised that rights are not absolute but rather relative in nature[7]. Reasonable limitations are imperative due to the need for a meticulous and discerning equilibrium to be established between the personal interests of individuals and the legitimate concerns of the State.

It is noteworthy that Articles 14 and 21 do not contain any specific limitations, thereby placing the responsibility on the judiciary to delineate the allowable scope of limits. The courts has formulated specific theories to ascertain rational limitations. From its inception, the Supreme Court has maintained that the right to privacy is not without limitations, as constraints may be imposed in instances involving national security, public interest, criminal investigations, innovation, and other relevant considerations.

II. Restrictions on Right to Privacy

The Indian Constitution guarantees fundamental rights, and the Supreme Court has specifically acknowledged several requirements that must be met in order to impose restrictions on these rights. In the current privacy case, the court has asserted that fundamental rights should not be interpreted in isolation, and any limitation on these rights must satisfy the established criteria outlined in Articles 21 and 14 of the Constitution. If Article 19 is interpreted, the limitations that apply to the right are enforceable.

According to Article 21 of the Constitution, it is imperative to establish a legally prescribed mechanism in the event of the deprivation of life or liberty. The first prerequisite for limiting the entitlement to privacy is the existence of a legal framework that substantiates the infringement. Article 21 of the constitution specifically stipulates that individuals cannot be deprived of their life or liberty, unless it is done in accordance with the legally established system. The curtailment of any right under Article 21 is contingent upon the presence of legal support. According to Fazal Ali J., the concept of 'law' extends beyond being a mere statute enacted by the Parliament. It encompasses fundamental concepts of justice that are embedded within it[8]. The term 'law' as referred to in Article 21 denotes a legally binding regulation, while the term 'procedure' signifies a set of specific and substantive regulations, rather than a mere superficial formality[9]. According to Article 21 of the Constitution of India, the laws must adhere to the several fundamental rights outlined in Part III of the Constitution. Chandrachud J. emphasised that the presence of a legitimate legal framework is a prerequisite for any curtailment of the fundamental right to privacy. It is important to acknowledge that the term 'law' encompasses both ordinances and rules established by the government[10].

In the context of law and arbitrary, according to the Black Law Dictionary, an arbitrary conduct is one that lacks any basis in prejudice or preference. Actions that are grounded in irrelevant facts and lack awareness of pertinent conditions demonstrate arbitrariness. Article 14 has been established by the judiciary to examine the validity of legislative acts and administrative actions. There are two methods by which development occurs: The initial step involves administering the classification test to see if the contested action is predicated on an unjustifiable categorization. Additionally, the judiciary examines the extent to which the action is equitable, impartial, and devoid of capriciousness. The law must possess a valid and justifiable objective. A legislation or activity is deemed legal or constitutional only if it meets the requirements of these two elements. When authorities pursue a procedure without providing reasons for their actions, it might be considered a clear manifestation of arbitrariness.

The acceptance of the arbitrariness test by the majority in the case of *E.P. Royappa v. State of Tamil Nadu*[11] marked its initial recognition. Bhagwati J., along with Chandrachud and Krishna Iyer, JJ., asserted that when an action is deemed arbitrary, it is seen to be unequal in accordance with both political reasoning and constitutional jurisprudence. It has been observed that there exists an inherent opposition between equality and arbitrariness, as these two concepts are fundamentally incompatible and diametrically opposed to one another. Arbitrariness is predicated only upon the arbitrary and capricious decisions made by an autocratic ruler. In the case of *Indira Gandhi v. Raj Narain*[12], the court invalidated clauses 4 and 5 of Article 329A based on the argument of arbitrariness. The court determined that any action intended to undermine the principles of the rule of law is considered arbitrary. According to Dr. L.M. Singhvi, it is permissible to impose limitations on the right, as long as such restrictions are not arbitrary or excessive. Additionally, Dr. Singhvi emphasises the importance of the Parliament exercising thoughtful consideration and prudence while determining the appropriate constraints[13].

In a subsequent case, *Maneka Gandhi v. U.O.I.*[14], the court made use of the aforementioned reasoning and concluded that reasonableness is a crucial aspect of both legal and philosophical dimensions of equality and non-arbitrariness. Articles 14 and 21 were consolidated, and it was asserted that "the principle of reasonableness, which is an indispensable component of both legal and philosophical notions of equality or non-arbitrariness, permeates Article 14 in a pervasive and all-encompassing manner. Additionally, any procedure outlined in Article 21 must satisfy the criterion of reasonableness in order to align with the provisions of Article 14".

The *Maneka Gandhi case* expanded the application of Article 21 and incorporated the principle of non-arbitrariness within its purview. The development of law concerning the interconnectedness of basic rights and the expansive interpretation of fundamental rights originated in India. The judiciary began to examine the law with a focus on its adherence to principles of justice, fairness, and the absence of arbitrariness, capriciousness, and oppression. If arbitrariness is evident in the actions of the State, whether legislative or executive, it is necessary to invalidate such actions[15]. The premise mentioned was further established in the cases of A.*L. Kalra v. Project and Equipment Corp*[16]*., Babita Prasad v. State of Bihar*[17], and *Dr. K.R. Lakshmanan v. State of Tamil Nadu*[18]. The statues in the *Mithu Singh*[19] and *Sunil Batra*[20] case were deemed unconstitutional due to their arbitrary nature. The principle of reasonableness and non-arbitrariness is pervasive throughout the entire constitutional framework.

In the case of *Shayara Bano v. U.O.I.*[21], the court made an observation that actions undertaken in a capricious, irrational, and inadequately principled manner, characterised by excessiveness and disproportionality, may be deemed obviously arbitrary and subject to being invalidated. The case of *Independent Thought v. Union of India*[22] also drew upon the prior decisions of the Supreme Court, which had determined that exemption 2 of section 375 of the Indian Penal Code is arbitrary and infringes upon the fundamental values outlined in Articles 14, 15, and 21 of the Indian Constitution.

III. Test of Proportionality in the case of Rights and Privacy

The principle of proportionality holds significant prominence within the realm of European Administrative law. The rule provides guidance to the judges in the process of determining and reconciling the conflict that arises from the competing rights of individuals. In the context of the proportionality test, the court will independently assess the merits and drawbacks of a government action, and will only validate said action if it is deemed beneficial for the general public. The courts are responsible for suppressing mischievous actions carried out by the government. The procedures employed by the Parliament ought to be commensurate with the objective intended to be accomplished by the legislation. The test serves as a safeguard against arbitrary actions by the State and is a legal framework developed by the judiciary to address conflicts.

The validity of a law is contingent upon fulfilling four conditions as prescribed by the proportionality test. There are several key considerations that should be taken into account when formulating a law. Firstly, it is important that the law is grounded in a legitimate State goal. Secondly, appropriate measures should be developed to effectively advance the objectives of the law. Thirdly, it is crucial to ensure that there are no equally effective alternatives available to achieve the desired outcomes. Lastly, it is essential to avoid any disproportionate impact on the rights of individuals affected by the law. During the fourth stage, known as the balancing stage, the court assesses the appropriate equilibrium between individual rights and competing rights or the general interest. Proportionality is a crucial element in safeguarding individuals from unjust and capricious actions by the State, as it serves to protect against undue infringement upon fundamental rights. The fundamental principle of the proportionality test is in the concept of reasonableness. The Supreme Court, in the case of *M.R.F. Ltd. v. Inspector, Kerala Government*[23], made an observation emphasising the requirement of a direct and proximate connection between the imposed constraints and the intended objective. Only provisions that meet this constitutional criterion are to be deemed valid.

Ranjit Thakur v. U.O.I.[24] is a notable Supreme Court case that addressed the concept of proportionality. The court stated that if a decision, including those made by a court-martial, is seen to be a flagrant disregard of logic, it may be subject to judicial scrutiny. The case of *Om Kumar v. U.O.I.*[25] provided a comprehensive analysis of the proportionality theory. In the current case, the division bench of the court observed three fundamental elements of the doctrine. The first requirement for an act formulated by the legislature is to effectively accomplish a specific rational aim. Additionally, it is imperative that the methods employed are logically linked to the desired outcome and minimise infringement upon individual liberties. Thirdly, it is crucial to establish a sense of proportionality between the action taken and the desired objective.

In the case of *Teri Oat Estates (P) Ltd. v. U.T. Chandigarh and Others*[26], Sinha J. emphasised the need of proportionality in evaluating the restriction of fundamental rights during the enactment of legislation or the implementation of administrative measures. It was argued that any limitations imposed should be both suitable and modest. It is imperative to keep a suitable equilibrium between the negative consequences of an action on individual rights and freedoms. According to Sikri J. (2016), jurisprudentially, the concept of proportionality encompasses a collection of principles that are utilised to establish the essential and adequate criteria for the restriction of constitutionally safeguarded rights[27]. The process of achieving equilibrium between two essential elements, namely individual rights and the restrictions imposed on those rights, is undertaken by considering the comparative societal values associated with each competing facet.

Justice Chandrachud has acknowledged the application of the proportionality test in situations when there is a restriction on basic rights. The assertion has been made that proportionality is a crucial aspect of safeguarding against arbitrary actions by the government, as it ensures that any infringement upon fundamental rights is not excessive in relation to the intended objectives of the law. In the case of *Justice K.S. Puttaswamy (Retd.) v. U.O.I*[28] (Aadhaar case), Sikri J. thoroughly examined the notion of proportionality and determined that a component of section 57 of the Aadhaar Act is deemed unlawful due to its failure to meet the proportionality test. The principle of proportionality necessitates that the State employ the least invasive means in order to accomplish the intended aim.

Justice Ashok Bhushan further emphasised the significance of proportionality as a crucial element in safeguarding individuals against arbitrary actions by the State. This concept aims to ensure that there is no excessive infringement on individuals' rights. Chandrachud J. elucidated the concept of the test of proportionality, which asserts that the State's encroachment on the exercise of fundamental rights, such as privacy, dignity, choice, and access to basic entitlements, should be commensurate with the objective it aims to accomplish[29].

The tripartite necessity arises from the reciprocal reliance between safeguarding individuals from arbitrary state actions and ensuring the preservation of life and personal liberty. The aforementioned regulations also safeguard the entitlement to privacy, derived from Article 21, as well as the liberties enshrined in Part III of the Constitution.

IV. Restrictions on the Right to Privacy

The right to privacy is not an absolute right and, like other fundamental rights, it can be subject to limitations and restrictions. The courts have the ability to impose various types of restrictions based on the nature and circumstances surrounding the claimed privacy interest. In order to interpret the right to privacy in conjunction with Article 19, it is necessary to consider the limitations outlined in Articles 19 and 21. The aforementioned principle is also valid when considering the intersection of the right to privacy with other fundamental rights, specifically Articles 14, 22, and 25. The limitations are to be construed narrowly, while the essential rights are to be interpreted more broadly. The court should exercise caution in adopting a strictly literal approach when interpreting the fundamental right enshrined in the Constitution, as it is necessary to afford broader protection to these rights. The phrase 'reasonable' refers to the exercise of informed care and judgement[30].

The application of limitations on rights should be done in an impartial manner, considering the overall benefit of society rather than focusing just on the individual subject to the limits or on theoretical grounds[31]. The reasonableness of a restriction cannot be only determined by its severity. It is not possible to establish a universal formula for imposing limits in all cases[32]. In the process of determining reasonable restrictions, courts take into account various elements, including the scope and duration of the restrictions, the manner and circumstances in which they are imposed, and the nature of the right being restricted[33]. The aforementioned three tests establish that the limitation on the right to privacy can be justified based on the following justifications.

(i) Public Interest

In the *Gobind case*, the court specifically articulated that the right to privacy is not an absolute right and may be subject to limitations in light of compelling public interest. The term 'public interest' is referenced in Article 19(6), however, the Gobind case included the modifier 'compelling' to precede it. The presence of a compelling state interest signifies the application of the strict scrutiny test and the requirement for narrow tailoring. This court exercises the authority of judicial review over the acts of the legislative and executive branches. The compelling nature of public interest is contingent upon its necessity rather than being a mere question of preference. Therefore, Article 21 imposes a more stringent threshold of scrutiny for the right to privacy compared to Article 19.

The concept of compelling state interest has been developed by the judiciary in the United States and holds significant importance in contemporary constitutional law. The use of a compelling state interest test is employed, in conjunction with a stringent scrutiny test. These examinations enhance the safeguarding of constitutional rights by subjecting the activities of governments to a heightened level of scrutiny by the judiciary. The compelling state interest test was first introduced during the late 1950s and early 1960s within the context of the First Amendment. The term 'exigent obviously compelling' was initially introduced in the case of *Sweezy v. New Hampshire*[34], when Justice Frankfurter emphasised the importance of political power refraining from interfering with the exercise of freedom, unless there are clearly compelling grounds to do so.

In the case of *X v. Z*[35], the court made an observation regarding the extent of privacy, emphasising that it should not be regarded as an absolute entitlement but rather as a right that is contingent upon considerations of public health, morals, and the rights of others. In the context of the COVID-19 pandemic, the relaxation of specific health regulations can be deemed justifiable based on considerations of public interest. The concept of public interest encompasses a wide range of factors, such as public health and safety. The court in the Aadhaar case affirmed the legality of the legislation, citing its potential to provide widespread public benefits.

(ii) National Security

Article 8 of the European Convention on Human Rights (ECHR) stipulates that the rights to privacy may be subject to interference by public authorities in cases pertaining to national security, public safety, or the economic well-being of the nation. The Indian judiciary has explicitly declared that national security falls under the purview of 'legitimate State interest'. Hence, the legitimacy of privacy-infringing provisions under different legislations is contingent upon their alignment with national security interests. An illustration of this can be found in section 5(2) of the Indian Telegraph Act, which grants the government the authority to intercept communication in situations of 'public emergency and public safety'. In a similar vein, Section 69 of the Information Technology Act, 2000, has the authority to impose justifiable limitations in situations where there exists a potential risk to the security of the nation, the integrity of the nation, and the security of the State. In the contemporary day, characterised by a proliferation of unlawful acts like as terrorism and cyber-attacks, it is arguable that an unconditional entitlement to privacy cannot be guaranteed.

The occurrence of the 9/11 incident in the United States and the 26/11 strike in India has significantly heightened concerns regarding security. The populace of the nation is willing to relinquish their privacy entitlements in exchange for the assurance of enhanced security measures. Professor Amitai Etzioni argues that individuals have a moral obligation to relinquish their privacy in order to benefit society as a whole. Privacy and security are fundamental rights that are crucial to individuals; nonetheless, they frequently find themselves in a state of conflict[36]. Policymakers should consistently strive to achieve communal objectives while minimising any infringement on privacy. Digital monitoring is a pertinent measure in safeguarding the populace from acts of terrorism and other forms of motivated criminal activities, particularly those directed towards individuals belonging to minority communities. In certain instances, the sacrifice of personal privacy might lead to an improved quality of life.

The relationship between national security and privacy is intricate, since they are closely interconnected. It is important to note that the preservation of national security does not necessarily entail the relinquishment of private. It is essential to maintain a suitable balance between the two. The transformation of the State into a surveillance State cannot be justified solely on the grounds of security. In light of the disclosures made by Edward Snowden, it becomes evident that the practise of worldwide surveillance is incompatible with the principles and ideals of a democratic society. It is imperative to establish a comprehensive privacy legislation prior to the gathering of data for the aim of ensuring security. The existence of the NSA and the scandal surrounding its Pegasus malware must not be tolerated inside the realm of national security. The prioritisation of privacy should always supersede the prominence of surveillance.

(iii)   Search and Seizure

In instances involving search and seizure, it is possible for the privacy of an individual to be lawfully infringed upon. The Income Tax Act[37] provides provisions for conducting searches to recover undisclosed income. Similarly, the Customs Act of 1962[38] outlines procedures for conducting searches to uncover goods that have been manufactured or imported in violation of relevant statutes. Additionally, the Narcotics Act allows for searches of individuals who have contravened the provisions of the act, with or without a warrant[39]. In a similar vein, the Code of Criminal Procedure (Cr.PC) authorises the execution of searches at locations suspected of containing stolen or counterfeit goods. Section 165 of the Cr.PC grants police officers the authority to conduct searches based on reasonable reasons[40].

The authority to conduct searches and seizures should not be unduly excessive or arbitrarily infringe upon individuals' rights to privacy. The case of *District Registrar and Collector v. Canara Bank*[41] involved the examination of the constitutionality of a specific clause of the Indian Stamp Act of 1899, which had been modified by the Andhra

Pradesh Act, 17 of 1986. The amendment granted the authority for any individual to lawfully access premises for the purpose of making written observations or excerpts, as well as to confiscate or secure registers, books, records, papers, documents, and proceedings. Several banks raised objections to the clause, citing a breach of confidentiality as the basis for their appeal. The court acknowledged the argument and determined that the provision in question was in violation of the fundamental right to privacy. According to the court's ruling, the current amendment grants the authority to both public officials and private individuals to enter and search a person's residence without the requirement of probable cause. The aforementioned action constitutes a breach of the fundamental right to privacy.

V. Other Fundamental Rights and Right to Privacy

The right to privacy often clashes with other fundamental rights that are protected under Part III of the Constitution. The right to privacy of an individual may come into conflict with the right to access information of others. In a similar vein, conflicts may arise between the right to privacy and other fundamental rights such as the right to security, right to food, right to progress, and various others. All of these aforementioned rights are fundamental human rights within contemporary civilization.

The court has recognised both privacy and the right to information as fundamental rights of significant importance. In the case of *R. Rajagopal v. State of Tamil Nadu*[42], the court provided an elucidation on the significance of striking a balance between the freedom of the press and the right to privacy. The court determined that privacy is a basic right; yet, it concluded that the intended publication cannot be subject to prior restraint or prohibition based on privacy protection. The petitioner possesses the legal entitlement to engage in the act of publication, whereas the privacy of the incarcerated individual is to be safeguarded unless explicit authorization has been granted. In the *N.D. Tiwari case*, the court similarly declined to uphold the private rights of N.D. Tiwari, citing the significance of establishing the paternity of a child.

In the recent Aadhaar case, the majority of the court affirmed the legitimacy of the Aadhaar scheme, citing its ability to provide individuals with access to essential resources such as food, subsidies, and other government benefits. This aspect of the scheme was seen to be in line with the principles enshrined in Article 21. If the government collects some data in order to provide benefits, it does not inherently infringe upon privacy rights.

VI. Conclusions

Similar to other essential rights, privacy is not an absolute right. In order for the State to infringe upon privacy rights, it must meet the criteria of acceptable limitations. The legislation that imposes limitations on individuals' rights should adhere to principles of fairness, justice, and reasonableness. In order to invade an individual's privacy, it is imperative to fulfil the three essential criteria of legality, legitimacy, and proportionality. The tests that have undergone development and gained Acceptance by judicial bodies provide enhanced safeguards for fundamental rights. However, it should be noted that these tests do not provide an infallible solution for every circumstance. The outcome of these tests may be subject to interpretation. By implementing these measures to safeguard the right to privacy, a diligent approach towards its preservation is demonstrated. Based on the rulings of the courts, it is evident that the primary objective of the State should be to safeguard the national interest, ensure security, prevent and investigate criminal activities, promote social welfare, and enhance administrative efficiency. In an era characterised by swift technological advancements and global transformations, novel challenges and limitations concerning privacy are anticipated to arise inside society. Consequently, courts are tasked with the responsibility of elucidating and resolving these complex legal matters. The judiciary is currently displaying significant interest in the suitability of tests aimed at safeguarding the right to privacy. The court's stance has undergone a noticeable shift in comparison to its previous rulings.

Endnotes

1   Justice K.S. Puttaswamy (Retd.) v. U.O.I., (2017) 10 SCC 1, AIR 2017 SC 4161.

2   Mattias Kumm, "Political Liberalism and the Structure of Rights: On the Place and Limits of Proportionality Requirement" in George Pavlakos (ed.), Law, Rights and Discourse: The Legal Philosophy of Robert Alexy 131, 133 (Oxford: Hart, 2007).

3   Dr. L.M. Singhvi, *Constitution of India* 655 (Modern Law Publication, New Delhi, 2nd edn. 2006).

4   Golak Nath v. State of Punjab, AIR 1967 SC 1643

5   VII, Constituent Assembly Debates, 40.

6   Charles A Beard, *American Government and Politics*, 23-24 (Macmillan, New York, 9th edn., 1947).

7   D.N. Banerjee , "Some Aspects of Our Fundamental Rights: Article 19" 11 *The Indian Journal of Political Science* 33 (1950).

8   1950 SC 27.

9   Ibid.

10 A.K. Roy v. U.O.I., AIR 1982 SC 710, Ratilal Mithani v. Asst. Collector of Customs, Bombay, AIR 1967 SC 1639.

11 AIR 1974 SC 555.

12 AIR 1975 SC 1590.

13 Dr. L.M. Singhvi, *Constitution of India* 655 (Modern Law Publication, New Delhi, 2nd edn. 2006).

14 AIR 1978 SCR (2) 621.

15 Ajay Hasia v. Khalid Mujib Sehravardi, (1981) 1 SCC 722.

16 (1984) 3 SCC 316.

17 1993 Supp (3) SCC 268.

18 (1996) 2 SCC 226.

19 Mithu Singh v. State of Punjab, AIR 1983 SC 473.

20 Sunil Batra v. Delhi Administration, (1978) 4 SCC 409.

21 W.P. (C) No.118 of 2016.

22 AIR 2017 SC 4904.

23 AIR 1999 SC 188.

24 J.T. 1987 (4) SC 93.

25 (2001) 2 SCC 386.

26 (2004) 2 SCC 130.

27 Modern Dental College and Research Centre v. State of Madhya Pradesh, (2016) 7 SCC 353.

28 W. P. (Civil) No. 494 of 2012.

29 Ibid.

30 Chintaman Rao v. State of M.P., AIR 1951 SC 118.

31 Dr. L.M. Singhvi, *Constitution of India* 655 (Modern Law Publication, New Delhi, 2[nd]edn. 2006).

32 Mohd. Hanif Quareshi v. State of Bihar, AIR 1958 SC 731.

33 M.P. Jain, *Indian Constitutional Law* 1426 (LexisNexis, Nagpur, 7[th]edn., 2018).

34 354 U.S. 234 (1957).

35 AIR 1999 SC 495.

36 Chiff Cheng, "The Limits of Privacy by Amitai Etzioni" 13 *The Academy of Management Executive* 121 (1999).

37 The Income Tax Act, 1961 (Act 43 of 1961), s.132.

38 The Customs Act, 1962 (Act 52 of 1962) Chapter XIII.

39 The Narcotic Drugs and Psychotropic Substance Act, 1985 (Act 61 of 1985), Chapter V.

40 The Code of Criminal Procedure, 1973, (Act 2 of 1974), ss. 94 and 165.

41 (2005) 1 SCC 496.

42 AIR 1995 SC 264.

Chapter 29

Origin of Data Protection Law: A Study with Emphasis on 1970s and 1980s

Stuti
B.A.L.L.B (Hons.) Student
SSOL , Sharda University
Greater Noida, U.P., India

Tahir Qureshi
Ph.D. Research Scholar
Centre for International Legal Studies
Jawaharlal Nehru University
New Delhi, India

Namit Kumar Srivastava
Assistant Professor
SSOL, Sharda University
Greater Noida, UP, India

Abstract

As more and more social and economic activities have place online, the importance of privacy and data protection is increasingly recognized. For any individual, privacy, as a value, is not absolute or constant; its significance can vary with time, place, age, and other circumstances. There is even more variability among groups of individuals. As organizations became increasingly reliant on computer systems to store and process data, the importance of protecting that data became evident. The 1970s set the stage for the ongoing evolution of data protection and cybersecurity, paving the way for more advanced security practices and technologies in subsequent decades. This paper sheds lights upon origin of data protection laws globally after US Department of Health, Education, and Welfare Report 1973 as basis of Privacy Principles and its Impact on Development of International Framework on Data Protection. The report was instrumental in highlighting the need to balance technological advancements and data processing with the protection of individual privacy rights. It called for the establishment of a comprehensive privacy framework that would govern the collection, use, and disclosure of personal information.

Keywords - Principle, international, U.S., Europe, 1973.

I. Introduction

Before the recognition of privacy as a human right, Courts also developed principles in the common law to allow suits for invasion of privacy in various situations involving financial or reputational injury of one person by another. But those decisions were ineffective to protect individuals against the potential adverse effects of personal-data record-keeping practice. Maintaining data bases is not as difficult as maintaining their confidentiality, so in the modern era, the most worried debate is to create an ideal system of data protection. The advancement of technological development resulted in a shift of the standard in crime. Most crimes in the modern era are committed by professionals using the most convenient medium, namely computers and electronic devices. Criminals are able to access the secured data with a single click. The desire for information is boosting the growth of cybercrime.

We have seen a remarkable increase in the usage of technology in recent years. Instead of using conventional methods, the current generation prefers internet buying, banking, communication, gaming, and so on. Furthermore, the tendency of keeping our social media accounts up to date exposes every individual's data to the world at large. Our data becomes vulnerable in the hands of the exploiters when we expose sensitive information such as credit card, debit card, and bank details, mobile number, geographic location, pictures, intimate, private, or business chats and calls, interests, financial, educational, family, and medical records, travel history, etc. Thus, data protection rules are essential to avoid exploitation and to monitor the efficient flow of data without infringing on rights of privacy of an individual, Organisation and Industry.

II. U.S. Department of Health, Education, and Welfare Report, 1973 as basis of Privacy Principles

The world became increasingly interconnected due to the growth of digital services and platforms. The governments recognized that online service delivery is a powerful vehicle for achieving policy objectives such as financial. inclusion and delivering cash transfers. In 1973, a Report of the U.S. Department of Health, Education, and Welfare proposed a set of principles that have been adopted in many countries' privacy frameworks. The "Records, Computers and the Rights of Citizens" report responded to rapid technological developments occurring in the 1970s, specifically computerization and automated processing by government and private firms. Subsequently, the main proposals of the report (such as, no data collection without consent, use limitations, transparency of data processing, and right to correction of data) were adopted by, among others, the Organization for Economic Co-operation and Development. Among privacy and computers (1972), the report of a task force established jointly by the Canadian department of communication and justice; Data and Privacy (1972), the report of the Swedish committee on automated personal systems; and Data banks in a free society in (1972), the report of the national academy of science project on computer data banks. It noted that under current law, a person's privacy is poorly protected against arbitrary or abusive record-keeping practices.

To establish standards of record-keeping practice appropriate to the computer age, the report recommended the enactment of a Federal "Code of Fair Information Practice" for all automated personal data systems. The Code rests on five basic principles that would be given legal effect as "safeguard requirements" for automated personal data systems. There must be no personal data record-keeping systems whose very existence is secret. There must be a way for an individual to find out what information about him is in a record and how it is used. There must be a way for an individual to prevent information about him that was obtained for one purpose from being used or made available for other purposes without his consent. There must be a way for an individual to correct or amend a record of identifiable information about him. Any organization creating, maintaining, using, or disseminating records of identifiable personal data must assure the reliability of the data for their intended use and must take precautions to prevent misuse of the data. 142 pages report concluded that a general right of privacy is not a reliable approach to achieving effective protection.

The Report noted that safeguarding privacy will not be without costs, and will vary from system to system. The cost to most organizations of changing their customary practices in order to assure adherence to our recommended safeguards will be higher in management attention and psychic energy than in dollars. These costs can be regarded in part as deferred costs that should already have been incurred to protect personal privacy, and in part as insurance against future problems that may result from adverse effects of automated personal data systems. From a practical point of view, we can expect to reap the full advantages of these systems only if active public antipathy to their use is not provoked.

III. Emergence of Domestic Laws for Data Privacy in 1970s onwards

The European Convention on Human Rights, 1953, with protocols and amendments laid the foundation for privacy laws in Europe, including principles related to the protection of personal data. The Convention played a significant role in shaping the data privacy landscape in European countries. Concern about the effects of computer-based record keeping on personal privacy appeared to be related to some common characteristics of life in industrialized societies. In the first place, industrial societies were urban societies. The effects of computer-based record keeping appears to have deep roots in the public opinion of each country, deeper roots than could exist if the issues were manufactured and merchandised by a coterie of specialists or reflected only the views of a self-sustaining group of professional Cassandras. The fragility of computer-based systems may account for some of the concern.

(i) West Germany

On October 7, 1970, the West German State of Hesse adopted the world's first legislative act directed specifically toward regulating automated data processing. This "Data Protection Act" applied to The Reaction in Other Countries to the official files of the government of Hesse; wholly private files are specifically exempted from control. The Act established a Data Protection Commissioner under the authority of the State parliament whose duty it is to assure that the State's files are obtained, transmitted and stored in such a way that they cannot be altered, examined, or destroyed by unauthorized persons. The Commissioner is also explicitly responsible for observing the effects of automated data processing on the operations of the State government and on its decision-making powers. He must take particular note 'Of whether computerization leads to any displacement in the distribution of powers among the governmental bodies of the State. Thus, the Data Protection Act of Hesse seems designed more to protect the integrity of State data and State government than to protect the interests of the people of the State. As a pioneer statute in the field of computer law, however, its exact practical effects could scarcely have been predicted, and in no, way diminish its usefulness as a guide for other jurisdictions that can learn from the Hesse experience. Concern about the effects of computer-based record keeping on personal privacy appeared to be related to some common characteristics of life in industrialized societies. In the first place, industrial societies were urban societies.

(ii) Sweden

After a notably thorough survey of personal data holdings in both public and private systems, the first national data protection law in Sweden came into existence. The commission issued a report containing draft legislation for a comprehensive statute for the regulation of computer-based personal data systems in Sweden in 1972. The Act was passed in 1973 for specifically the protection of data personal privacy. It came into effect on July 1st, 1973, and mandated that information systems that handled personal data obtain licenses from the Swedish Data Protection authority.

(iii) France

The 1972 Annual Report of the Supreme Court of Appeals went considerably out of its way, after reviewing a case of literary invasion of privacy, to comment on the subject of computers and privacy. In the report, a relatively small group of experts has given serious attention to the subject of computers and privacy, but that group has far outweighed its numbers in government.

(iv)   Britain

Report of the Committee on Privacy. Rt. Hon. Kenneth Younger as Chairman restricted in its terms of reference to private, rather than public, organizations that might threaten privacy, the committee's report is a model of clarity and concern. In brief, the Committee found that both the customs of society and the Common law had evolved defences against the traditional intrusions of nosey neighbours, unwelcome visitors, door-to-door salesmen, and the like. Against the new threats of technological intrusions-wire-taps, surveillance cameras, and, of course, computerized data banks-the Committee recognized that the traditional defences are inadequate. To help deal with the threat of the computer, the Committee recommended specific safeguards to be applied to automated personal data systems, although it left the method of application up to the government to decide. The Younger Committee also considered proposing specific legislation for automated personal data systems based on draft Bills submitted to Parliament prior to the formation of the Committee. After concluding that the proposed laws were too onerous to be justified by the level of threat as perceived by the Committee, the Committee reserved the option of recommending legislation at a later date, and limited its current recommendation to urging the data-processing industry to adopt the safeguards voluntarily as a code of good practice. This has now been accomplished through the adoption of a professional code by the British Computer Society.

(v)   Cananda

In April 1971, the Departments of Communications and Justice jointly established a Task Force on Privacy and Computers, growing out of earlier work in the Department of Communications on issues concerning the use of computers in communications. The Task Force was given broad terms of reference to consider the rights and values of the individual that cluster about the notion of privacy, and to examine present and foreseeable effects on those rights and values of computerized information systems containing personal data about identifiable individuals. In its report, published in late 1972, the Canadian Task Force concluded that computer invasion of privacy is still far short of posing a social crisis. However, the rapidly rising volume of computerized personal data and the equally rapidly rising public expectation of a right to deeper and more secure privacy threaten to converge at the crisis level. To forestall that Crisis, the Task Force recommends that a commissioner or ombudsman be established in a suitable administrative setting.

IV. Impact of 1973 Report on Development of International Framework on Data Protection

The developments discussed in the previous part affected the framing of a number of International Instruments. Several countries and regions drew upon the principles outlined in the 1973 Report when crafting their own privacy legislations and regulations.

(i) Convention 108

The Council of Europe's Convention for the Protection of Individuals with regard to Automatic Processing of Personal Data, often referred to as "Convention 108," is an international treaty that aims to safeguard the privacy and data protection rights of individuals in the context of automated data processing. It establishes principles and standards for the responsible handling of personal data by both public and private entities. The Convention provides for free flow of personal data between states party to the Convention. This free flow may not be obstructed, for personal data protection reasons, unless Parties derogate from this provision, which they might do in two explicit cases: when personal data are transferred to a third state that is not a Party to the Convention or when personal data protection in the other Party is not "equivalent."

(ii) OECD Guidelines,1976

The OECD Guidelines for Multinational Enterprises were first established in 1976 and subsequently updated in 1980. These guidelines provide recommendations to multinational enterprises on various aspects of responsible business conduct, including employment and industrial relations, human rights, environmental protection, and corruption prevention. The guidelines aim to promote responsible behavior among businesses operating internationally. In any revision of the Guidelines, the OECD should focus on the governance models of Multinational Enterprises. It should be a company centric model moving away from the shareholder centric model. Multinational Enterprises should be guided to focus on the long-term health of their enterprises rather than maximization of shareholder wealth, which will be in the long-term better interests of all their stakeholders including their shareholders. Keep in mind that these guidelines have evolved over the years in 2013.

(iii) The Asia-Pacific Economic Cooperation (APEC) Privacy Framework ,2005

The Asia-Pacific Economic Cooperation (APEC) Privacy Framework is a set of principles and guidelines developed by the APEC member economies to promote and facilitate the protection of personal information in the Asia-Pacific region. It was established to address the challenges posed by the rapid growth of digital technology and the increasing flow of personal data across borders.

APEC Cross-Border Privacy Rules (CBPR) system is a government-backed data privacy certification that companies can join to demonstrate compliance with internationally-recognized data privacy protections. The CBPR system implements the APEC Privacy Framework endorsed by APEC Leaders in 2005 and updated in 2015. The CBPR system benefits consumers and business alike by ensuring that regulatory differences do not block businesses' ability to deliver innovative products and services.

(iv)   General Data Protection Regulation

The General Data Protection Regulation (GDPR) originated as a result of the need to update and harmonize data protection laws within the European Union (EU). It was adopted on April 14, 2016, and became enforceable on May 25, 2018. The GDPR established strict rules for the processing of personal data, including consent requirements, data subject rights, data breach notification, and substantial fines for non-compliance. It significantly impacted how organizations handle personal data and set a new benchmark for data protection worldwide. The GDPR's goal is to protect individuals and the data that describes them, as well as to ensure that organizations that collect that data do so responsibly. The GDPR also requires that personal data be stored securely; the regulation states that personal data must be protected against "unauthorized or unlawful processing, as well as accidental loss, destruction, or damage."

V. Conclusions

The 1970s marked a significant period in the history of computing leading to the increased use of computers and their integration into various aspects of daily life, business, education, healthcare, entertainment, and more. Computer record protection became a growing concern. The development of legal principles comprehensive enough to accommodate a range of issues arising out of pervasive social operations, applications of a complex technology, and conflicting interests of individuals, record-keeping Organizations, and society will have to be the work of legislative and administrative rule-making bodies. The 1973 report by the U.S. Department of Health, Education, and Welfare provided foundational privacy principles that have had a lasting impact on the development of data protection frameworks, both in the United States and internationally. These principles have guided the evolution of privacy laws and regulations to adapt to the challenges posed by advancing technologies and the growing importance of safeguarding personal data. Today, computer record protection involves a complex array of measures, including advanced encryption algorithms, multi-factor authentication, cybersecurity training, compliance with data privacy laws and continuous monitoring for threats and vulnerabilities.

References

1   Report of the Secretary's Advisory Committee on Automated Personal Data Systems U.S Department of Health, Education and Welfare (1973)| https://www.justice.gov/opcl/docs/rec-com-rights.pdf

2   APEC Privacy Framework (2005) | https://www.apec.org/docs/default-source/Publications/2005/12/APEC-Privacy-Framework/05_ecsg_privacyframewk.pdf

3   GDPR (2018) | https://gdpr.eu/what-is-gdpr/

4   OECD Guidelines (1976) | https://www.oecd.org/investment/mne/1903291.pdf

5   Convention 108 | https://www.coe.int/en/web/data-protection/convention108/background

6   Swedish Data Act (1973) | https://www.ojp.gov/pdffiles1/Digitization/49670NCJRS.pdf

7   https://www.computerhistory.org/timeline/1970/.

Chapter 30

Unveiling the Imperatives for Ethical Hacking in the Digital Age

Prof. (Dr.) Pradeep Kulshreshtha
Dean, School of Law
Bennett University
Greater Noida, U.P., India

Nehru
Assistant Professor
ICFAI University Law School
The ICFAI University
Jaipur, Rajasthan, India

Abstract

The evolution of ethical hacking is intertwined with the need to secure computer systems and networks against unauthorized access and malicious activities. Ethical hacking has become an integral component of modern cybersecurity strategies, allowing organizations to proactively identify vulnerabilities, mitigate risks, and enhance their overall security posture. The concept of ethical hacking gained more structure and recognition in the late 20th and early 21st centuries, leading to the establishment of formal certifications, programs, and guidelines for individuals interested in becoming ethical hackers.

Today, ethical hacking is a vital component of cybersecurity, helping organizations secure their digital assets and protect against cyber threats. The CEH certification equips professionals with the skills needed to assess an organization's security posture and help in improving its defenses. Ethical hackers play a vital role in cybersecurity by identifying vulnerabilities before malicious actors can exploit them.

Keywords - Certification, Hacking, ethical, Security, Weakness.

I. Introduction

The Digital Age is characterized by the widespread use of computers, the internet, mobile devices, and other advanced technologies that have transformed the way we communicate, work, learn, and conduct various aspects of our lives. One of the earliest documented cases of ethical hacking dates back to the 1970s and involved a group of students and researchers at the Massachusetts Institute of Technology (MIT). These individuals are considered pioneers in the field of ethical hacking. The TMRC members, including Peter G. Neumann, discovered vulnerabilities or "bugs" in the computer systems of that era. Rather than exploiting these bugs maliciously, they responsibly reported them to system administrators and helped fix the issues. They set the foundation for a culture that would later evolve into ethical hacking as a recognized and legitimate field focused on improving cybersecurity.

Hacking is identifying weakness in computer systems or networks to exploit its weaknesses to gain access. Ethical hacking plays a role in identifying and addressing vulnerabilities to safeguard privacy. A Hacker is a person who finds and exploits the weakness in computer systems and/or networks to gain access. Hackers are usually skilled computer programmers with knowledge of computer security.

One of the earliest and most well-known cases of unethical hacking is the case of Kevin Mitnick, one of the most notorious hackers in the

1980s and 1990s. Mitnick's hacking activities ranged from computer intrusions to wire fraud, and he became infamous for his prolific hacking skills and evading law enforcement for an extended period. gained unauthorized access to computer systems of major corporations, including IBM, Nokia, Novell, and Motorola, among others. He also hacked into various government systems, including the FBI and the Pentagon. hacking spree came to an end in 1995 when he was arrested by the FBI. His capture was the result of a joint effort by multiple law enforcement agencies.

In India, a hacker or a group identifying themselves as "The Indian Cyber Army" claimed responsibility for hacking several high-profile websites in India in 2010. The targeted websites included those of the Central Bureau of Investigation (CBI), the National Informatics Centre (NIC), and some private organizations. In 2011, the Central Bureau of Investigation (CBI) announced the arrest of a person believed to be the key individual behind the "Indian Cyber Army" pseudonym, known as "Geekboy." The arrested individual was identified as a software engineer.

II. Classification of Hackers

The hackers are of different kinds, depending upon intention of the hackers.

(i)   White Hat Hackers (Ethical Hackers)

White hat hackers are ethical hackers who use their skills to help organizations and individuals by identifying vulnerabilities and securing systems and networks. They often work to improve security measures and protect against potential cyber threats.

(ii)   Black Hat Hackers (Malicious Hackers)

Black hat hackers are malicious individuals who exploit vulnerabilities in systems and networks for personal gain, financial motives, or to cause harm. They engage in illegal activities, including stealing sensitive information, distributing malware, or disrupting services.

(iii) Grey Hat Hackers

Grey hat hackers fall between white hat and black hat hackers. They may hack into systems without permission, but their intentions are not necessarily malicious. They may disclose vulnerabilities to the affected organization without causing any harm.

(iv) Hacktivists

Hacktivists are hackers who use their skills to promote a social or political agenda. They may target government, corporations, or organizations to raise awareness about specific issues, express dissent, or protest against policies they disagree with.

(v) Script Kiddies

Script kiddies are individuals with limited technical skills who use pre-written scripts or tools to exploit vulnerabilities. They often lack a deep understanding of hacking and rely on existing tools to carry out their activities.

(vi) State-Sponsored Hackers (Advanced Persistent Threats - APTs)

State-sponsored hackers are individuals or groups employed or supported by governments or state organizations to conduct cyber-espionage, sabotage, or gather intelligence. Their actions are typically highly sophisticated and well-funded.

(vii) Hacktivist Groups

Hacktivist groups are organized collectives of individuals with a common cause or ideology. They engage in cyber-attacks to advance their agenda, often promoting political, social, or ideological goals.

(viii) Cybercriminals

Cybercriminals engage in various illegal activities for financial gain, such as identity theft, credit card fraud, ransomware attacks, and other forms of cybercrime. Their primary motivation is monetary profit.

(ix) Hacktivist for Hire

These are individuals or groups that provide hacking services for a fee, often hired by other malicious entities to carry out cyber-attacks. They may conduct targeted attacks, distribute malware, or engage in other malicious activities on behalf of their clients.

(x) Crackers

Crackers engage in bypassing software or digital copyright protection to access and distribute copyrighted material without authorization. It's often associated with pirated software or media.

(xi) Distributed Denial of Service (DDoS) Attackers

DDoS attackers involve overwhelming a target server or network with a flood of traffic from multiple sources, causing a temporary or permanent disruption of services.

(xii) Malware Authors

Malware authors create malicious software like viruses, worms, Trojans, and ransomware to gain unauthorized access to systems, steal data, or disrupt services.

(xiii) Phreakers

Phreakers manipulate telecommunications systems to make free calls, exploit services, or gain unauthorized access. While less common today, their legacy influences modern hackers.

III. Components of Ethical Hacking

Ethical hacking, also known as penetration testing or white hat hacking, involves authorized and legal attempts to identify vulnerabilities and weaknesses in computer systems, networks, applications, and infrastructure. The primary purpose of ethical hacking is to uncover potential security flaws and provide recommendations to enhance the security posture of the target systems.

(i) Authorized Testing

Ethical hacking involves authorized testing of systems and networks to discover vulnerabilities that could be exploited by malicious hackers. The key difference is that ethical hackers have explicit permission from the system owners to perform these tests.

(ii) Security Improvement

Ethical hackers identify security weaknesses and potential threats, allowing organizations to proactively address these issues before malicious actors exploit them. This proactive approach helps in improving the overall security of systems and networks.

(iii) Risk Mitigation

By identifying vulnerabilities and weaknesses, ethical hacking helps organizations assess their level of risk. This information enables them to prioritize security measures and allocate resources to areas that need the most attention, ultimately reducing the risk of cyber-attacks.

(iv) Compliance and Regulations

Many industries and sectors are subject to regulatory requirements that mandate regular security testing. Ethical hacking assists organizations in complying with these regulations by ensuring that security measures are up to the required standards.

(v) Awareness and Training

Ethical hacking also plays a crucial role in creating awareness and educating employees and stakeholders about cybersecurity risks. It helps in fostering a security-conscious culture within an organization.

IV. Need for Ethical Hacking

Prevent Unauthorized Access Ethical hacking helps in identifying vulnerabilities that could allow unauthorized access to systems and data. By addressing these vulnerabilities, organizations can prevent unauthorized individuals from gaining access and potentially causing harm.

(i) Protect Sensitive Data

Companies often handle sensitive and confidential data, such as customer information, financial data, and intellectual property. Ethical hacking helps ensure that this data remains secure and protected from unauthorized access and theft.

(ii) Maintain Business Continuity

Cyber-attacks can disrupt business operations, leading to downtime and financial losses. Ethical hacking helps in identifying potential weaknesses that could be exploited to disrupt services and allows organizations to strengthen their defenses to maintain business continuity.

(iii) Build Customer Trust

Demonstrating a commitment to cybersecurity through ethical hacking and proactive vulnerability assessments can enhance customer trust. Customers are more likely to trust an organization that takes security seriously and invests in protecting their data.

(iv) Compliance with Regulations

Many regulatory frameworks, such as GDPR, HIPAA, and PCI DSS, require organizations to maintain specific security measures. Ethical hacking helps ensure compliance with these regulations by identifying and addressing security gaps.

(v) Stay Ahead of Cyber Threats

The threat landscape is constantly evolving, with new vulnerabilities and attack vectors emerging regularly. Ethical hacking helps organizations stay ahead of these threats by proactively identifying and addressing potential security risks.

In summary, ethical hacking is a crucial practice for organizations to ensure the security and integrity of their systems, protect sensitive data, maintain compliance with regulations, and build trust with stakeholders. It's an essential component of a comprehensive cybersecurity strategy in today's digital age.

V. Development of Ethical Hacking

Ethical hacking has gone through a number of phases over the past decades.

(i) Phases

1960s - 1970s witnessed Early Hacking Culture. 1970s - 1980s saw emergence of Ethical Hacking. 1990s - 2000s was the phase of Formalization and Professionalization. Post 2000, modern ear of Hackings started.

a. MIT Hacking Culture (1960s)

Hacking culture emerged at the Massachusetts Institute of Technology (MIT) in the 1960s, where students and researchers explored the limits of computer systems, software, and hardware. They sought to understand and enhance the capabilities of computers.

b. Phone Phreaking (Early 1970s)

Phone phreaking was an early form of hacking that involved manipulating telephone networks and systems to make free or unauthorized calls. This culture contributed to the development of early hacker ethics.

c. The First Ethical Hackers (1970s)

The term "hacker" initially had a positive connotation and referred to individuals who were passionate about exploring and understanding computer systems. Some of these early hackers recognized the need to secure systems against unauthorized access and became the first ethical hackers.

d. Progress in Computer Security (1980s)

During the 1980s, computer security concerns grew as more systems were connected to networks. This led to the establishment of the first computer security teams within organizations, often comprised of individuals with hacking backgrounds who focused on protecting systems.

e. Formation of Hacker Groups (1990s)

The 1990s saw the formation of formal hacking groups, such as the L0pht and Cult of the Dead Cow, which advocated for responsible hacking and security research. These groups played a significant role in advancing the ethical hacking movement.

f. Certified Ethical Hacker (CEH) Program (2003)

The International Council of E-Commerce Consultants (EC-Council) introduced the Certified Ethical Hacker (CEH) certification program in 2003. The CEH program provided a structured framework for individuals to learn ethical hacking skills and gain certification.

g. Legal and Regulatory Frameworks (2000s)

During the 2000s, governments and organizations recognized the importance of ethical hacking in bolstering cybersecurity. Legal and regulatory frameworks were established to guide ethical hacking activities and ensure compliance with laws and standards.

h. Increased Demand for Ethical Hackers (2010s)

The rapid growth of cyber threats and cyber-attacks in the 2010s led to an increased demand for skilled ethical hackers. Organizations began to invest more in cybersecurity and ethical hacking services to secure their assets.

i. Professionalization and Specialization (2010s - Present)

The field of ethical hacking has evolved into a well-established profession with various certifications, specialized roles (e.g., penetration testers, vulnerability assessors), and a strong emphasis on professional ethics, standards, and continuous education.

j. Role of The International Council of E-Commerce Consultants

The International Council of E-Commerce Consultants (EC-Council) is a globally recognized organization that offers a variety of certifications related to cybersecurity, including ethical hacking. One of their most well-known and sought-after certifications in the field of ethical hacking is the Certified Ethical Hacker (CEH) certification.

The Certified Ethical Hacker (CEH) certification is designed for individuals who want to become ethical hackers or penetration testers. The certification focuses on teaching individuals to think and act like a malicious hacker, in an ethical and legal manner, to identify vulnerabilities and weaknesses in computer systems and networks.

(ii) Instances of Ethical Hacking for larger good

This part highlights different instances of ethical hacking.

a. Aberdeen Proving Ground Network Test (1970s)

In the 1970s, the United States Department of Defense (DoD) authorized a group of computer experts to conduct a legal hacking exercise, known as the "Aberdeen Proving Ground Network Test," to identify vulnerabilities in its computer systems and network security.

b. NSA's RED TEAM (1990s - Present)

The National Security Agency (NSA) established a "Red Team" program composed of skilled cybersecurity professionals. These teams simulate adversarial attacks to identify security weaknesses and help strengthen the NSA's cybersecurity measures.

c. Microsoft's Bounty Programs (2003 - Present)

Software giant Microsoft initiated various legal hacking programs, offering monetary rewards to ethical hackers who discover vulnerabilities in their software products. These programs encourage responsible disclosure and help Microsoft enhance the security of their products.

d. Pwn2Own Contest (2007 - Present)

The Pwn2Own contest is an annual hacking competition that invites ethical hackers to find and exploit vulnerabilities in popular software and operating systems. Participants are rewarded for their successful exploits, and the vulnerabilities discovered are shared with the respective vendors for patching.

e. Bug Bounty Programs Across Companies (Ongoing)

Many organizations, such as Google, Facebook, Amazon, and PayPal, run bug bounty programs. These programs encourage ethical hackers to identify security flaws in their systems and reward them for responsibly reporting these vulnerabilities, promoting a safer online environment.

f. HackerOne's Hack the Pentagon Program (2016)

The U.S. Department of Defense (DoD) collaborated with HackerOne to launch the "Hack the Pentagon" initiative. It invited ethical hackers to test the cybersecurity of select DoD websites and report vulnerabilities. This program aimed to strengthen the DoD's security posture.

g. European Union Agency for Cybersecurity (ENISA) - EU Bug Bounty Program (2019)

The European Union Agency for Cybersecurity (ENISA) initiated a bug bounty program, inviting ethical hackers to discover vulnerabilities in several open-source software solutions used by the EU institutions. This program aims to enhance the security of critical open-source software.

These cases demonstrate how legal hacking, when conducted responsibly and with authorization, can contribute to improving cybersecurity and enhancing the resilience of systems, networks, and applications against potential cyber threats.

VI. Ethical Hacking in India

The Government of India has recognized the critical need for cybersecurity and has taken several initiatives to promote and regulate ethical hacking activities. Ankit Fadia, Saket Modi, Sunny Vaghela, Trichet, Arun Yadav (known as "JackH4xor") , Rahul Tyagi Amit Dubey, (known as "the Mentor"), Onkar Sonawane and so on have made substantial contributions to the cybersecurity landscape in India and globally, emphasizing the importance of ethical practices, responsible disclosure, and cybersecurity education.

(i) National Cyber Security Policy (NCSP)

The Indian government introduced the National Cyber Security Policy in 2013 to address the challenges of cybersecurity and enhance the country's resilience to cyber threats. The policy emphasizes the importance of promoting research and development in cybersecurity, including ethical hacking, to strengthen the nation's cybersecurity posture.

(ii) Certification and Training Programs

Various government bodies and organizations offer certification and training programs in ethical hacking. For instance, the Ministry of Electronics and Information Technology (MeitY) and the National Institute of Electronics & Information Technology (NIELIT) offer courses and certifications in cybersecurity and ethical hacking to develop skilled professionals in this field.

(iii) Cybersecurity Competitions and Challenges

The government, along with various private organizations, often organizes cybersecurity competitions, challenges, and hackathons to identify and nurture cybersecurity talent. These events encourage ethical hacking and provide platforms for individuals to showcase their skills and knowledge in a competitive environment.

(iv) Incident Response and Coordination

The government establishes mechanisms for incident response and coordination to handle cybersecurity incidents effectively. Ethical hackers may collaborate with these incident response teams to identify vulnerabilities and assist in mitigating potential threats.

(v) Cybersecurity competitions

Competitions are organized to promote awareness, education, and skills development in the field of cybersecurity. These competitions provide a platform for participants to demonstrate their knowledge, problem-solving abilities, and practical skills in various aspects of cybersecurity. Here are some notable cybersecurity competitions in India.

a. InCTF (Indian Capture the Flag)

InCTF is one of the largest and most prestigious national-level cybersecurity competitions in India. It is organized annually by Amrita Vishwa Vidyapeetham University and sees participation from students and professionals from across the country. The competition focuses on various aspects of cybersecurity, including web security, cryptography, reverse engineering, and more.

b. CTF (Capture the Flag) Competitions

Numerous CTF competitions are organized by universities, cybersecurity organizations, and communities across India. These competitions typically involve challenges related to network security, web application security, digital forensics, and cryptography. Participants work in teams to solve these challenges and "capture the flag" to earn points.

c. Defcon Capture the Flag (DCG)

India Qualifiers DEFCON is a well-known international hacking and cybersecurity conference. India hosts its own DEFCON Capture the Flag (DCG) qualifiers, where participants compete for a chance to represent India in the DEFCON Capture the Flag finals held in the USA.

d. Nullcon HackIM

Nullcon is a popular security conference held annually in Goa, India. It includes the HackIM CTF competition, where participants can showcase their skills in areas such as cryptography, reverse engineering, exploitation, and more. It attracts cybersecurity enthusiasts from India and around the world.

e. Cyber Security Challenge UK-India

This competition is a joint initiative between the UK and India, focusing on fostering cybersecurity talent and collaboration between the two countries. It aims to bring together students, professionals, and cybersecurity enthusiasts to compete in various challenges and gain recognition for their skills.

f. CodeVita by TCS (Tata Consultancy Services)

CodeVita is an annual global coding competition organized by TCS. While not solely focused on cybersecurity, it includes algorithmic challenges that test participants' programming and problem-solving skills, which are essential for a strong foundation in cybersecurity.

Many universities and corporations in India organize hackathons and cybersecurity challenges to encourage students and professionals to develop innovative solutions, tools, or defenses against cybersecurity threats.

VII. Duties of Ethical Hackers

Ethical hackers ensure that their operations are handled responsibly and ethically by acquiring proper authorisation, operating within legal boundaries, and complying with applicable rules and regulations.

(i) Do's

The hacker ethics were first written down by Steven Levy in his book "Hackers: Heroes of the Computer Revolution" in 1984. The Book delved into the early days of computing and the pioneers who shaped the culture and technology of the computer revolution.

a. Responsibility, timely communication, and openness are vital ethical principles to abide by, and clearly distinguish a hacker from a cybercriminal.

b. Ethical hackers should establish a clearly defined scope of engagement with the organisation or individual they are working. This includes defining the systems, networks, or applications to be tested, the methodology employed, and the engagement's restrictions. Staying within the agreed-upon scope keeps the hacker's efforts legal and focused.

c. Ethical hackers must consult with legal professionals who specialise in cybersecurity and privacy laws to ensure compliance with applicable regulations.

d. Keeping membership in professional organisations, such as the EC-Council or the International Council of Electronic Commerce Consultants (EC-Council), can provide ethical hackers access to resources, guidelines, and best practices related to legal and ethical hacking. *Code of Ethics and Conduct outlined by organizations like ISC2, EC-Council, and ISACA emphasize integrity, honesty, transparency, and respect for privacy.*

(ii) Don'ts for Ethical Hackers

An ethical hacker must not indulge in certain specified activities such as -

a. Unauthorized Hacking

Do not engage in any hacking activities without proper authorization, regardless of intent or motivation.

b. Data Theft or Misuse

Never steal, modify, or misuse data accessed during the testing process. Respect privacy and confidentiality at all times.

c. Exceed Scope of Work

Do not go beyond the defined scope of the engagement or attempt to access systems or data outside the authorized boundaries.

d. Damage Systems or Networks

Avoid actions that could cause damage to systems, networks, or data, even if unintentional. Prioritize the safety and integrity of the systems being tested.

e. Public Disclosure without Authorization

Refrain from publicly disclosing any vulnerabilities or sensitive information without proper authorization from the affected organization.

f. Misleading or False Information

Do not provide false or misleading information to the organization or misrepresent the results of the assessment.

g. Reckless Behavior

Avoid reckless actions or testing methodologies that could potentially disrupt or harm the organization's operations or infrastructure.

h. Ignoring Legal Obligations

Do not neglect legal obligations or violate any laws or regulations during the testing process.

VIII. Legal Framework for Hacking

The legal framework for hacking varies from country to country and is influenced by national laws, international agreements, and regional regulations. Hacking, in the context of unauthorized access to computer systems, networks, or devices, is generally considered illegal and punishable by law in most jurisdictions. In India, Sections 43, 66, 66 C and 66D of Information Technology Act, 2000 are relevant. Section 43 deals with unauthorized access to computer systems, computer networks, or any other device. It covers unauthorized downloading, extraction, copying of data, or introduction of computer contaminants. Section 66 deals with computer-related offenses such as hacking, identity theft, and introducing malware. It prescribes penalties for unauthorized access, introduction of computer viruses, and denial-of-service attacks. Section 66C deals with identity theft and imposes penalties for fraudulent use of electronic signatures, passwords, or any unique identification features. Section 66D addresses cheating by impersonation using a computer resource and prescribes penalties for such acts.

Computer Fraud and Abuse Act, 1986 (CFAA) enacted in 1986 and amended over the years is the primary federal law addressing unauthorized access to computer systems and related activities in the U.S.A. It criminalizes unauthorized access to protected computers and networks. Computer Misuse Act, 1990 criminalizes unauthorized access, unauthorized access intending further offenses, and unauthorized modification of computer material. Penalties can include imprisonment and fines in U.K. Section 342.1 of Criminal Code addresses unauthorized access to computer systems, networks, or data in Canada. It criminalizes unauthorized use or interception of computer functions and data. Section 478.1 of Criminal Code Act, 1995 addresses unauthorized access to, or modification of, restricted data held in a computer in Australia. It prescribes penalties for unauthorized access, including imprisonment.

IX. Conclusions

Hacking can encompass a wide range of activities, from benign ethical hacking (authorized testing to uncover vulnerabilities for security improvement) to malicious hacking (cyberattacks, data breaches, etc.). Understanding the classification of hackers helps in assessing and addressing cyber threats effectively, as each type of hacker requires a different approach for cybersecurity and risk mitigation. Ethical hacking or white hat hacking, involves authorized and legal activities aimed at identifying vulnerabilities in computer systems, networks, applications, or any digital asset. The purpose of ethical hacking is to strengthen cybersecurity by proactively identifying weaknesses that malicious hackers could exploit. Ethical hackers, also known as white-hat hackers, are professionals who apply their skills and knowledge to uncover vulnerabilities, analyze security measures, and assist organizations in fortifying their defenses against cyber threats.

Ethical hackers continually update their skills and knowledge to keep pace with evolving technologies and emerging cyber threats. They contribute to a culture of continuous improvement in cybersecurity practices. Ethical hacking should always be conducted within legal and ethical boundaries, with proper authorization and consent from the target organization. Collaboration between ethical hackers, government agencies, private sectors, and academia is essential to ensure a holistic approach to cybersecurity. By integrating ethical hacking practices into the defense strategy, countries can better prepare and respond to evolving cyber threats, ultimately strengthening the cybersecurity landscape. As the cybersecurity significance continues to evolve, ethical hacking will play an even more critical role in safeguarding digital assets and sensitive information.

References

1   https://www.scmagazine.com/perspective/three-principles-ethical-hackers-can-adopt-as-a-code-of-conduct

2   Computer Fraud and Abuse Act, 1986 (U.S.A.).

3   Computer Misuse Act, 1990 (U.K.).

4   Criminal Code, 1985 (Canada).

5   Information Technology Act, 2000 (India).

6   Dr. Allen Harper, Daniel Regalado, Ryan Linn, Stephen Sims, Branko Spasojevic, Linda Martinez, Michael Baucom, Chris Eagle and Shon Harris, *Grey Hat Hacking: The Ethical Hackers Handbook* (McGraw Hill, 2020).

7   https://supervisorbullying.com/ethics-of-ethical-hackers/.

8   https://www.darkreading.com/vulnerabilities-threats/should-hacking-have-a-code-of-conduct-.

9   Markus Christen, Bert Gordijn & Michele Loi, *The Ethics of Cybersecurity* (Springer Open, 2020).

10 https://timebusinessnews.com/the-ethical-hackers-code-of-conduct-what-you-need-to-know/.

11 Dan Verton, *The Hacker Diaries: Confessions of Teenage Hackers* (McGraw-Hill, New York, 2002.)

12 Johnny Long, Bill Gardner and Justin Brown, *Google Hacking for Penetration Testers* (Syngress, 2011).

13 Steven Levy, *Hackers: Heroes of the Computer Revolution* (Doubleday, 1984).

Chapter 31

Global Legal Framework vis-à-vis Data Protection

Dr. Ramesh Verma
Assistant Professor
Department of Law
Himachal Pradesh University
Shimla, Himachal Pradesh, India

Madhav Sharma
Associate, Unitedlex
Gurugram, Haryana, India

Vani Chaudhary
B.A.L.L.B.(Hons.) Student
SSOL, Sharda University
Greater Noida, U.P., India

Abstract

The long existence and flourishing of data privacy laws have helped to build a language of data privacy and a set of ethical standards. Nearly two thirds of the world's countries with data privacy laws are in Europe. These regulations are the most common. The U.S.A. cannot be regarded to have a comprehensive national data privacy legislation. There are noteworthy cases of jail terms or suspended sentences and huge fines for breaking data privacy rules in other nations, while legislation in certain jurisdictions offers considerable potential penalties. The law in the field of data-protection can be seen catching up with technological and technical advancements, but the vacuum is created with unwillingness of law makers to make dynamic laws. Some countries who lacked any law to regulate manual data protection are making data-subjects' friendly laws. While national measures could certainly be adopted, a coordinated and harmonized approach through the world may be more successful in strengthening personal data protection.

Keywords - Comparison, Comprehensive, Data protection, Europe, Privacy.

I. Introduction

There are vast disparities how data is protected under different jurisdictions across the world. Starting from scattered legal provisions, developments have taken place as to the protection of online data in various countries. Data protection is a critical aspect of managing and safeguarding data to ensure its confidentiality, integrity, and availability. Each country has its own process for creating laws related to data protection. However, there are some common factors that may influence the development of such laws. In many cases, laws related to data protection emerge in response to public demand for greater privacy and security of personal information. This can be particularly true in countries where there have been high-profile data breaches or other incidents that have raised public awareness of the need for greater protection. The present paper seeks to examine various legal provisions in selected countries.

II. Regulatory Position Regarding Data Protection in France

A legislation was introduced by France relating to personal data and computer files in 1978, making provision for French Data Protection Authority. The Act on Data Processing, Data Files, and Individual Liberties, 1978 popularly known as the French Data Protection and Freedoms Act (DPA) requires organizations implementing data processing or holding data files to guarantee their security and privacy. This security must be considered for all operations involving this data, including its production, usage, backup, archiving, and destruction. It relates to its availability, confidentiality, integrity, and authenticity. In 2004, the legislation was amended. The current law applies to the following situations: the processing of personal data by a data controller that is established in France or conducts its business there, regardless of its legal structure; and the data controller that, despite not having an establishment in France or another EU member state, uses means of processing that are located there, with the exception of processing that is only done for transit through France or another EU member state.

III. Regulatory Position Regarding Data Protection in Australia

Federal Privacy Act, 1988 applies to Federal and ACT Government Agencies and many private sector organizations. The Privacy Act of 1988 is overseen by the Office of the Federal Privacy Commissioner. In respect to organizations and agencies that are required to preserve privacy under the Privacy Act, the office offers information and guidance, as well as assistance with managing complaints and policy-related issues. It has created different set of requirements for different kind of organization handling data. Most Australian and Australian Capital Territory Government agencies must comply with a set of eleven standards, known as the Information Privacy Principles (IPPs), when handling personal information. When processing personal information, private sector firms, including all private health service providers, are required to adhere to the ten National Privacy Principles (NPPs) or a recognized privacy code. Credit providers and credit reporting agencies are required to abide with Part IIIA of the Act's requirements regarding consumer credit information as well as the Credit Reporting Code of Conduct, which was established by the Australian Information Commissioner (formerly the federal Privacy Commissioner). Everyone who handles tax file numbers must comply with the Tax File Number Guidelines, issued by the Australian Information Commissioner.

IV. Regulatory Position Regarding Data Protection in U.K.

Among the Council of Europe Member States UK was one of the nations that considered data protection at this initial stage of development of the law.

(i) Maiden Reports laid in U.K.

The development of data protection policies witnessed three major discrepancies at the end of the 1980s concerning scope (public-private sector), nature of data (automated or manual) and lastly choice of policy instruments. The two Committees formed in U.K. laid the foundation for development of information privacy. The administration of accounting and other numerically based records was the exclusive application of the older idea of electronic data processing, which was expanded into other areas of information processing by the younger Committee Report.

(ii) Younger Committee Report

Younger Committee Report identified ten principles which were intended as guidelines to computer users in the private sector. It can be concluded that a broad statement of a universal right to privacy could only be made by legislation, leaving it to the courts to formulate a general policy through specific rulings. In an area where fast change was expected, it was acknowledged that this lengthy approach ran the danger of failing to reflect the perspectives of modern society. The principles detailed in the Younger Report formed the foundation for future reports and the Data Protection Act.

(iii) Lindop Report

The Lindop Report on Data Protection studied computer systems in the public and commercial sectors and recommended a flexible legal framework with a set of overarching principles to direct a data protection authority in the creation of codes of practice targeted at different economic sectors. It proposed an independent authority to formulate Codes of Practice.

(iv) Current Legal Regulation in U.K.

Data Protection Act, 1984 recognized that organizations must use personal data but preserve the balance by providing individuals with several levers to control that use. To protect against the unintentional or unlawful processing of personal data, accidental loss, deletion, or damage to personal data, appropriate technological and organizational measures must be implemented." Data Protection Act, 1998 applies both to manual data and data processed by computers, provides eight principles for collecting, holding, using, processing, disclosure, and protection of personal data. Section 4, Electronic Communications Act 2000 restricts disclosure of information which has been obtained under Sections 1-6. Part III titled Investigation of electronic data protected by encryption etc. of the Regulation Of Investigatory Powers , 2000 deal with power to require disclosure ,contributions to costs and Safeguards.

The Data Protection Act 2018 supplements GDPR and provides additional provisions specific to the UK.

V. Regulatory Position regarding Data Protection in Germany

Public and private bodies processing personal data either on their own behalf or on behalf of others shall take the technical and organizational measures which are necessary to ensure the implementation of the provisions of this Act, in particular the requirements set out in Annexure to this Act. Where personal data are processed automatically, measures suited to the type of personal data to be protected shall be taken to prevent unauthorized persons from gaining access to data processing systems with which personal data are processed (access control), to prevent storage media from being read, copied, modified or removed without, authorization (storage media control), to prevent unauthorized input into the memory and the unauthorized examination, modification or erasure of stored personal data (memory control), to prevent data processing systems from being used by unauthorized persons with the aid of data transmission facilities (user control), to ensure that persons entitled to use a data processing system have access only to the data to which they have a right of access (access control), to ensure that it is possible to check and establish to which bodies personal data can be communicated by means of data transmission facilities (communication control), to ensure that it is possible to check and establish which personal data have been input into data processing systems by whom and at what time (input control), to ensure that, in the case of commissioned processing of personal data, the data are processed strictly in accordance with the instructions of the principal (job control) , to prevent data from being read, copied, modified or erased without authorization during the transmission of personal data or the transport of storage media (transfer control), and to arrange the internal organization of authorities or enterprises in such a way that it meets the specific requirements of data protection (organizational control).

All countries, except France have one person at the Apex to look after matters relating to data protection, whereas France's Commission consists of seventeen members. The functions of all authorities are more or less same.

VI. Regulatory Position Regarding Data Protection in Canada

In the adoption of data protection laws affecting both the public and private sector, Canada is the nation which has emerged as a very significant leader. The Privacy Act of 1985 expands the existing rules of Canada that guarantee individuals' right to access personal information about them that is stored by a government institution while also protecting their right to privacy about such information. The Act of 2001 regarding Personal Information Protection and Electronic Documents balances an individual's right to the privacy of personal information with the need of organizations to collect, use or disclose personal information for legitimate business purposes. It regulates the private sector.

VII. Conclusions

The General Data Protection Regulation (GDPR) adopted by the European Union (EU) significantly impacted data protection laws and practices not only within the EU but also globally due to its extraterritorial reach. While most of the countries offer protections for personal information, whether directly or indirectly nominative, the interpretation of when information relates to an identifiable person is not uniform. The laws collectively establish comprehensive frameworks for data protection, emphasizing privacy, rights of individuals, and responsibilities of organizations across various sectors within each respective country. The GDPR had a substantial impact on countries within the EU, including the UK and France, due to its direct applicability and the need to align national laws with its provisions. Outside the EU, while GDPR doesn't have direct legal authority, its principles have influenced discussions and prompted some countries to consider updates or reforms to their own data protection laws.

References

1   The Privacy Act, 1988.
2   Chris Pounder and Freddy Kosten, Managing Data Protection, (Butterworth-Heinemann Ltd, Oxford,1992).
3   Federal Data Protection Act, 1990.
4   Younger Committee Report.
5   Lindop Report on Data Protection.
6   The Privacy Act,1985.
7   The Act on Data Processing, Data Files, and Individual Liberties, 1978.
8   The Electronic Communications Act, 2000.
9   The Data Protection Act, 2018.
10 The General Data Protection Regulation.

Chapter 32

Emergence of Data Centers: Enabling Sustainable Digitalization with Special Emphasis on Circular Economy Practices

Towseef Ahmad Dar

Advocate on Record

Supreme Court of India

New Delhi, India

Aishwarya Balodi

Assistant Professor

The NorthCap University

Gurugram, Haryana, India

Abstract

Today, data centers are essential for businesses governments, research institutions, and various other organizations to store, process, and manage vast amounts of data and applications. Given the data center sector's recent recognition as critical infrastructure in India and the ongoing digital transformation, coupled with the necessity for localized data storage, this industry is anticipated to experience rapid growth in the foreseeable future. However, this expansion is also predicted to result in a substantial rise in greenhouse gas emissions. Consequently, it is imperative for the data center industry to prioritize "environmental sustainability" with increasing vigor. Considering the high growth phase of the Indian data center industry, it is well-positioned to incorporate environmentally friendly technologies in both existing and upcoming data centers. The critical focus at this juncture should be to direct forthcoming investments in a sustainable direction, guided by policy directives and collaborative efforts within the industry. This approach ensures that new data centers are inherently designed with environmentally responsible features, and concurrently, existing data centers are retrofitted to align with green standards. While it's encouraging to witness certain data center developers voluntarily embracing sustainable practices, implementing a regulatory framework mandating the use of green energy in data center operations holds immense promise. This becomes especially significant as India gears up to become a pivotal data center hub, as it can provide clear guidelines for developers and substantially promote the integration of sustainable practices within data center operations. While the regulatory framework for Data Protection is emphasised, focus on sustainability of Data Centers is simultaneously inevitable matter.

Keywords - India, Data Center, Cloud, environment, Sustainable.

I. Introduction

With the rapid growth of digital industries such as banking, Fintech, Healthtech, Edtech, e-commerce, and telemedicine, there is an anticipated surge in data generation. This surge is expected to drive an increased demand for data centers. Data centers serve as the fundamental foundations of a digital economy. They represent physical spaces utilized by organizations to house critical Information Technology (IT) applications, along with supporting network and security operations. Every activity conducted over the internet is directed to and from a central data center where the IT infrastructure resides, responsible for storing and processing online transactions. These facilities require a substantial amount of energy and need to operate continuously to sustain servers and uphold the ideal temperature within the data center using robust air conditioning systems. As banks and financial institutions progressively integrate IT solutions into their fundamental business processes, there's a growing need for specialized data centers equipped with cutting-edge infrastructure.

Between 2010 and 2018, global data centers escalated by 26 times, while traffic instances surged by 11 times, and compute instances grew by 6.5 times. The data center services market had a valuation of $48.9 billion in 2020. Projections suggest that by 2026, this figure is expected to grow significantly to reach approximately $105.6 billion. This trajectory is projected to continue and possibly intensify, indicating a continuous upward trend in data center usage. Data center workloads are on the rise due to a heightened need for enhanced application performance, increased storage demands, higher mobile data utilization driven by a surge in applications, and a growing reliance on the internet. Consequently, organizations worldwide are increasingly transitioning to cloud-based data storage, consequently amplifying the demand for data center services.

Till very recently, Singapore expressed concerns and considered restrictions on the development of new data centers due to their substantial environmental impact. Likewise, there have been reports of cities like Amsterdam in the Netherlands imposing bans. Additionally, the Irish government is contemplating implementing restrictions on data center construction to align with emissions and renewable energy targets. We need to reconsider our approach to constructing, conceptualizing, and managing data centers. Data centers function within an international market, and the challenges of minimizing their environmental impact are not bound by location. Given this reality, collaboration on a multinational level holds significant potential.

Governments across globe can unite their resources to amplify their influence and achieve more impactful sustainability initiatives.

II. Environmental Challenges associated with Data Centres

Data centers consume significant amounts of electricity and necessitate a continuous and reliable power supply to ensure operational stability, especially during power fluctuations or outages to prevent data unavailability or loss. Moreover, the setup of data center facilities involves the use of electronic and electrical equipment (EEE), resulting in the creation of electronic waste, commonly known as e-waste. Therefore, it's crucial to establish a system that can assess, oversee, and mitigate the detrimental environmental effects caused by data centers. Data centers, while vital for modern digital operations, pose significant environmental challenges. These challenges stem from their high energy consumption, electronic waste generation, and water usage, among other factors:

(i) High Energy Consumption

Data centers are known for their substantial energy consumption to power and cool the multitude of servers and related infrastructure. The demand for electricity continues to rise as data centers grow in number and size. This high energy usage contributes to a significant carbon footprint and increased greenhouse gas emissions.

(ii) Greenhouse Gas Emissions

The massive energy consumption of data centers primarily comes from fossil fuel-based power sources, leading to a notable contribution to greenhouse gas emissions, including carbon dioxide (CO_2), methane (CH_4), and nitrous oxide (N_2O). These emissions further exacerbate climate change and global warming.

(iii) E-Waste Generation

The rapid turnover of electronic and electrical equipment in data centers leads to the generation of electronic waste or e-waste. Disposing of this e-waste poses a serious environmental threat due to the hazardous materials it contains, including heavy metals and toxic chemicals that can contaminate soil and water sources.

(iv) Water Usage

Data centers require a significant amount of water for cooling purposes. High water usage can strain local water supplies, especially in regions facing water scarcity. Moreover, the discharge of heated water from cooling systems into nearby water bodies can disrupt aquatic ecosystems.

(v) Heat Generation and Urban Heat Islands

The heat generated by data centers, especially during cooling processes, contributes to the urban heat island effect. This effect raises local temperatures, impacting the environment and human health in the surrounding areas.

(vi) Land Use and Habitat Disruption

Data centers often necessitate large plots of land for construction and operation. This can lead to deforestation, habitat disruption, and loss of biodiversity. The construction of data centers can also alter natural landscapes and affect wildlife habitats.

(vii) Supply Chain Environmental Impact

The production, transportation, and disposal of equipment and components used in data centers have associated environmental impacts. This includes the extraction of raw materials, manufacturing processes, and the carbon footprint of transportation throughout the supply chain.

III. Initiatives and Policies of Government of India for Data Centers

The Indian Government had been focusing on various initiatives and policies to manage and promote the development of data centers in the country. Government has a range of potential strategies to monitor and regulate the sustainability impact of data centers, drawing from global examples and leveraging current technology.

(i) Recognition as Critical Infrastructure

The Indian government acknowledged data centers as critical infrastructure, signifying their importance for the nation's digital economy and development. This recognition underscores the need for strategic planning and investment in the sector.

(ii) Inclusion in Harmonized List of Infrastructure

Data centers were included in the harmonized list of infrastructure sectors in the Union Budget, allowing them to receive benefits and support similar to other critical infrastructure sectors.

(iii) State-Specific Policies and Incentives

Several Indian states, including Tamil Nadu, Telangana, Uttar Pradesh, Maharashtra, West Bengal, and Karnataka, announced state-specific policies and incentives to attract data center investments. These policies often include tax incentives, subsidies, and streamlined regulatory procedures to encourage data center development.

(iv) Minimum Capacity Criteria

The government announced a minimum capacity criterion of 5 MW of IT load for data centers to be included in the harmonized list of infrastructure. This measure aimed to promote the establishment of larger, more efficient data centers.

(v)   Data Localization Requirements

The government emphasized data localization, mandating that certain types of sensitive data should be stored within India. This has driven the demand for data centers within the country, particularly to cater to sectors like finance, healthcare, and government services.

(vi)   Digital India and National e-Governance Plan

The broader vision of the Indian government, encapsulated in initiatives like Digital India and the National e-Governance Plan, emphasizes the importance of data centers for digital transformation and efficient delivery of public services.

(vii)   Promotion of Renewable Energy Use

There has been an increasing emphasis on the use of renewable energy sources to power data centers. Incentives and policies promoting the adoption of solar and other forms of renewable energy have been encouraged to minimize the environmental impact of data centers.

IV. Need for Green Data Centers

A green data centre is one in which mechanical, electrical and computer systems are designed for maximum energy efficiency and minimum environmental impact. A green data centre ensures sustainability in its operations by enabling improvement of power usage effectiveness (PUE) by addressing energy consumption, reduction in data processing, precision and comfort air conditioning systems, lighting & building envelop; encouraging the use of water in a sustainable manner through reduce, recycle and reuse strategies; substituting conventional power consumption through onsite and offsite renewable energy sources; collecting and handling e-waste in an environmentally safe manner; and provides adequate ventilation, daylight and occupant well-being facilities for the staff. A significant decrease in both energy and water usage. On the other hand, improved air quality, optimal natural lighting, and the well-being of the personnel operating within these establishments are the possible advantages.

V. Challenge of Water Cooling

Water conservation is still not considered significant while discussing the challenges related to data centers. A single data center can consume power equivalent to that of a small city and demands a substantial amount of water for its cooling systems. As an extreme illustration, a 15-megawatt data center in the USA could utilize up to 360,000 gallons of water daily. Typically, this water is stored in cooling towers, a situation associated with notable environmental and health risks. Conventional cooling systems demand immense amounts of energy, facing technical limitations such as unexpected downtimes, surges in energy usage, and reduced efficiency during heatwaves.

Legionella is acknowledged as one of the top three or four microbial culprits behind community-acquired pneumonia, with cooling towers being the primary source of this illness. According to the Institute Pasteur, approximately 8,000 to 18,000 individuals contract Legionella annually. In the United States, the Centre for Disease Control pinpoints cooling towers as a primary origin of legionnaires' disease. In France, a malfunctioning cooling tower in the Nord-Pas-de-Calais region led to the infection of 86 individuals and the unfortunate demise of 18 individuals in 2003. Unfortunately, numerous data centers opt for cooling towers due to their lower upfront costs compared to closed circuit alternatives, and they also offer a space-efficient solution.

Data center operators bear the responsibility of safeguarding both our planet by adopting genuinely energy-efficient technologies and the health of nearby residents by making deliberate decisions to utilize technologies that eliminate the need for cooling towers.

An adiabatic cooling system, mimicking how the human body sweats to cool down is used by **Scale way**. By evaporating a few grams of water into the air, a few hours per year, the air coming from the outside can be cooled by nearly 10°C. Google, Facebook, and Amazon have acquired land in Sweden for constructing data centers, leveraging the cooler climate to reduce the resources needed for server cooling. In contrast, Microsoft is exploring an unconventional yet highly promising approach with Project Natick, researching the deployment of subsea data centers. Placing them on the seabed, protected from corrosive elements, moisture, and physical disturbances, allows data centers to operate effectively.

VI. Circular Economy Practices for Green Data Centers

Implementing circular economy practices in data centers involves designing, operating, and managing the data center's lifecycle in a way that maximizes resource efficiency, minimizes waste, and promotes sustainability. There is a need to reconsider the approach to constructing, conceptualizing, and managing data centers. Here are several key aspects and practices related to circular economy in data centers:

(i) Equipment Reuse and Refurbishment

Extend the lifespan of data center equipment by refurbishing and reusing hardware components or entire systems. Re-purposing servers, storage devices, and networking equipment can significantly reduce e-waste and resource consumption.

(ii) Recycling and Responsible Disposal

Establish procedures for responsible recycling of electronic waste (e-waste) generated within the data center. Ensure that e-waste is properly sorted, recycled, or disposed of in compliance with environmental regulations and best practices.

(iii) Non-volatile memory Express

Non-Volatile Memory Express (NVMe) refers to a host controller interface designed to optimize the data transfer process between enterprise and client systems. NVMe is extensively employed for solid-state storage, main memory, cache memory, or backup memory. It maximizes performance through parallel processing and is fully utilized by both the host application and hardware. It contributes to sustainability in the IT industry by promoting energy efficiency and reducing the environmental footprint of data centers and storage systems.

(iv) Materials Recovery and Reutilization

Extract valuable materials from decommissioned or obsolete equipment and reuse them in the production of new hardware. This approach helps reduce the demand for raw materials and minimizes environmental impact.

(v) Energy Efficiency Optimization

Continuously optimize the data center's energy efficiency through measures such as advanced cooling techniques, efficient power distribution, and energy-saving hardware. Reducing energy consumption ensures that fewer resources are needed to power and cool the data center.

(vi) Renewable Energy Integration

Prioritize the use of renewable energy sources, such as solar, wind, or hydroelectric power, to meet the energy needs of the data center.

This reduces reliance on fossil fuels and lowers the carbon footprint associated with data center operations.

(vii) Modular Design and Scalability

Adopt a modular and scalable design approach for the data center infrastructure. This allows for the efficient scaling up or down of resources based on demand, reducing the need for complete infrastructure overhauls and minimizing waste.

(viii) Waste Heat Recovery

Implement waste heat recovery systems to capture and repurpose excess heat generated by data center operations. Redirect this heat for heating purposes in nearby buildings, thereby reducing the overall energy consumption of the community.

(ix) Water Recycling and Efficiency

Implement water recycling and reuse systems within the data center for non-potable uses like cooling. Additionally, optimize cooling systems to minimize water usage and reduce the strain on local water resources.

(x) Sustainable Procurement

Prioritize the purchase of energy-efficient, eco-friendly, and recyclable products when procuring new equipment or components for the data center. Consider the environmental impact throughout the entire lifecycle of the product.

(xi) Education and Awareness

Educate data center staff and stakeholders about the principles and benefits of the circular economy. Encourage sustainable practices, resource efficiency, and waste reduction throughout the data center's operations.

VII. Conclusions

As society becomes more dependent on the internet and mobile usage, data is no longer merely viewed as information but rather as a vital service and a fundamental public utility. There are billions of Internet-connected devices, and this number continues to increase. A significant portion of these devices generates substantial volumes of data, necessitating efficient recording, processing, storage, assessment, and retrieval mechanisms to manage and utilize this data effectively. It is quite likely that the environmental impact in terms of energy consumption and emissions generated by data centers will become significantly more crucial in the future. The evolution of data centers is likely to continue with a focus on energy efficiency, sustainability, and leveraging emerging technologies like artificial intelligence and edge computing to meet the evolving needs of the digital age.

By incorporating the circular economy practices, data centers can reduce their environmental footprint, decrease waste generation, and contribute to a more sustainable and responsible IT infrastructure. Addressing these environmental challenges requires innovative approaches such as improving energy efficiency, adopting renewable energy sources, enhancing waste management and recycling processes, optimizing water usage, and implementing sustainable data center designs. Additionally, focusing on circular economy principles and incorporating environmentally conscious practices throughout the data center lifecycle is essential for mitigating the adverse environmental impacts of data centers. Data centers will expand and enhance their utilization of data compression, deduplication, and other efficiency-boosting techniques.

Suggestions

Establishing precise and resilient standards and metrics is crucial to facilitate meaningful comparisons of data center operations and effectively regulate their environmental impact. No single policy approach will address the sustainability impact of large and growing data centre use on account of the wide range of sizes, operational models and usage. Instances of policies implemented in various regions can serve as a reference for contemplation by governments elsewhere. While a single policy may be impactful, most policies tend to complement one another, implying that the cumulative adoption of policies can amplify the impact.

(i) At a national scale, collecting data on existing data center performance via a comprehensive industry-wide survey on data center energy consumption would offer insights into how data center performance aligns with publicly available data from other regions. This approach would establish a foundational benchmark for future adjustments grounded in empirical evidence.

(ii) Data centers might consider obtaining recognized accreditations for environmentally sustainable data centers, such as IGBC certification for green data centers and LEED certification for green data centers.

(iii) Organisations can devise a comprehensive strategy to achieve carbon neutrality for their data centers, offsetting greenhouse gas emissions through careful measurement and balancing.

(iv) Organisations can incorporate eco-friendly practices into their data center operations, encompassing elements like design, materials, construction, energy usage, and waste management. For their upcoming data centers, sustainability considerations can be integrated from the very initial stages, well before the construction phase.

(v) The electronic waste produced during data center refresh activities can be directed to government-approved recyclers to ensure responsible and sustainable decommissioning. Embracing the principles of 3Rs -Reduce, Recycle, and Reuse within data center operations can be helpful.

(vi) Data centers have the potential to achieve a net-zero emissions status by procuring renewable energy from suppliers and compensating for any non-renewable power used by purchasing green credits from these suppliers.

(vii) It is imperative to make the adoption of renewable energy and sustainable waste management, among other eco-friendly practices, mandatory in a phased approach for high power consumption operations such as data centers.

(viii) Given that cooling systems represent a significant portion of total energy consumption in data centers, implementing AI-driven smart systems could be a viable approach to optimize power usage and enhance airflow within these facilities.

(ix) It is critical for data centers to gain a comprehensive understanding of the urgent challenge regarding water consumption and its side-effects by considering Dry technologies like the dry coolers.

(x) Incorporating NVMe into storage infrastructure can be a step toward achieving more sustainable data center operations, promoting energy efficiency, and minimizing the ecological impact of IT operations.

References

1   Greenpeace International, *How dirty is your data? A Look at the Energy Choices That Power Cloud Computing*, available at https://www.greenpeace.org/static/planet4-international-stateless/2011/04/4cceba18-dirty-data-report-greenpeace.pdf.

2   *Greenpeace International, How Clean is your Cloud*, April 2012, available at https://www.greenpeace.org/static/planet4-international-stateless/2012/04/e7c8ff21-howcleanisyourcloud.pdf.

3   *Greenpeace International, Clicking Clean: Who is Winning the race to build a Green Internet?* 2017, available at https://www.greenpeace.de/publikationen/20170110_greenpeace_clicking_clean.pdf.

4   Brij Raj and Deepak Verma, *Green Data Centres: Pathway to Sustainable Digitalisation*, Nov. 2022, available at https://rbi.org.in/scripts/BS_ViewBulletin.aspx?Id=21402.

5   India to Be A Cloud Computing And Data Centre Hub, available at https://static.pib.gov.in/WriteReadData/specificdocs/documents/2022/dec/doc2022128141601.pdf.

6   https://stl.tech/blog/the-rise-of-data-centers-in-india/.

7   *Policy Initiatives Fueling Data Centres in India, 2021, available at https://stl.tech/blog/policy-initiatives-that-could-fuel-data-centre-growth-in-india/.*

8   Data Centre, Energy Storage System to be included in the list of infrastructure: FM, Business Line, 1ˢᵗ Feb, 2022, available at https://www.thehindubusinessline.com/info-tech/data-centre-energy-storage-system-to-be-included-in-the-list-of-infrastructure-fm/article64961155.ece.

9   Karnataka's new data centre policy aims Rs 10,000 cr investments in 5 years, The Economic Times, 19 April 2022, https://economictimes.indiatimes.com/news/india/karnatakas-new-data-centre-policy-aims-rs-10000-cr-investments-in-5-years/articleshow/90934419.cms?from=mdr.

10   T.R.A.I. Consultation Paper on Regulatory Framework for Promoting Data Economy Through Establishment of Data Centres, Content Delivery Networks, and Interconnect Exchanges in India, 2021, available at https://www.trai.gov.in/sites/default/files/CP_16122021_0.pdf.

11   Tamil Nadu Data Centre Policy 2021, available at https://cms.tn.gov.in/sites/default/files/documents/TN_Data_Centre_Policy_2021.pdf.

12   Telangana Data Centres Policy 2016, available at https://www.telangana.gov.in/PDFDocuments/Telangana-Data-Centres-Policy.PDF.

13   U.P.Data Center Policy 2021, available at https://www.uplc.in/docs/Data%20Center%20Policy%202021_English.pdf.

14   Maharashtra's State Data Center , *available at* https://it.maharashtra.gov.in/sites/default/files/Maharashtra%20State%20Data%20Center_0.pdf.

15   *West Bengal Data Centre Policy*, 2021, available at https://sahayata.itewb.gov.in/pages/data-centre-policy-2021.

16   Andreja Velimirovic, Data Center Tiers Explained, Nov. 2021, available at https://phoenixnap.com/blog/data-center-tiers-classification.

17 Steps by Government of India to promote renewable energy in the country, July 2022, available at https://pib.gov.in/PressReleaseIframePage. aspx?PRID=1843538.

18 Union Budget *2022-23, available at* https://www.indiabudget.gov.in/ budget2022-23/.

19. ICICI Bank's Data Centre is the first IGBC Platinum rated data centre in India, available at ICICI Bank Annual Report FY2018-19 - https://www. icicibank.com/annual-report-microsite/ICICI_AR_2019.pdf.

20. PayPal Sets Up India Tech Center For Data Localization, July 2019, https://www.pymnts.com/news/international/2019/paypal-sets-up-tech-center-in-india-for-data-localization/.

21 MeitY, Draft Data Centre Policy 2020, available at https://www.meity. gov.in/writereaddata/files/Draft%20Data%20Centre%20Policy%20 -%2003112020_v5.5.pdf.

22 Global warming: Data centres to consume three times as much energy in next decade, experts warn, Jan. 2016, available at https://www. independent.co.uk/climate-change/news/global-warming-data-centres-to-consume-three-times-as-much-energy-in-next-decade-experts-warn-a6830086.html.

23 Axis Bank Annual Report 2020-21, available at https://www.axisbank. com/docs/default-source/annual-reports/for-axis-bank/annual-report-for-the-year-2020-2021.pdf.

24 NASSCOM, India - The Next Datacenter Hub, 2021.

25 Aligning Data Centres with Sustainable Development Goals, July 2022, available at https://www.cyrusone.com/resources/blogs/aligning-data-centres-with-sustainable-development-goals.

26 Data Centers: can they stay cool and go green?, available at https:// www.tcs.com/insights/topics/sustainability-topic/article/data-centre-temperature-monitoring.

27 George Kamiya and Oskar Kvarnström , Data centres and energy - from global headlines to local headaches? Dec. 2019, available at https://www. iea.org/commentaries/data-centres-and-energy-from-global-headlines-to-local-headaches.

28 Central Electricity Authority, Ministry of Power, Government of India, *Report On Optimal Generation Capacity Mix for 2029-30, Version 2.0*, April

2023, available at https://cea.nic.in/wp-content/uploads/irp/2023/05/Optimal_mix_report__2029_30_Version_2.0__For_Uploading.pdf.

29 https://www.iba.org.in/depart-res-stcs/key-bus-stcs.html.

30 Eric Masanet, Arman Shehabi, Nuoa Lei , Sarah Smith and Jonathan Koome, *Recalibrating global data center energy-use estimates* , Mar. 2022, available at https://www.science.org/journal/science.

31. Caroline Donnelly, *Why water usage is the datacentre industry's dirty little secret*, Sept. 2021, https://www.computerweekly.com/blog/Ahead-in-the-Clouds/Why-water-usage-is-the-datacentre-industrys-dirty-little-secret.

32. Caroline Donnelly, *Climate change and datacentres: Weighing up water use,* 2021, available at https://www.computerweekly.com/feature/Climate-change-and-datacentres-Weighing-up-water-use.

33 Expert Market Research titled, 'Global Non-Volatile Memory Express Market Size, Report and Forecast 2022-2027, available at https://www.expertmarketresearch.com/reports/non-volatile-memory-express-market.

34 Yan Jianfeng, *Our Green Future: Why a Digital-First World needs Zero-Carbon Data Centers*, 2022, available at https://blog.huawei.com/2022/06/02/green-future-digital-first-world-zero-carbon-data-centers/.

35 *Zhang Fan, Green data centers in four steps, 2015 , available at* https://www.huawei.com/en/huaweitech/publication/75/green-data-centers-in-four-steps.

36 *Paul Mah, How China is going green with data centers*, 2019, available at https://www.datacenterdynamics.com/en/analysis/how-china-going-green-data-centers/.

Chapter 33

The Personal Data Protection Bill, 2019: Challenges to Good Governance for The Banking Sector

CS (Dr.) Yogesh Sharma
Ph.D. Research Scholar
School of Law, Maharaja Agrasen University
Barotiwala, Baddi, Solan, H.P., India

Yatika Gupta
Advocate
Supreme Court of India
New Delhi, India

Abstract

To ensure growth of digital economy while keeping personal data of citizens secure and protected, the Government of India appointed an expert Committee headed by the then Supreme Court Judge, Justice B.N. Srikrishna. This Committee submitted its draft along with draft legislation on data protection in July 2018. The government set up a Joint Parliamentary Committee to review the Personal Data Protection Bill. On 16th December 2021, the Joint Parliamentary Committee submitted its report to the Indian Parliament after deliberating on the Personal Data Protection Bill, 2019. The much awaited Bill is the culmination of a series of extensions provided to the Joint Parliamentary Committee and will pave the way for a strong data protection law in the world's largest democracy. The Committee visited data centres and processing centres in India and also consulted various stakeholders with an objective of understanding as to how personal and sensitive data can be processed in real time with implementation of actual data protection safeguards preventing data leakages. The status of client data protection and privacy in India seems to be in a good state because of a strong legal framework such as The Banking Regulations Act, and the Information Technology Act, industry watchdog like BCSBI and compliance by national networks of banks i.e. IBA and MFIs. With an increasing focus and investment in e-governance and technology based services in India, the topic of privacy rights and data security is very crucial to national development and appropriate functioning of these services.

Key Words - Personal Data Protection, Stakeholders, Banking companies, Corporate Governance.

I. Introduction

The foundation of good governance in banks rests on a comprehensive set of principles and practices that ensure transparency, accountability, integrity, and effective risk management. the Banking Regulation Act is an important legislation in India that provides a framework for the regulation and supervision of banking companies in the country. However, the act primarily focuses on aspects related to banking operations, governance, capital requirements, licensing, and regulation by the Reserve Bank of India (RBI). It does not directly address data protection or privacy matters.

The intersection of the Banking Regulation Act and data protection primarily comes into play when banks handle personal data, which is a significant aspect of their operations. Banks are required to adhere to data protection and privacy laws, including any future legislation while conducting their business and handling customer data. The Information Technology Act, 2000, and its associated Rules establish legal guidelines for data protection, including sensitive personal data or information. It outlines reasonable security practices and procedures for handling sensitive personal data and imposes penalties for data breaches.

II. Legal Framework for Regulation of Banking Sector

The Reserve Bank of India, being the central regulatory authority for banks, issues guidelines and circulars related to data protection and cybersecurity in the banking sector. The Banking Regulation Act, 1949 empowers the R.B.I. to inspect and supervise commercial banks. It is empowered under a number of legislations to supervise the functioning of banks in India.

(i) Payment and Settlement Systems Act, 2007

This Act provides the legal framework for the regulation and supervision of payment systems in India. It includes provisions related to data security and confidentiality in electronic payment transactions.

(ii) Cybersecurity Framework for Banks

The RBI has set up a comprehensive cybersecurity framework that mandates banks to implement security measures to protect customer data and information systems. The RBI continuously updates and revises its cybersecurity framework to adapt to evolving cyber threats and technology advancements. Banks are required to stay updated with the latest guidelines and implement necessary measures to maintain a high level of cybersecurity and protect customer data. Compliance with RBI's cybersecurity framework is essential to safeguard the integrity and trustworthiness of the banking system in India.

(iii) Payment Card Industry Data Security Standard (PCI - DSS)

PCI-DSS is a global data security standard applicable to organizations that handle credit and debit card information. Its compliance is mandatory for all entities that process payment card transactions. This includes banks, financial institutions, merchants, payment gateways, and service providers involved in handling payment card data. PCI DSS outlines 12 high-level security requirements organized into six categories namely Build and Maintain a Secure Network; Protect Cardholder Data; Maintain a Vulnerability Management Program; Implement Strong Access Control Measures; Regularly Monitor and Test Networks; Maintain an Information Security Policy. It categorizes entities into different compliance levels based on the number of transactions they process annually. Level of compliance requirements varies based on these levels.

(iv) Financial Sector Computer Emergency Response Team (FINCERT)

FINCERT is an initiative by the RBI to enhance the cybersecurity posture of the Indian banking and financial sector. It plays a crucial role in monitoring and mitigating cyber threats.

(v) The Banking Codes and Standards Board of India (BCSBI)

Following the proposal outlined in the Annual Policy Statement for the fiscal year 2005-06, the Banking Codes and Standards Board of India (BCSBI) was established on February 18, 2006, through a collaborative initiative involving the Reserve Bank of India (RBI) and various banks. This establishment was modeled after a similar system in the UK and aimed to oversee the Banking Code, a voluntary set of guidelines devised by the British Bankers Association (BBA) and adopted by all banks in the UK. The decision to create the BCSBI was influenced by a recommendation from the Committee on Procedures and Performance

Audit on Public Services addressed the issues of Benchmarking, ISO Certification, and Performance Audit. The BBA's "Banking Code" establishes standards of exemplary banking practices for financial institutions to adhere to when interacting with personal customers in the UK. This code serves as a vital means of safeguarding customers in their day-to-day transactions and during financial challenges. Its coverage extends to savings deposits, current accounts, card products and services, loans, overdrafts, and payment services, including foreign exchange.

In June 2004, the Indian Banks Association (IBA) introduced the "Bankers' Fair Practice" code, which was voluntarily adopted by all its member banks. This code represented a pledge to uphold fairness and transparency in interactions with individual customers. Additionally, the IBA independently developed the "Fair Practice Code for Credit Card Operations" and the "Model Code for Collection of Dues and Repossession of Security" to specifically address customer concerns regarding banking practices in these domains. BCSBI enhanced the existing Code and introduced the "Code of Bank's Commitment to Customers" outlining essential banking practice standards for interactions with individual customers. The unveiling of the "Code of Bank's Commitment to Customers" occurred during an inaugural event at the Reserve Bank of India (RBI) on July 1, 2006, with Dr. Y.V. Reddy, the Governor of RBI.

(vi) Draft Bill

The Personal Data Protection Bill, 2019 is an Indian proposed law aimed at regulating the collection, storage, processing, and transfer of personal data. The Bill aims to provide individuals with greater control over their personal data and establish obligations on entities that process this data. The Personal Data Protection Bill, 2019 presents several challenges for the banking sector, including compliance costs, data security, data localization, consent management, and operational adjustments to adhere to the Bill's provisions. Effective implementation and adaptation will require significant effort, investment, and strategic planning by the banking industry to ensure compliance with the Bill's requirements while maintaining good governance.

III. Challenges for the Banking Sector

In implementing and adhering to new proposed law on Data Protection, the banking sector can face many challenges -

(i)   Data Privacy Compliance Costs

The banking sector may face significant financial burdens in ensuring compliance with the stringent data protection requirements of the Bill. Implementing necessary technical and organizational measures to protect personal data can require substantial investments in infrastructure, training, and technology.

(ii)   Data Localization

The Bill proposes the concept of data localization, which mandates that certain categories of sensitive personal data must be stored within Indian borders. This could be a challenge for international banks or those with a global presence, as it may disrupt their existing data storage and processing mechanisms.

(iii)   Data Security Measures

The Bill imposes strict requirements for data security and breach notification. Ensuring robust data security measures can be a challenge for banks, especially considering the constantly evolving cyber threats and the need for ongoing investment in cybersecurity technologies and practices.

(iv) Consent Management

The Bill emphasizes obtaining explicit consent from individuals for processing their personal data. Banks may find it challenging to manage and track individual consent at various stages of the customer relationship, especially in cases where multiple services are provided.

(v) Data Processing Limitations

The Bill imposes restrictions on the purposes for which personal data can be processed. Banks often process data for various purposes, including risk assessment, fraud detection, marketing, and compliance. Adhering to the prescribed limitations while fulfilling their operational requirements may present a challenge.

(vi) Cross-border Data Transfer

The Bill imposes restrictions on the transfer of personal data outside India, requiring explicit consent from individuals and adherence to certain conditions. This could pose difficulties for international banking operations and collaborations with global partners.

(vii) Data Subject Rights

The Bill grants various rights to data subjects, such as the right to access, rectify, and erase personal data. Ensuring compliance with these rights while managing the vast amount of data typically held by banks can be operationally challenging.

(viii) Third-Party Data Sharing and Contracts

Banks often collaborate with third parties for various services. Ensuring that these third parties comply with the data protection requirements and establishing robust contracts to manage data sharing and processing can be complex.

IV. Significant Factors to ensure protection of personal information in the Banking Sector

It's essential for banking organizations operating internationally to understand and comply with the data protection laws applicable in the regions they operate to ensure good governance and data security. The best protection of personal information for bank customers is typically achieved through a combination of robust data protection laws, stringent industry regulations, secure technological measures, employee training, and a strong culture of privacy and security within the banking sector. The key elements that contribute to effective protection of personal information in the banking sector are -

(i)   Compliance with Data Protection Laws and Regulations

Banks must strictly comply with relevant data protection laws and regulations specific to their jurisdiction. These laws often mandate how customer data should be handled, stored, accessed, and shared.

(ii)   Privacy Policies and Consent Mechanisms

Banks should have clear and transparent privacy policies that outline how they collect, use, and share customer data. Obtaining explicit consent from customers for data processing activities is crucial.

(iii)   Data Encryption and Security Measures

Implement strong encryption techniques to protect sensitive customer data during storage, transmission, and processing. Use advanced security measures, including firewalls, multi-factor authentication, access controls, and regular security assessments.

(iv) Regular Security Audits and Risk Assessments

Conduct regular security audits and risk assessments to identify vulnerabilities, assess potential risks, and implement measures to mitigate them effectively.

(v) Incident Response Plan

Develop and maintain a comprehensive incident response plan to handle data breaches or security incidents effectively. This plan should include steps for containment, investigation, communication, and resolution.

(vi) Data Minimization and Purpose Limitation

Collect and retain only the data necessary for legitimate banking purposes. Avoid unnecessary data collection and ensure that customer data is only used for its intended purposes.

(vii) Employee Training and Awareness

Train employees on data protection policies, security protocols, and privacy best practices. Create a culture of awareness and accountability within the organization.

(viii) Regular Customer Education and Communication

Educate customers about data protection practices, their rights, and how their data is being used. Keep them informed about security measures in place to protect their information.

(ix) Third-Party Vendor Management

Establish clear guidelines and contracts with third-party vendors to ensure they comply with data protection standards when handling customer data on behalf of the bank.

(x) Secure Data Disposal and Retention Policies

Implement secure methods for disposing of customer data when it's no longer needed. Establish clear data retention policies to ensure data is not kept longer than necessary.

(xi) Access Controls and Monitoring

Implement strong access controls to limit access to customer data only to authorized personnel. Regularly monitor access and detect any unauthorized activities promptly.

(xii) Regular Security Training and Drills

Conduct regular training sessions and simulated drills to ensure that employees are prepared to respond to security incidents effectively.

By implementing these measures and continually adapting to evolving threats and regulatory requirements, banks can provide strong protection for their customers' personal information and maintain a high level of trust and confidence.

V. Data Protection and Good Governance in the Banking Sector

Data protection in the banking sector plays a critical role in ensuring good governance by establishing a framework that promotes ethical conduct, trust, transparency, compliance with laws and regulations, and the responsible handling of customer information. Here's how data protection contributes to good governance in the banking sector

(i) Compliance with Regulations and Legal Requirements

Adherence to data protection laws and regulations is fundamental for good governance. Banks must comply with relevant laws governing data privacy, security, and confidentiality. Compliance helps in avoiding legal liabilities and penalties, ensuring transparency, and building trust with customers and regulators.

(ii)   Customer Trust and Confidence

Effective data protection fosters trust and confidence among customers. When customers believe that their personal and financial information is handled securely and in compliance with privacy laws, they are more likely to engage with the bank, use its services, and maintain long-term relationships.

(iii)   Ethical Handling of Customer Data

Good governance requires ethical conduct. Data protection ensures that customer data is handled ethically, with consent and for legitimate purposes. Respecting customer privacy and maintaining the confidentiality of their information demonstrates ethical business practices within the banking sector.

(iv)   Accountability and Transparency

Data protection policies and practices promote accountability and transparency. Banks need to clearly communicate their data handling practices to customers, employees, and stakeholders. Transparent disclosure of data collection, processing, storage, and sharing practices builds credibility and fosters good governance.

(v)   Risk Mitigation and Resilience

Implementing robust data protection measures helps in mitigating risks associated with data breaches, unauthorized access, or cyber-attacks. A secure data environment ensures resilience against potential threats, enhancing operational stability and contributing to good governance by minimizing risks.

(vi) Operational Efficiency and Effectiveness

Properly managing customer data in compliance with data protection principles enhances operational efficiency. Efficient data processing, secure storage, and streamlined data access facilitate smooth operations, contributing to good governance by optimizing processes and minimizing inefficiencies.

(vii) Preventing Data Abuse and Fraud

Data protection measures, such as access controls, authentication, and encryption, help in preventing data abuse, fraud, and identity theft. By protecting customer data from unauthorized access and misuse, banks maintain their reputation and trustworthiness, vital for good governance.

(viii) Stakeholder Confidence and Reputation Management

Effective data protection practices bolster stakeholder confidence and protect the bank's reputation. Stakeholders, including investors, partners, and the public, have confidence in banks that demonstrate a strong commitment to safeguarding customer data, positively impacting governance and reputation management.

(ix) Data Governance Frameworks

Implementing structured data governance frameworks, including data quality, data lifecycle management, and data stewardship, ensures that customer data is handled in a controlled and responsible manner. This contributes to good governance by establishing clear roles, responsibilities, and processes for data management.

(x) Continuous Improvement and Adaptability

Data protection encourages a culture of continuous improvement and adaptability. Banks need to stay updated with evolving data protection requirements and technological advancements, ensuring that their data protection strategies evolve to meet changing threats and regulatory landscapes.

(xi)   Strong Board Oversight and Independence

A well-structured board with independent directors provides strategic guidance, oversight, and holds management accountable. The board should represent a diverse range of skills, experiences, and perspectives to ensure effective decision-making.

(xii)   Robust Risk Management Framework

Banks need a comprehensive risk management framework to identify, assess, and mitigate risks effectively. This includes credit, market, operational, and compliance risks. Risk management should be integrated into the bank's strategy and operations.

(xiii) Transparency and Disclosure

Banks must maintain transparency in their operations and disclose relevant financial and non-financial information in a timely and accurate manner. This promotes accountability, builds trust, and helps stakeholders make informed decisions.

(xiv) Ethical Conduct and Corporate Social Responsibility (CSR)

Banks should adhere to a high standard of ethical conduct, promoting integrity, fairness, and honesty in all interactions. Engaging in CSR activities that benefit the community and environment demonstrates social responsibility and good governance.

(xv)   Compliance with Laws and Regulations

Strict adherence to local and international laws and regulations is essential. Compliance ensures that the bank operates within legal boundaries, avoids penalties, and maintains its reputation.

(xvi) Customer-Centric Approach

A customer-centric approach involves placing customer interests at the core of business operations. Banks should prioritize customer satisfaction, fair treatment, and provide transparent, easily understandable products and services.

(xvii) Sustainable Business Practices

Incorporating sustainability into business strategies is important for long-term success. Banks should consider environmental, social, and governance (ESG) factors, aiming for sustainable growth that benefits both the institution and society.

(xviii) Sound Internal Controls and Audit Functions

Effective internal controls, including regular audits and assessments, ensure compliance, detect fraud, and provide insights for process improvement. Internal audit functions should be independent and report directly to the board.

(xix) Talent Management and Succession Planning

Attracting, developing, and retaining talent is critical for organizational success. Banks should have robust talent management strategies, leadership development programs, and succession plans to ensure continuity and excellence in leadership.

(xx) Technological Innovation and Cybersecurity

Embracing technological advancements is essential for efficiency and competitiveness. However, banks must ensure robust cybersecurity measures to protect sensitive data and maintain the integrity and trustworthiness of digital systems.

(xxi) Fair Treatment of Shareholders

Banks should treat all shareholders fairly and equitably, respecting their rights and providing access to information. Shareholder engagement and involvement in key decisions should be encouraged.

(xxii) Adaptive Strategy and Agility

Banks need to adapt their strategies in response to changing market dynamics, customer needs, and technological advancements. Agility allows banks to seize opportunities and manage challenges effectively.

VI. Conclusions

Strong data protection practices are critical for maintaining the security, privacy, and trust of customers in the banking sector, regardless of the location or jurisdiction. Different countries have their own data protection laws, regulations, and industry best practices that guide banks in ensuring data security and privacy. data protection within the banking sector is a prerequisite for good governance. It ensures compliance with laws, ethical handling of data, trust-building with customers, risk mitigation, operational efficiency, and overall enhancement of the bank's reputation and stakeholder confidence. Integrating strong data protection practices into the governance structure is essential for the sustainable success of banking institutions

The Banking Regulation Act primarily focuses on regulating the banking sector, while data protection laws regulate the handling of personal data. Banks must comply with sector specific and general law to ensure lawful and responsible handling of customer data. The specific implications may evolve as new data protection legislation is enacted or amended in the future.

Suggestions

(i)   Collaboration and Engagement with Regulatory Authorities

Establish channels for collaboration and open communication with regulatory authorities, seeking guidance and clarification on data protection matters to ensure compliance

(ii)   Data Mapping and Classification

Understand what personal data is collected, processed, and stored within the organization. Classify the data based on sensitivity and importance to ensure appropriate security measures and compliance with data protection laws.

(iii)   Data Transfer Safeguards

Ensure secure data transfers, especially for cross-border data transfers, by using appropriate safeguards and adhering to legal requirements for international data transfers.

(iv)   Data Backups and Recovery Plans

Implement regular data backups and maintain well-defined recovery plans to ensure business continuity and minimize data loss in case of a disaster or security incident.

(v)   Regular Updates and Patch Management

Keep all software, applications, and systems up to date with the latest security patches and updates to mitigate vulnerabilities that could be exploited by malicious actors.

References

1   Rachita Gulati et al. , Governance, Efficiency and Soundness of Indian Banks, 2022, available at https://rbi.org.in/Scripts/PublicationsView.aspx?id=21083.

2   RBI Guidelines for Fintech: The Role of Data Protection and Data Backup, available at https://nimesa.io/blogs/rbi-guidelines-for-fintech-the-role-of-data-protection-and-data-backup/#:~:text=The%20RBI%20guidelines%20also%20place,updating%20security%20software%20and%20systems.

3   RBI guidelines on Information Security, Electronic Banking, Technology Risk Management and Cyber Frauds (G.Gopalakrishna Committee) vide Circular DBS.CO.ITC.BC.No.6/31.02.008/2010-11 dated April 29, 2011.

4   RBI's Cyber Security Framework in Banks, 2016 , available at https://www.rbi.org.in/Scripts/NotificationUser.aspx?Id=10435&Mode=0.

5   PCI DSS Quick Reference Guide: Understanding the Payment Card Industry Data Security Standard version 3.2.1. , 2018, available at https://listings.pcisecuritystandards.org/documents/PCI_DSS-QRG-v3_2_1.pdf.

6   https://www.pcisecuritystandards.org/about_us/.

7   FSDC Secretariat, Department of Economic Affairs, Ministry of Finance, Government of India Report Of The Working Group For Setting Up Of Computer Emergency Response Team In The Financial Sector (CERT-Fin) , 2017 , available at https://dea.gov.in/sites/default/files/Press-CERT-Fin%20Report.pdf.

8   National Cyber Security Policy 2013, available at https://www.meity.gov.in/writereaddata/files/downloads/National_cyber_security_policy-2013%281%29.pdf.

9   https://financialservices.gov.in/cybersecurity-and-fintech-section.

10 https://dea.gov.in/fsdc.

11 Shaktikanta Das, Excellence in Customer Service in the Changing Paradigm of Financial Services, Nov. 2022, available at https://rbi.org.in/scripts/BS_ViewBulletin.aspx?Id=21393.

12 Banking Regulation Act, 1949.

13 RBI Act, 1934.

14 Payment and Settlement Systems Act, 2007.

15 Master Circular - Corporate Governance, 2014, available at https://www.rbi.org.in/commonperson/English/Scripts/Notification.aspx?Id=1421.

16 Codina Sabau, 5 Ways in Which Banks Secure Their Data, June 2022, available at https://www.endpointprotector.com/blog/ways-banks-secure-data/.

17 Information Technology (Reasonable Security Practices and Procedures and Sensitive Personal Data or Information) Rules, 2011.

18 NGData, Banks and Financial Institutions: A Data Privacy Guide, 2020, available at https://www.ngdata.com/data-privacy-guide-for-banks-and-financial-institutions/.

19 Tobias M.C. Asser, Legal Aspects of Regulatory Treatment of Banks in Distress, 2001, International Monetary Fund.

20 https://www.rbi.org.in/commonman/English/Scripts/DeptofBS.aspx#:~:text=The%20Banking%20Regulation%20Act%2C%201949,inspection%20and%20off%20site%20surveillance.

21 Fintech Laws and Regulations 2023, available at https://www.globallegalinsights.com/practice-areas/fintech-laws-and-regulations/india.

22 https://www.iba.org.in/.

23 GlobalData Thematic Research, Data Privacy in Banking – Thematic Research, 2022, available at https://www.globaldata.com/store/report/data-privacy-in-banking-theme-analysis/.

24 ISO/IEC 27701:2019, Security techniques - Extension to ISO/IEC 27001 and ISO/IEC 27002 for privacy information management - Requirements and guidelines.

25 Credit Information Companies (Regulation) Act, 2005.

26 Companies Act, 2013.

27 Foreign Exchange Management Act, 1999.

28 Securitisation and Reconstruction of Financial Assets and Enforcement of Security Interest Act, 2002.

Chapter 34

Right to be Forgotten: A Study with Special Reference to India

Swati Pandita
Assistant Professor
School of Law, FIMT-GGSIPU
New Delhi, India

Lovely Sharma
Lecturer in Law
Divine Law College
Meerut, UP, India

Shivani Johri
Assistant Professor
S.S.O.L., Sharda University
Greater Noida, UP, India

Abstract

In recent years, data privacy issues have dominated the discussion among media and academic circles, more so after the Right to be forgotten was included in European Union's Data Protection Regulation. The "Right to be Forgotten" is a legal concept that allows individuals to request the removal of personal information from online platforms and search engine results under certain circumstances. The concept is primarily associated with privacy and data protection laws, particularly in the European Union (EU). It is a right under Article 17 of GDPR. It means the data subject has the right to get his personal data erased or delisted from any internet database or public platform if the information is no longer necessary. As of now, this right is not backed by any statute in India but courts on multiple occasions have ruled in its favor. The right to be forgotten is not an absolute right and must be balanced against other fundamental rights, such as freedom of expression and the public's right to access information. This paper critically analyses the Personal Data Protection Bill, 2019 and the report of Justice Srikrishna Committee. The author also considers RTBF in the context of the Right to know and examines the stand of the judiciary.

Keywords: Right to be forgotten, Data Privacy, Google Spain Case, Data Protection Bill, right to Know, Right of Erasure.

I. Introduction

The conceptualization and subsequent formalization of the right to be forgotten can be traced back to a significant legal case in the European Court of Justice (ECJ) involving Google, a leading search engine. In 1998, La Vanguardia, a Spanish newspaper published an announcement of bidding for two properties being sold due to social security debt. Mario Costeja González, owner of one of the properties was named in the newspaper. In 2009, he requested the paper to erase the data from its online edition. After his request was denied, he contacted Google Spain to remove the links from the search engine as he objected to the processing. Google Spain forwarded the request to its registered office in California, which denied the request again. Desperate to seek redressal, Mario filed a complaint with AEPD (Agencia Española de Protección de Datos), a Spanish data protection agency asking both bodies to remove his personal data from the internet. AEPD rejected the complaint against the newspaper but asked Google Inc. and Google Spain to remove the data. They separately brought an action against the decision before the National Court (Audiencia Nacional) and argued that it does not come under the ambit of EU Directives nor they can be considered controllers. They further added that uploading data cannot be called processing. The court held that Google Inc. is an establishment, and it collects, retrieves, records, and discloses data hence engaging in processing and can be called a controller. This case was decided by the European Court of Justice on May 13, 2014. The ECJ ruled that individuals have the right to request the removal of outdated, inaccurate, or irrelevant information about them from search engine results. This ruling was based on the EU's Data Protection Directive and privacy rights. The case emphasized the responsibilities of search engines and data controllers in managing and processing personal data.

Article 17 of GDPR allows the data subject to ask for the erasure of his personal data if it's no longer necessary for the purpose it was collected, he withdraws his consent and the controller has no legal standing to keep it, he objects to the collection, or the data collection was illegal. European Union's bid to protect an individual's privacy right, although imperfect, is crucial in the times of rapidly growing penetration of the Internet.

In July 2017, the Ministry of Electronics and Information Technology formed a ten member Committee to identify loopholes in data protection and measures to address them. After due deliberations, the committee submitted a report titled "A Free and Fair Digital Economy – Protecting Privacy, Empowering Indians" and draft data protection Bill.

II. Salient Features of the Personal Data Protection Bill, 2021

Personal Data Protection Bill, 2021 has following features:

(i) Data

The 2018 Bill dealt with the processing of the personal data of an individual by the government or any private establishment, whether incorporated in India or abroad. In 2019, the Bill was amended to include non-personal data, as it could also affect privacy and it would be difficult for the government to create authorities for two different types of data that are more or less similar. Scholars believe that this provision is too far overreaching as now it shall include data from the military and companies too.

It allowed processing on the following grounds:

(i)   "any function of Parliament or state legislature, or if required by the State for providing benefits to the individual,

(ii) if required under law or for compliance with any court judgment,

(iii) to respond to a medical emergency or a breakdown of public order,

(iv) purposes related to employment, such as recruitment, or,

(v)   for reasonable purposes specified by the Data Protection Authority concerning activities such as fraud detection, debt recovery, credit scoring, and whistle-blowing."

(ii) Obligations

The Bill laid down certain obligations on the data fiduciary like "processing personal data fairly and reasonably, notifying the data principal of the nature and purposes of data collection, and their rights, among others, and collecting only as much data as is needed for a specified purpose, and storing it no longer than necessary."

(iii) Exceptions

However, certain data processing activities were not subject to the obligations specified. Data principle cannot claim right if the data is for national security (pursuant to a law), prevention, detection, investigation, and prosecution of contraventions to law, legal proceedings, personal or domestic purposes, and journalistic purposes.

(iv) Child Protection

The Bill specially mentioned the need to have stringent provisions to protect the privacy of children. It has prohibited companies from processing data like behavioral monitoring, tracking, targeted advertising, and any other processing harmful to the child. BN Srikrishna committee had recommended creating division in the personal data of children. Websites specially created for children would fall under the category of guardian data fiduciary and they will not be allowed to collect data. However, websites for adults will not fall into this category. The Bill did away with this distinction. This implies that every single data fiduciary has to ensure whether they are dealing with an adult or child, which is a bit tedious task to execute. It also recommended that minors should be allowed to revalidate their consent on attaining majority.

(v) Surveillance

Another problematic provision of the Bill states that any agency that processes personal data in the interest of the sovereignty, security, or public order or for preventing a crime can be exempted by the Central Government from all or any provisions of this Bill if the oversight is fair, reasonable and proportionate. Privacy advocates criticized this provision as it allows the central government to override the cyber security standards and not enough balusters have been created by the committee to protect an individual against state vendetta. Gautam

Bhatia, an advocate working on privacy and data protection, remarked "without addressing surveillance head-on, a data protection law will remain incomplete and ineffective."

The Committee recommended that data principles have the right to be forgotten. They can apply to adjudicators appointed by the government.

III. Overview of Right to be Forgotten

Right to be forgotten is defined as "the ability of individuals to limit, de-link, delete, or correct the disclosure of personal information on the internet that is misleading, embarrassing, irrelevant, or anachronistic." B N Krishna Committee also recommended that this right must be granted but after ensuring that freedom of speech and expression is not compromised in an unjustified manner.

Deputy Information Commissioner of the UK When you unpick it, much of what is there of the right to be forgotten is just a restatement of existing provisions-data shouldn't be kept for longer than is necessary; if it has been processed in breach of the legal requirements it should be deleted, which goes without saying.

In the GDPR, the right to be forgotten does not exist alone, but with the right to erasure. Article 17(1) of the Regulations says, 'the data subject shall have the right to obtain from the controller the erasure of his/her personal data without undue delay, and the controller, on the other hand, shall have to do the same without undue delay if a certain condition exists.

Clause 20 of the Personal Data Protection Bill 2019 allows the data principle to restrict the disclosure of his personal data if:

a. "It has served the purpose for which it was collected or is no longer necessary for the purpose.

b. It was made with the consent of the data principal under section 11 and such consent has since been withdrawn; or

c. It was made contrary to the provisions of this Act or any other law for the time being in force"

The data principal has to apply on the above-mentioned grounds showing that his right to be forgotten overrides the freedom of speech and expression of the data fiduciary.

The application shall be filed before the Adjudicating officer appointed by the central government. He shall before adjudge, have regard to the sensitivity of the data in question, the scale of disclosure by data fiduciary and the extent to which information is asked to be curtailed, the role of the individual in public life, and the importance of the data to the people. Any person can file for the review of the order if he has a reason to believe that the order did not satisfy the abovementioned grounds. An aggrieved party can also appeal to the Appellate Tribunal if not satisfied with the decision.

Recently, the government withdrew the data protection Bill intending to introduce a new draft in early 2023 after due public consultation. Union Minister for Information and Technology stated that "The Personal Data Protection Bill, 2019 was deliberated in great detail by the Joint Committee of Parliament. 81 amendments were proposed, and 12 recommendations were made toward a comprehensive legal framework for the digital ecosystem. Considering the report of the JCP, a comprehensive legal framework is being worked upon."

IV. Role of Indian Judiciary

No legislative or constitutional provision expressly provides for the right to privacy. However, the judiciary has time and again stepped up and in the **Puttaswamy case** declared the right to privacy as well as the right to be forgotten. It enabled data principals to control the information related to them in the physical and virtual worlds. However, it imposed certain restrictions on grounds of the fulfillment of legal responsibilities, execution of duty in the public interest, protection of information in the larger public interest, for scientific or historical studies, for matters related to any legal claim, and the exercise of freedom of expression and information.

Justice Kaul stated that the "right of an individual to exercise control over his personal data and to be able to control his/her own life would also encompass his right to control his existence on the Internet." The judges recognized that individuals are prone to making mistakes and the digital footprints should not hinder their ability to reform.

In **Laksh Vir Singh Yadav V Union of India, the** petitioner has requested the court to erase information pertaining to his wife and mother from an online legal case law database. He claimed that despite his non-involvement in the case, he faced bias and discrimination because the information was in the public domain. The Delhi High Court dismissed his plea.

Kerala High Court in **Civil Writ Petition No 9478 of 2016** gave interim order to a legal database website- Indian Kanoon, to remove the name of the rape victim from the documents. On the other hand, in **Dharamraj Dave vs. State of Gujarat**, the Court rejected the plea for permanent restraint on display and indexing of the non-reportable judgment on the internet on grounds of lack of legal provisions establishing his right to erasure.

Karnataka High Court in **Sri Vasunathan vs. The Registrar General and Ors** gave a similar order. The petitioner requested the erasure of his daughters' name from a case posted online as the parties had later entered into an agreement. The court gave relief but only to the extent of deletion of copies of the case from an internet search. The certified copies of the case on the website of the High Court were allowed.

In **Zulfiqar Ahman Khan vs. Quintillion Business Media Pvt. Ltd. and Ors.**, Delhi High Court held that the right to be left alone is a part of the right to privacy and ruled in favor of the plaintiff and asked the respondent to take down #metoo posts alleging sexual harassment from the internet. In another recent case of **Jorawer Singh Mundy vs. Union of India** the petitioner, an American citizen approached the Delhi High Court requesting it to order the erasure of information related to his case under the NDPS Act. The court emphasized the need for the right to be forgotten and stated that "freedom allows an individual to silence earlier events in his life."

The Madras High Court in **Karthick Theodore vs. The Registrar General**opined that given the lack of legislation regulating data protection, the burden is on the judiciary to fill the gap. It ruled that people who have been accused of offenses and later acquitted by the court have a right to deletion of information from the internet.

V. Balancing the Right to be Forgotten, Right to Information, and Freedom of Speech and Expression

It is to be noted that the genesis of the right to information is in the fundamental right of freedom of speech and expression as a person can be informed only through speech. Hence, in this part, the author shall discuss the anomalous relationship between the right to be forgotten and freedom of speech and expression. In this age, personal data is no less than a digital currency.

Jeffery Rosen in his paper argues that the Right to be Forgotten interferes with someone's right to receive information. When CJEU declared that individuals have the right to be forgotten, advocates of free speech claimed it was nothing but censorship.

This issue was discussed in **Olivier G v. Le Soir** also. A Belgian newspaper was sued for making public its archives which included a newspaper report disclosing the full name of a truck driver involved in an accident. The driver in his plea claimed that he was convicted and duly rehabilitated and requested the information to be removed from the internet. The court ruled in his favor by establishing the right of erasure in cases of a gap of a significant number of years, or disproportionate damage.

The General Comment No 34 of UNHRC has reaffirmed that protection given under Article 19 of ICCPR extends to the online platform also. In India, the same is said under Article 19 of the Constitution. Srikrishna committee emphasized the need to balance both these rights. They proposed a test where the principal's right to the erasure of information is checked on the altar of the right of speech and information of others.

Ashwinee Kumar in his paper argues that in this case, the committee has not considered the rights of the public but of tech companies against the right of the data principal. He states that "All the related fundamental rights should be tested one on one by keeping 'informational self-determination', of, first, on one hand, and 'collective interest' of the rest on the other. Here, 'collective interest' does not mean the society as a whole but concentrates on the digital service providers. The real fight is between the tech moguls and a common man. A war must be fought between equals having almost analogous weapons."

VI. Criticism of the Right to be Forgotten

With the judgment in Google V Spain case, criticism of the Right to be Forgotten had initiated. The consequences of information being permanently accessible over the internet had just started gaining traction.

(i) State as an Exception

Privacy rights activists are criticizing the exception granted to the government under section 35 of the PDP Bill, 2019. Justice B.S. Srikrishna Committee recommended exception to statutory requirements on only one ground, i.e. security of the state. Under the new Bill, the Central government is authorized to any of the privacy safeguards if it believes it to be in the interest of the sovereignty and integrity of India, the security of the state, and friendly relations with foreign countries. It can refuse to comply with the statutory requirements to prevent any offense related to the above-mentioned issues. It is of great concern that the government can use the data of common people without them being aware of it in the name of security and order. It gives undue power to the central government which can be easily misused. Jairam Ramesh said, "separate privileged class whose operations and activities are always in the public interest."

(ii) Dilution of User's Rights

Consent for the processing of personal data can be withdrawn by the data principal at any point. However, the citizen might have to face legal consequences if the state thinks that the withdrawal was devoid of any valid reason. Here, what constitutes a valid reason is not defined. The Bill, on one hand, preaches that consent should be easy to give and withdraw and on the other hand puts the burden of proof on the data principal. Also, the data fiduciaries are permitted under the PDP Bill to charge fees from the principals for processing their requests. These one-sided provisions favor the state heavily and would deter the common masses from exercising their right to withdraw consent under the act. Activists have been claiming that the "document has become prerogative of the rich."

(iii) Non-Consensual Data Processing

The Bill states that the data can be processed without the knowledge and consent of the data principles. The central government can direct fiduciaries to provide it with any non-personal data or anonymous personal data, whereby the government cannot identify the person but still improve the target area for services. It is of major concern, especially in a country like India, where multiple data breaches have happened in the past. Citizens have raised concerns that this provision can be arbitrarily used by the government to discriminate against the people on political and religious grounds.

Further, employers are also exempted from having consent to process data of its employees. They can process it for the purpose of appointment or termination, attendance verification or assessment of performance. It is feared that it would be arbitrarily used by employers to monitor the workers and invade their privacy. Recently, a legal notice was issued to BECIL by Internet Freedom Foundation for floating a tender for a 'personnel tracking GPS.'

(iv) Data Localization

Data Localization can stand for any mandate on the free flow of data across borders. The committee required mirroring of sensitive and personal data in the base country. However, the new Bill allows the transfer of personal data outside India except for sensitive data. This categorization of data is a good move, but it shall lead to increased operational costs for businesses. There are many companies, especially startups, that rely on cloud services for storage. This policy has the potential to cause negative business sentiment. National Institute of Public Finance and Policy in their paper argued that privacy protection needed to depend on the location of the data. Security of the data depends more on technical measures, skills, and cybersecurity protocols.

(v)   Social Media User Verification System

The Bill makes it mandatory for all social media companies to provide users an option to verify their identity. In case of failure to do the same, these companies shall lose the status of intermediaries, which shall make them legally liable for the content posted by their unverified users. Regulation of social media is necessary in times when it can lead to an increase in hate crimes or misinformation. However, building a massive verification system would be possible only for a few giant companies, leading to monopolizing the sector. Another issue is the data principals would have to submit official documents like Aadhar, PAN Card, Ration Card, etc. for verification. There is a possibility of this data being used for advertising agendas of the company without explicit consent of the user and also of a data breach or identity theft happening. Also, it is believed that anonymity on social media has its benefits too. It has helped people in past to blow the whistle against their organizations and ensure identity protection and personal safety.

(vi)   Power Imbalance

A cursory reading of the Bill will make the reader realize that there is an imbalance of power in the hands of the central government. The Bill provides for an independent Data Protection Authority to safeguard the rights of data principals. However, the draft Bill authorizes the central government to appoint members of the data protection authority on basis of recommendations of an outside committee and to remove them according to the law. It also lacks the presence of any judicial member on the panel as suggested by the Srikrishna Committee. It is clear from the draft that this DPA is nothing but a toothless tiger, too fragile to take to task any violation committed by the government.

Also, an adjudicating officer is appointed by the central government to deal with the requests of data principals to avail the right to be forgotten. These officers are to be appointed by the members of the DPA, who are appointed by the state. This violates the doctrine of separation of power.

VII. Conclusions and Suggestions

The Google Spain case had a profound and lasting impact on data protection and privacy laws, not only in Europe but also globally. It emphasized an individual's right to request the removal of outdated, inaccurate, or irrelevant personal information from search engine results and other online platforms. The concept of the Right to be Forgotten was later codified in the EU's General Data Protection Regulation (GDPR), which became enforceable in May 2018. The concepts introduced in the Google Spain case and the GDPR influenced data protection laws and regulations globally. Several countries and regions adopted or updated their data protection laws to align with the principles of the GDPR, including the right to erasure or similar concepts. Search engines are now required to assess and respond to requests for erasure based on specific criteria, as outlined in the GDPR.

The Right to be Forgotten is still in an early stage in India, though recognized by the judiciary. A comprehensive Bill needs to be drafted keeping in mind the suggestions offered by privacy advocates and JPC members. Along with these regulations, a stable accountability mechanism also needs to be formulated. The legislature also needs to propose a test to decide the conflict between the right to be forgotten, the right to information, and freedom of speech and expression.

Srikrishna committee while drafting the framework for the regulation of data collection did not deliberate about the issue of mass surveillance by the state. In the past few years' complaints of targeted surveillance by political activists and journalists have increased. In such a situation it becomes imperative to include provisions regulating the powers of government to collect data from the principals. Even if data needs to be collected or stored, there have to be some constitutional safeguards to protect the interest of the citizens. This is where the role of legislators becomes more prominent than ever. The law should put limitations on the state to ensure the data collection was directly linked with the objective sought to be achieved by the government and if it was collected to the extent it was necessary. In case of non-compliance, the victim should be adequately compensated by a fast-track court. Data Protection Authority should have judicial members on the panel to limit the power of the executive and ensure independence in its functioning. The role of the central government in appointment and termination should be a bare minimum.

The right to be forgotten, freedom of speech and expression, and right to information need to be balanced. The legislature should draft laws and the court should implement them in such a manner as not to give prominence to any single right over the other. A clearly defined test should be laid out to determine what information can be made public and what should remain private. Dr. Mozika in her paper advocates for a twofold assessment. The first is on lines with ECHR, which protects private information like race, religion caste, health status, bank details, or contact information. And the second one evaluates the actual harm caused to the individual because the information is in the public domain. In the end, the concern of data protection should not be left to the ad hoc judicial protection of the court, but a comprehensive law needs to be drafted with a rights-based approach to regulating the collection, storage, and processing of data.

References

1   GDPR Article 17: Right to erasure ('right to be forgotten') - GDPR Software Solutions

2   Theresa M. Payton & Theodore Claypoole, Privacy in The Age of Big Data: Recognizing Threats, Defending Your Rights, And Protecting Your Family 1 (2015).

3   Sushovan Sircar & Vakasha Sachdeva, "Key Highlights from Srikrishna Committee Report on Data Protection", The Quint, 27 July 2018 available at: Srikrishna Committee: Key Highlights From Srikrishna Committee Report on Data Protection and Draft Bill (thequint.com) (last visited 15 October, 2022)

4   Recommendation No. 2, p. 26 available at:https://prsindia.org/Billtrack/the-personal-data-protection-Bill-2019 .

5   Draft Personal Data Protection Bill, 2018 (prsindia.org) (last visited 15 October 2022).

6   "Data Fiduciary" means any person, including the State, a company, any juristic entity or any individual who alone or in conjunction with others determines the purpose and means of processing of personal data.

7   "Data Principal" means the natural person to whom the personal data relates.

8   Committee of Experts under the Chairmanship of Justice B.N. Srikishna, Report on a Free and Fair Digital Economy: Protecting Privacy, Empowering Indians (2018) available at: https://meity.gov.in/writereaddata/files/Data_Protection_Committee_Report.pdf (last visited 15 October, 2022).

9   Dhananjay Dhonchak, "Righ to be Forgotten: Privacy v Freedom", Indian Express, August 2021.

10 Michael J. Kelly and David Satola, "The Right to be Forgotten", University of Illinois Law Review (2017).

11 Ashwinee Kumar, "The Right to be Forgotten in Digital Age: A Comparative Study of the Indian Personal Data Protection Bill, 2018 and the GDPR", Shimla Law Review, 2020.

12 Kazim Rizvi, "New Data Protection Bill must Enable a Progressive Data Governance System", The Times of India, October 9 2022 available at: https://timesofindia.indiatimes.com/blogs/voices/new-data-protection-

Bill-must-enable-a-progressive-data-governance-framework/ (last visited on 22 October 2022)

13 Justice K.S. Puttaswamy (Retd.) & Anr. vs. Union of India & Ors., (2017) 10 SCC 1.

14 Laksh Vir Singh Yadav v Union of India, WP(C) 1021/2016).

15 Dharamraj Bhanushankar Dave v. State of Gujarat, 2015 SCC OnLineGuj 2019.

16 Name Redacted v. The Registrar General, 2017 SCC OnLineKar 424.

17 Zulfiqar Ahmad Khan v. Quintillion Business Media Pvt. Ltd., CS (OS) 642/2018.

18 Jorawar Singh Mundy vs Union of India, MANU/DE/0954/2021.

19 Karthick Theodore v. Madras High Court., 2021 SCC OnLine Mad 2755.

20 Giovanni Sartor, "The Right to be forgotten: Balancing Interest in the Flux of Time", International Journal of Law and Information Technology, 24, 72–98 (2016).

21 Jeffrey Rosen, "The Right to Be Forgotten", Stanford Law Review Online (2012) at p.88

22 Marcus Wohlsen, "For Google, the 'Right to Be Forgotten' Is an Unforgettable Fiasco", Wired, July 3, 2014, available at http://www.wired.com/2014/07/google-right-tobeforgotten-censorship-is-an-unforgettable-fiasco (visited on 19 August, 2022).

23 P.H. v. O.G., Cour de Cassation Belgique, Apr. 29, 2016, N° C.15.0052.F (Belg.).

24 Hiroshi Miyashita, "The 'Right to Be Forgotten' and Search Engine Liability" Brussels Privacy Hub (2016).

25 Available at: https: //brusselsprivacyhub.eu/BPH-Working-Paper-VOL2-N8 (last visited Jan. 18, 2020).

26 Vakasha Sachdeva, "Why is the JPC Report on the Personal Data Protection Bill being Criticised", The Quint, December 17, 2020 available at: https://www.thequint.com/tech-and-auto/tech-news/jpc-report-personal-data-protection-Bill-criticism-govt-exemption-dpa-control-social-media-verification (last visited on October 22, 2022).

27 The Wire Staff, "Data Protection Bill: Congress MPs File Dissent Notes over JPC Report," The Wire, 22 November,2002. available at: https://thewire.in/government/jpc-report-pdp-Billjairam-ramesh-dissent-note-

unbridled-exemptions-government-agencies (last visited on 20 October 2022).

28   Barik, Soumyarendra and Aashish Aryan, "US Bodies Push Back on Data Protection Bill, Seek New Working Group," Indian Express, 3 March 2022, available at: https:// indianexpress.com/article/india/us-bodiespush-back-on-data-protection-Bill-seek-newworking-group-7798193 (last visited on 20 October 2022)

29   Rishabh Bailey, "The issues around data localization", The Hindu, February 25, 2020 available at: https://www.thehindu.com/opinion/op-ed/the-issues-around-data-localisation/article62108462.ece (last visited on 20 October 2022)

30   Dasgupta, Surajeet, "India's Data Localisation Rules to be a Barrier to Digital Trade: US," Business Standard, 11 April 2022, available at: https://www.business-standard.com/article/economy-policy/india-s-datalocalisation-rules-to-be-a-barrier-to-digital-tradeus-122041100008_1.html (last visited on 20 October 2022)

32   Mandavia, Megha, "Personal Data Protection Bill Can Turn India into 'Orwellian State': Justice B N Srikrishna," Economic Times, 12 December 2022, available at: https://economictimes.indiatimes.com/news/ economy/policy/personal-data-protection-Billcan-turn-india-into-orwellian-state-justice-bnsrikrishna/articleshow/72483355.cms?. (Last visited on 24 October 2022).

33   Dr Jyoti Mozika, "Integrating the Right to be Forgotten in the Indian Legal Framework in the Light of Experiences from the European" Indian Journal of Law and Justice, Vol. 12 Issue No. 1, available at: https://ir.nbu.ac.in/bitstream/123456789/4138/1/Integrating%20the%20Right%20to%20be%20Forgotten%20in%20the%20Indian%20Legal%20Framework%20in%20the%20Light%20of%20Experiences%20from%20the%20European%20Union.pdf (last visited on 22 October 2022).

Chapter 35

A Critical Analysis of Privacy Rights and Data Protection Laws in India

Vinita Singh

Ph.D. Research Scholar

SSOL, Sharda University

Greater Noida, U.P., India

Dr. Ritu Gautam

Assistant Professor

SSOL, Sharda University

Greater Noida, U.P., India

Abstract

Privacy is the right to keep personal information, thoughts, and activities secluded from public view or unauthorized access. It encompasses an individual's ability to control what information about themselves is collected, shared, and used by others. This control allows people to maintain a sense of autonomy, protect their identity, and determine when and how their personal data is disclosed or utilized. Various sectors in India, such as banking, healthcare, and telecommunications, have their own regulations and guidelines pertaining to data protection and privacy. Privacy even though now no longer expressly furnished beneath the Constitution; it implicitly takes into it the proper to privacy as non-public liberty assured beneath Article 21. There is an inherent battle between proper to privacy and records safety. The records safety can also additionally encompass monetary details, fitness information, commercial enterprise proposals, highbrow belongings, and touchy records. Data safety and privacy were dealt in the Information Technology (Amendment) Act, 2008 however now no longer in an exhaustive manner. The IT Act is not enough in safety of records and consequently, a specific law in this regard is required. The purpose of this paper is to provoke a critical debate on proper privacy and records safety with inside the Indian perspective.

Keywords: Article 21, Constitution, Information Technology, India, privacy.

I. Introduction

The goal of this Chapter is to begin a genuine discussion on the privilege to security and information insurance from the Indian purpose of perspectives. Security, despite the fact that it is not explicitly accommodated in an Indian Constitution, certainly suggests the benefit of protection as a flexibility under Article 21. There is an inborn clash between the privilege to security and information insurance. Information assurance may incorporate money related subtleties, wellbeing data, strategic plans, protected innovation, and classified information. Information assurance and protection have been tended to inside the IT Law of 2008, however not thoroughly. The IT Law is not adequate in information assurance and along these lines separate enactment is required in such manner.

II. The Right to Privacy

Article 21 secures the privilege to protection and promotes respect for the person. As if late there has been growing concern about huge measure vast amount of data on individual found in a computer document. The security privilege refers to a person's particular option to control the variety, use, and disclosure of individual data. Individual data might be as close to home interests, propensities and exercises, family and instructive records, interchanges (counting postal and phone records), clinical records, and money related records. An individual could without much of a stretch be hurt by the presence of electronic information about him/her that is off base or deluding and that could be moved to an unapproved outsider at rapid and at exceptionally ease. This development in the utilization of individual information has numerous advantages yet could likewise create numerous issues.

III. Telephone Tapping and Privacy

The privilege to security is influenced by new advancements. The privilege to security identified with an individual's correspondence has gotten a matter of discussion because of innovative turns of events. There have been instances of capture attempt of messages and phone correspondence from political rivals, just as employment candidates. Indian Post Office law, section 5 (2) and Indian Telegraph law, section 26 (1) engage the focal and State gov.to catch transmitted and post linter changes in a crisis open in light of a legitimate concern for open wellbeing, would be an infringement of Articles 21 and 19 (1) (a) of the Constitution.

The Popular Civil Liberty vs. UOI that the option to have a telephone discussion in the security of Office without obstruction can positively be asserted as a privilege to protection. For this situation, the SC set up certain procedure rules for directing lawful block attempts, and furthermore settled a significant level audit council to research the pertinence of such captures. However, such alert has been tossed into the air in late government office mandates as obvious by wiretapping episodes that have become known.

IV. Freedom and Privacy of Women

Assault is not only a wrongdoing against the individual of a lady, but also a wrongdoing against the entire of society. As a casualty of a sex wrongdoing, I would accuse nobody however the guilty party. The attacker not just disregards the security and individual trustworthiness of the person in question, however, unavoidably causes genuine mental and physical damage simultaneously. Assault is not just an ambush: it is regularly damaging to the casualty's whole personality. The privilege to protection is a basic prerequisite for the human character to grasp inside itself the high feeling of ethical quality, respect, tolerability, and direction towards esteem.

The topic of the connection between benefit to and conjugal right initially emerging Sareetha vs. Vankta Subbaih, where the Andhra Pradesh High Court maintained the arrangements of Section 9 of the Hindu Marriage Act of 1955, that is, the compensation of conjugal rights, as unlawful as it abuses Article 21 of Indian Constitution corresponding to one side to protections.

The inquiry is whether the privilege to security envelops the lady's choice. A lady's right to decide on regenerative choices is also an element of individual flexibility as contemplated in Article 21 of Indian Constitution. Conceptual choices can be made to reproduce and avoid multiplication. The urgent thought is that ladies' nobility and substantial uprightness must be regarded. Conceptive rights incorporate the privilege of ladies to convey the full-term pregnancy, conceiving an offspring and thus bring up children. A lady's entitlement to end her pregnancy is not outright and might be constrained somewhat by authentic interests of the state in defending ladies. insurance likely human life.

V. Press, Electronic Media and Privacy

The freedom of the press has not been explicitly referenced in the article 19 of Indian Constitution yet has been deciphered to be verifiable in it. In Rajgopala vs. Tamil Nadu express, the SC held the candidates save the benefit to disperse what they case to be auto Shankar's memoir/life record to the extent that it appears visible to everyone records, even without their consent or endorsement. In any case, if they go further and circulate their memoir, they may be assaulting your privilege to insurance. Constitution exhaustively records the suitable clarifications behind constraining chance of enunciation in Article 19 (2), it would be difficult for the court to incorporate security as one more inspiration to drive reasonable impediments. Thusly, a woman who is the loss of assault, getting, seizing or a similar bad behavior should not be needy upon the shock of her name and the scene that is conveyed in the media.

VI. Information Privacy

An information assurance or data security is the association between the arrangement and dispersing of data advancement, the open want for insurance, and the legal and strategy driven issues including them. The degree to which privacy ought to be shielded could be comprehended from certain cases.

The inquiry emerges how much a voter has an option to think about a competitor's protection. The voter's entitlement to think about a competitor's security can be ensured and flourished by disposing of the traps of laws identified with voters' entitlement to data. Protection suggests the choice to control the correspondence of before long conspicuous information about any person. It requires a mentality of equalization, a reasonable intrigue. Hence, at last, it requires a solid and charming interrelation between social great and individual opportunity. In this manner, it is reasoned that one must keep up a harmony between the privilege to data of a resident and the privilege to protection of an up-and-comer looking for races.

VII. Health and Privacy

As indicated by clinical morals, a specialist is required not to reveal the mystery data about the patient, as the divulgence will contrarily influence or imperil the lives of others. In Mr. ''X'' vs Medical Clinic ''Z'' the SC held that the specialist persistent relationship, albeit essentially business, is expertly a matter of trust and, along these lines, pros are morally and ethically dedicated to take care of security. In such a condition, open introduction of even clear private events can a portion of the time lead to confrontation of 1 person's privilege to be dismissed with another person's qualification to be taught. For another situation, the preliminary court said the emergency clinic or specialist was available to reveal the data to people identified with the young lady he expected to wed and that she reserved an option to think about the appealing party's HIV status. The court likewise held that the litigant's privilege was not influenced at all by uncovering her HIV status to the family members of her life partner.

In Selvi vs. Karnataka, SC held that the microanalysis, lie discovery and beep test coincidentally disregard endorsed protection limits. A clinical assessment can not legitimize the weakening of sacred rights, for example, the privilege to security. On the off chance that DNA testing is prominently important to get to reality, the court must exercise the dissector of an individual's clinical assessment. In this manner, the Supreme Court was of the supposition that despite the fact that the privilege to the individual freedom has been perused in Article 21, it cannot be tasted as a flat-out right. To permit the court to arrive at a reasonable determination, an individual could be exposed to a test despite the fact that it would attack your entitlement to security.

Article 21 of the Indian constitution: 'No individual will be denied of her life or individual freedom, aside from as per the methodology set up by law. The target of this essential right is to forestall the intrusion of individual freedom'. In spite of the fact that "security of the privilege to protection, property and information" is not determined in the article, the Supreme Court of India deciphers it and incorporates it under "individual freedom". The IT Act of 2000 was acquainted with order the development of electronic exchanges, to give lawful acknowledgment to electronic business and electronic exchanges, to encourage electronic government, to forestall cybercrime, and to guarantee safe practices and strategies being used of the data. innovation around the world. In any case, India has gotten mechanically impartial because of the appropriation of electronic marks as a lawfully substantial method of mark execution. The article talks about the developing significance of electronic and computerized media, including necessities for copyright proprietors to give more consideration to electronic and advanced rights issues. Contrasted with the laws of created countries, since the laws of India expect information to be characterized dependent on utility and significance, the article prescribes changes to the standards identified with information extraction, assurance and obliteration, just as corporate shield approaches and support of consistence commitments.

VIII. Data Protection Law in India

Data protection implies the game plan of security law and philosophy that mean to restrict interference into a person's insurance realized by the arrangement, amassing and dissipating of individual data. Singular data generally suggests information or data that relates to a person who can be recognized from that information or data, whether or not it is assembled by any lawmaking body or any private affiliation or association.

The constitution does not give the major right to protection. In any case, the court have perused the privilege of society in the existing other basic right, that is, opportunity of articulation and articulation under article 19 (1) (a) and the privilege to life and personal freedom in under article 21 of the Indian Constitution. In any case, these principal right under the Indian constitution are dependent upon sensible limitations set out in article 19 (2) might be forced by the state. As of late, the recorded instance of justice KS Putaswamy vs UOI., The constitution bank of the SC maintained the privilege to protection as a central right.

The IT Act 2000 arrangements with the issue relating to portion of compensation (civil) and discipline (criminal) if there ought to emerge an event of ill-advised presentation and maltreatment of individual data and encroachment of lawfully restricting terms in respect of individual data.

Under Section 43 A of the IT law of 2000, overseeing or managing any delicate individual data or information, and imprudent in realizing and keep up reasonable security takes a shot at achieving uncalled for adversity or inappropriate expansion to any person, by then such body corporate may be held in danger to pay damages to individual so impacted. It is basic to observe that, there is no most extreme limit showed for the compensation that can be ensured by the impacted party in such conditions.

The IT Rules provide sound security practices and methodology, which the corporate body or any person who, for the benefit of the corporate body, collects, obtains, stores, manages or manages data, must follow in administering "individual sensitive data or information". If an occurrence arises from any penetration, the corporate body or some other individual that tracks for the benefit of the corporate body, the corporate body could be subject to damages for the highly influenced individual.

According to popular demand, you may orchestrate any fitting government association to catch, screen, or unravel or cause any information made, transmitted, got, or set aside on any computer resource for be blocked or checked or unscrambled. This zone draws in the gov. to catch, screen or translate any information, recollecting information of an individual sort for any computer resource. Exactly when the information is with the ultimate objective that it should be revealed in the open intrigue, the gov. may require the exposure of such information. Information related to antagonistic to national activities that are against national security, violates of law or genuine commitment or blackmail can fall into this class.

IX. Information Technology Act, 2000

IT Act of 2000, is a demonstration to lawful acknowledgement to exchange made through the electronic trade of information and other electronic methods for correspondence, ordinarily known as "electronic business ", which incorporate the use of elective strategies for paper corresponding and the limit of information to support the electronic account of chronicles with government associations.

(i)   Reasons Why the Government Can Interfere with The Data

As demonstrated by Section 69 of the IT Law, any individual approved by the government or any of its specialists remarkably affirmed by the government if they are convinced that it is significant or fitting to do as such considering a genuine worry for the influence or uprightness of India, the opposition of India, state security, heartfelt relation with solicitation or to avoid incitation to the commission of any unquestionable bad behavior related to the previously mentioned or for the assessment of bad behavior, for inspirations to be recorded as a printed version , according to popular demand, they can direct any organization association to catch, screen or decipher or catch to be gotten or watched or unscrambled any information made, transmitted, got or set aside in any PC resource. The extent of Section 69 of the IT law incorporates both block attempt and checking alongside unscrambling to explore cybercrime. The government has likewise told the IT Rules (Procedure and Safeguard for the Interception), 2009, in the past area.

X. Conclusions

Data Protection is an essential human right. Sections IX and XI of the Information Technology Law characterize the obligations regarding the infringement of classification and protection of information identified with unapproved access to the PC, PC framework, system or PC assets, change, cancellation, expansion, alteration, decimation, duplication or unapproved transmission of information, PC database, and so on. Information security may incorporate monetary subtleties, wellbeing data, strategic agreements, licensed innovation, and secret information.

In any case, today any data identified with anybody can be gotten to from anyplace whenever, yet this speaks to another danger to private and classified data. Globalization has grasped innovation around the world. As indicated by the expanding prerequisite, various nations have presented diverse legitimate structures Like Data Protection Act of 1998 (UK), Electronic Privacy Act of 1986 (of America), etc. every once in a while. In the US, there are some uncommon security laws to ensure understudy instruction records, youngsters' online protection, people's clinical records, and private money related data. In the two nations, self-administrative endeavors encourage the meaning of improved security conditions.

The privilege to security is perceived in the Constitution, yet its development and advancement is absolutely helpless before the legal executive. In the present associated world, it is hard to keep data from getting away into the open area on the off chance that somebody is resolved to get it out without utilizing amazingly abusive techniques. Information insurance and protection have been tended to inside the Information Technology (Amendment) Act, 2008, yet not thoroughly. The IT Law must set up explicit guidelines identified with the techniques and reason for acclimatizing the privilege to security and individual information. To finish up, it would do the trick to state that the IT Law faces the issue of information insurance, and that different enactment is required for information security that accomplishes a compelling harmony between individual flexibilities and protection.

References

1   IT Act, 2000.

2   Data Protection Act.

3   Electronic Communications Privacy Act, 1986.

4   https://www.nap.edu/read/5756//7.

5   http://www.oecd.org/env/country-review.pdf.

6   Universal Declaration of Human Rights, 1948.

7   International Covenant on Civil and Political Rights, 1966.

8   Convention on the Rights of the Child, 1989.

9   Indian Penal Code, 1860.

10 Constitution of India.

Chapter 36

Covid 19 Pandemic: A Need for Robust Cybersecurity Response

Rishabh Singh Randhawa
B.B.A.L.L.B. (Hons.) Student
SSOL, Sharda University
Greater Noida, UP, India

Utkarsh Paliwal
B.B.A.L.L.B. (Hons.) Student
SSOL, Sharda University
Greater Noida, UP, India

Abstract

Since March 2020, the world has witnessed a significant increase in digital connectivity, making it more susceptible to cyber threats than ever before. The surge in online communications and the widespread adoption of remote business operations have amplified the risk of cyberattacks to an unprecedented level. This digital transformation has also introduced a plethora of new vulnerabilities. Organizations are now facing the threat of breaches in their perimeter security, necessitating continuous surveillance and real-time risk assessment for both physical and digital entry points. In this evolving landscape, cybersecurity professionals must proactively address these risks. Initially, they should promptly raise awareness among their company's remote workforce about potential scams and provide them with training to avoid falling victim to these threats. E-learning and web-based training platforms are valuable tools in this regard. Additionally, IT security experts must adopt a forward-thinking approach, recognizing that remote work might persist long after the pandemic has subsided. Security leaders should evaluate existing security solutions for scalability, provisioning speed, cloud-based availability, and remote management capabilities. Furthermore, they should proactively collaborate with trusted partners to plan for dynamic scalability and the delivery of necessary services and solutions. Leaders in the field of cybersecurity are facing an increasing need to embrace innovative approaches and consider new operational technologies. To support these initiatives, security leaders should leverage digital technology and service models that are optimized to achieve more with fewer resources.

Keywords: Covid, Cyber, Security, Risk, Global.

I. Introduction

The COVID-19 pandemic has sent shockwaves through our world, introducing unprecedented levels of uncertainty and anxiety. Alongside the health crisis, it has triggered a radical shift in the way organizations operate, with a sudden and widespread demand for remote work. Many businesses found themselves grappling with the need to swiftly adapt to this new reality, prompting hasty adjustments in both physical office spaces and hastily crafted remote work policies.

This rapid transition to remote work has posed numerous challenges, especially in the realm of cybersecurity. A glaring issue is the lack of preparedness among most organizations. Indeed, research indicates that a mere 38% of businesses had established cybersecurity policies prior to the pandemic, leaving the majority ill-equipped to navigate this digital transformation securely. As the world embraces remote work as the new normal, it has brought about an alarming increase in potential attack vectors and data security risks.

One of the most notable aspects of this remote work shift is the reliance on employees' personal devices and home networks. These environments are inherently less secure than corporate settings and often lack the robust security measures required to safeguard sensitive information. Even in cases where businesses provided their staff with company devices, these devices are typically configured with minimal administrative rights. Consequently, when employees require additional software installations or permissions, it can become a logistical nightmare.

This evolving landscape has forced organizations to grapple with the need for more realistic solutions. While empowering employees with greater rights and flexibility can enhance productivity, it also opens up a Pandora's box of potential security vulnerabilities. Striking the right balance between convenience and security is a delicate task that requires careful planning and implementation.

Furthermore, the pandemic has given rise to a troubling trend in cybersecurity. As the world became preoccupied with the health crisis, cybercriminals seized the opportunity to exploit vulnerabilities in both individuals and systems. Security incidents and cyber threats have surged, posing a significant and ongoing challenge to businesses and institutions worldwide. The need for robust cybersecurity measures has never been more critical.

In light of these pressing concerns, it is imperative for organizations to address cybersecurity comprehensively. This includes not only establishing robust policies but also investing in employee training and awareness programs. Ensuring that employees are well-versed in cybersecurity best practices can significantly reduce the risk of security breaches. Additionally, companies must take proactive steps to secure remote work environments, such as implementing multi-factor authentication, encryption, and regular security audits. The COVID-19 pandemic has ushered in a new era of remote work, significantly impacting the way organizations operate. This shift has exposed vulnerabilities in cybersecurity preparedness, leading to increased risks and threats. To navigate this new landscape effectively, businesses must prioritize cybersecurity, striking a balance between employee empowerment and data security. Failure to do so could leave them exposed to a range of cyber threats that continue to evolve in the wake of the global pandemic. As we move forward, addressing these challenges remains paramount to safeguarding our digital future.

II. Challenges in Ensuring Cybersecurity Amid the COVID-19 Pandemic

The COVID-19 pandemic has laid bare the vulnerabilities of our current institutions when it comes to safeguarding human health and well-being. In this unprecedented global crisis, two critical issues have come to the forefront: the deficiency of timely and accurate data and the rampant spread of misinformation. These shortcomings have exacerbated the harm caused by the pandemic and exacerbated the tensions between public health concerns and data privacy. As a result, the COVID-19 crisis is not only a health crisis but also an information crisis and a crisis of trust. One of the most glaring weaknesses exposed by the pandemic is the inadequacy of our data infrastructure. Timely and accurate data are the lifeblood of effective public health responses, yet the pandemic has revealed significant gaps in our ability to collect, analyze, and disseminate such information. These gaps have hindered our ability to make informed decisions and allocate resources where they are most needed. The lack of reliable data has hampered efforts to track the spread of the virus, identify hotspots, and implement targeted interventions. This has resulted in a slower and less effective response to the pandemic, leading to more infections and deaths than might have been preventable with better data systems in place. Moreover, without accurate data, it has been challenging to assess the impact of various public health measures, making it difficult to fine-tune our strategies as the situation evolves. In addition to the data deficit, the COVID-19 crisis has highlighted the alarming prevalence of misinformation. Misinformation has flourished in the absence of authoritative and accurate sources of information. This has led to confusion, fear, and mistrust among the public, further undermining our ability to mount an effective response to the pandemic. The spread of false information about COVID-19 has ranged from misleading claims about miracle cures to conspiracy theories about the origins of the virus. Such misinformation has not only sowed doubt about the safety and efficacy of vaccines but has also fueled public resistance to public health measures such as mask-wearing and social distancing. This has created a dangerous disconnect between scientific consensus and public opinion, making it harder to implement evidence-based policies. Moreover, the pandemic has strained the delicate balance between public health concerns and data privacy. Efforts to track the virus's spread and trace contacts have raised legitimate concerns about individual privacy. In some cases, these concerns have resulted in delays and limitations in contact tracing efforts, potentially allowing the virus to spread unchecked. The erosion of trust in institutions during the COVID-19 crisis is another critical issue that cannot be ignored. People's faith in governments, public health authorities, and the media has been shaken by mixed messages, changing guidance, and political polarization. This has made it more difficult to rally public support for necessary measures and has contributed to the patchwork response observed in many countries. Furthermore, supply chain failures have been glaringly evident during the pandemic, particularly concerning personal protective equipment (PPE) and life-saving ventilators in clinics and hospitals. The inability to source these critical items quickly and reliably has put healthcare workers and patients at risk. This highlights the need for resilient supply chains and better preparedness for future health crises.

The healthcare sector has emerged as a prime target for cyberattacks, particularly during the ongoing pandemic. These relentless hacking attempts have exposed the glaring vulnerabilities and pervasive cybersecurity challenges plaguing the healthcare industry, encompassing a wide spectrum of entities, from healthcare providers and institutions to pharmaceutical companies and research organizations. The ramifications of these cyber threats have been substantial, exemplified by incidents such as the devastating WannaCry ransomware attack that crippled the National Health Service (NHS) in 2017. A pivotal factor contributing to the susceptibility of healthcare organizations to such attacks is their often constrained budgets, a reality dictated by their reliance on funding from cities or countries that typically subject them to stringent fiscal controls.

The widespread adoption of the **Internet of Things (IoT) in the healthcare sector** is hindered by a significant barrier: the looming threat of cyber risks. Central to this challenge is the imperative to safeguard patients' privacy while simultaneously preventing unauthorized tracking and verification of their sensitive data. IoT, with its promise of enhanced healthcare services, introduces novel opportunities for cyberattacks and the improper capture of personal information, necessitating vigilant security measures. One primary reason why applications based on IoT are vulnerable to cyberattacks lies in the wireless nature of many IoT communications. This wireless environment makes eavesdropping relatively easy, particularly when robust encryption is not employed.

This vulnerability underscores the critical need for robust encryption protocols to protect the confidentiality and integrity of healthcare data transmitted via IoT devices. Additionally, the low energy consumption characteristic of most IoT components adds complexity to the security landscape. While the energy efficiency of IoT devices is a fundamental advantage, it poses challenges in implementing robust security schemes. Balancing energy efficiency with security measures remains an ongoing concern for healthcare practitioners and IoT developers alike. Interoperability across diverse IoT platforms emerges as another crucial factor in ensuring the accessibility and safety of healthcare services. In a fragmented IoT landscape, achieving seamless data exchange and collaboration among various devices and systems becomes imperative. This interoperability can enhance the efficiency of healthcare delivery, reduce costs, and ultimately improve patient care. A globalized healthcare environment also introduces concerns related to data sharing across international borders. While data security, confidentiality, and privacy are typically subject to federal regulations within individual countries, international hosts or suppliers may not always adhere to domestic laws and regulations. This misalignment underscores the need for international cooperation and agreements to establish a comprehensive framework for data protection in cross-border healthcare scenarios. One approach to addressing these challenges is the implementation of federal regulations that mandate data security, confidentiality, and privacy standards across IoT-based healthcare applications. These regulations can serve as a foundation for safeguarding patient information and ensuring its integrity. However, achieving international consensus on such regulations remains a complex and ongoing endeavor. To mitigate the inherent cyber risks associated with IoT in healthcare, a multi-faceted strategy is required. This strategy should encompass the following key elements: Robust Encryption: Implementing strong encryption protocols for IoT communications is paramount to protect against eavesdropping and data breaches. Energy-Efficient Security Solutions: Developing security solutions tailored to the energy constraints of IoT devices is essential. This involves striking a balance between security and energy efficiency. Interoperability Standards: Establishing industry-wide interoperability standards ensures that IoT devices and platforms can seamlessly exchange data, promoting accessibility and safety in healthcare services. International Collaboration: Encouraging international cooperation and agreements to harmonize data protection regulations across borders can address the challenge of data sharing in a global healthcare ecosystem. In conclusion, while IoT holds immense potential for revolutionizing healthcare, it also introduces significant cyber risks that must be addressed comprehensively. Patients' privacy and data security should remain paramount, with a focus on encryption, energy-efficient security solutions, interoperability, and international collaboration. By proactively addressing these challenges, the healthcare industry can harness the transformative power of IoT while ensuring the safety and privacy of patients' information on a global scale.

A glaring illustration of this financial constraint is the prevalent use of outdated software and unsupported operating systems within healthcare institutions, including the archaic Windows 7 and Windows XP, which persist in managing critical medical devices throughout hospitals. In this landscape, Europol has asserted that healthcare facilities present an alluring and lucrative target for ransomware attackers. The modernization of healthcare has led to an extensive reliance on computers and the Internet of Things (IoT) in contemporary hospitals. These technologies are harnessed extensively to store and monitor patient data, oversee the operation of medical devices such as intensive care units (ICUs), and manage vital equipment like ventilators.

Amidst this cybersecurity crisis, a joint advisory report and guidelines issued by the United Kingdom's National Cyber Security Centre (NCSC) and the United States Department of Homeland Security (DHS) Cyber Security and Infrastructure Security Agency (CISA) have provided invaluable insights and recommendations. These guidelines comprehensively address issues ranging from the ever-present threat of phishing attacks to the security challenges associated with remote work tools, including platforms like Zoom.

Despite efforts to combat cyber threats, it is ominously predicted that Advanced Persistent Threat (APT) groups will persist in targeting healthcare and essential services on a global scale. Recent collaborative assessments by the NCSC and Canada's Communications Security Establishment (CSE) have strongly indicated the involvement of Russian intelligence services, specifically APT29, colloquially known as "Cozy Bear," in cyber-attacks against organizations involved in the development of COVID-19 vaccines. These malicious endeavors are primarily aimed at pilfering critical information pertaining to COVID-19 vaccines.

To achieve their nefarious objectives, APT29 employs an array of sophisticated techniques, including vulnerability scanning, the deployment of publicly available exploits, and the insidious practice of phishing. These tactics enable them to infiltrate target networks covertly. Additionally, APT29 utilizes custom-designed malware, going by the monikers 'WellMess' and 'WellMail,' to propagate further damage within compromised systems.

In summation, the healthcare sector's vulnerability to cyber threats has been magnified during the pandemic, underscoring the pressing need for robust cybersecurity measures. Budgetary constraints, outdated technology, and the increasing reliance on interconnected digital systems have rendered healthcare organizations easy prey for cybercriminals.

Collaborative efforts and advisory reports from cybersecurity agencies offer a glimmer of hope, but the persistence of APT groups and nation-state actors in targeting healthcare services worldwide remains a formidable challenge that demands ongoing vigilance and enhanced cybersecurity practices.

III. Key Technologies employed in the battle against COVID-19

The emergence of tactile edge technology, particularly in the context of 5G and beyond 5G (B5G) networks, has ushered in a new era of possibilities in the fight against the COVID-19 pandemic. This innovative technology represents a significant leap forward compared to its predecessor, 4G-enabled technology, and promises to revolutionize our approach to controlling novel diseases like COVID-19. At its core, the integration of edge computation with 5G wireless networks forms the backbone of this transformative approach.

One of the key advantages of this paradigm shift is the establishment of a hierarchical system of edge computing. This hierarchical structure offers several benefits, including scalability, low latency, and enhanced protection for training model data. These advantages are crucial in addressing the unique challenges posed by a pandemic, where rapid response and data security are paramount.

Pervasive edge computing plays a pivotal role in augmenting security measures. By extending the computational capabilities closer to the data source, this approach minimizes the vulnerabilities associated with central data processing. In the context of healthcare, especially during a pandemic, ensuring the highest level of security for sensitive medical data is non-negotiable.

A noteworthy development in this regard is the creation of a B5G-based healthcare framework, tailored specifically to combat pandemics like

COVID-19. This framework comprises three integral layers: the cloud layer, the edge layer, and the stakeholder layer. This layered approach allows for a holistic and coordinated response to the pandemic's challenges.

The integration of this framework with a surveillance system is a game-changer. It empowers authorities to monitor crucial parameters such as mask-wearing, social distancing, and body temperature testing on a large scale. These surveillance measures are pivotal in curbing the spread of the virus and ensuring public safety.

One of the most remarkable outcomes of this technological revolution is the development of a highly efficient COVID-19 diagnostic method. This method enables the identification of patients without COVID-19 infection, a crucial step in preventing overcrowding in hospitals and conserving vital healthcare resources. Furthermore, this diagnostic method operates with the utmost sensitivity and can process sensitive personal data securely, adhering to stringent privacy standards.

In parallel, an authentication scheme tailored for cloud-assisted vehicular ad hoc networks (VANETs) has been developed. This innovation extends the reach of the technology beyond healthcare and into other critical domains, such as transportation. It allows for the timely measurement of passengers' physical status without the need for direct contact, leveraging the power of vehicular cloud (VC).

Moreover, the integration of a blockchain-based vehicle recording mechanism enhances the security and transparency of data management in VANETs. This blockchain technology ensures the immutability and integrity of data, making it an invaluable asset in maintaining the integrity of vital transportation systems, especially during a pandemic. The convergence of tactile edge technology with 5G and B5G networks has opened up a world of possibilities in the fight against COVID-19 and similar pandemics. From enhanced surveillance and diagnostic methods to secure data handling and transportation innovations, these technological advancements are paving the way for a safer and more resilient future. As we continue to harness the potential of these technologies, the collective effort to combat COVID-19 remains steadfast, empowered by the tools of the digital age.

The COVID-19 pandemic has reshaped the way we work, and the concept of working from home has become the "new normal" for many individuals. As we look towards the post-pandemic era, it's evident that remote work is here to stay. Even as vaccines are distributed and life begins to return to some semblance of normalcy, a significant portion of the workforce is expected to continue working from home. This paradigm shift has not only impacted the way we work but has also brought about a series of security and reputation risks, particularly for businesses handling sensitive data. The pandemic forced organizations to adapt rapidly to the new reality of remote work. In doing so, they had to grapple with the sudden surge in network activities as employees connected to company servers from their home offices. This transition came with its own set of challenges, including the need to implement robust security measures to protect sensitive information. Businesses had to quickly pivot their strategies to ensure that their remote workforce remained secure and productive. One of the primary concerns that emerged during this shift to remote work was the increased vulnerability to cyberattacks. As employees accessed company networks from various locations, hackers adapted their tactics accordingly. Instead of solely targeting corporate offices, they began focusing on individuals' home networks and the platforms they used for remote communication, such as Zoom and Netflix. This shift in focus posed a significant threat to both businesses and individuals. The security risks associated with remote work are multifaceted. Employees working from home often use personal devices, which may not have the same level of security as their office counterparts. This creates potential entry points for cybercriminals to exploit. Additionally, the use of unsecured Wi-Fi networks at home can expose sensitive data to interception and hacking attempts. Furthermore, the blurred lines between personal and professional life in a remote work setting can lead to inadvertent security breaches. Employees may be more relaxed about security protocols when working from the comfort of their homes, potentially exposing confidential information. The absence of physical supervision and face-to-face interactions can also make it harder for organizations to monitor and enforce security measures. In addition to security concerns, the shift to remote work has also raised questions about the reputational risks faced by businesses. A breach of sensitive data can have far-reaching consequences, damaging a company's reputation and eroding trust among customers and partners. The public disclosure of a data breach can lead to a loss of confidence, financial losses, and legal ramifications. To mitigate these risks, organizations have had to invest in advanced cybersecurity measures, including encryption, multi-factor authentication, and employee training on security best practices. Many have also implemented strict policies to govern remote work, emphasizing the importance of maintaining the same level of security diligence at home as in the office.

IV. Utilizing Blockchain Technology to enhance Cybersecurity

Block chain technology is well-suited for supply chain management as it provides immutable and distributed ledgers with auditable records. It offers secure and unchangeable record-keeping, making it valuable for tracking assets in various sectors, especially in healthcare during the COVID-19 pandemic. Block chain can help governments and hospitals identify COVID-19 cases, locations with reported cases, and high-risk areas. It enhances healthcare data security and facilitates patient tracking and symptom analysis in affected countries. The lack of data for assessing COVID-19 risks led to its rapid spread, especially considering asymptomatic cases. Permissioned block chain offers advantages in such scenarios by allowing members of a medical consortium to verify transaction and information details, and any attempts to block information would result in a change in the hash, preventing tampering. This ensures the secure transmission of personal records. For COVID-19 sequences, a closed hub controls access and prevents unauthorized redistribution, addressing commercial concerns that hinder data sharing. Blockchain also provides proof of the existence and content of specific data objects. The Centers for Disease Control and Prevention (CDC) in the United States use blockchain to manage medical data, identify symptom patterns, and track medical supply chains, ultimately enhancing diagnostic accuracy and treatment effectiveness. The "Social Internet of Things" (SIoT) integrates people and smart devices within IoT, often through IoT platforms, but faces security and privacy issues, particularly in cloud IoT ecosystems. Blockchain can enhance the security and privacy of digital health systems. A blockchain-based system was proposed for secure home quarantine management. A novel blockchain-based framework is being developed to facilitate global tracking of COVID-19 cases and testing. Blockchain's security features, including tamper resistance and immutability, bridge supply chain visibility gaps, especially during the pandemic. IBM's "Rapid Supplier Connect" is an example of a blockchain-based network that strengthens the medical supply chain. The combination of AI and blockchain is proposed for self-testing and tracking systems to increase testing rates and risk stratification. Blockchain is effective for data sharing between groups. Its applications in healthcare encompass health data analytics, remote patient monitoring, pharmaceutical supply chain management, and electronic medical records (EMRs) management. Blockchain can reduce clinical bias when managing electronic health records (EHR). Interoperability challenges among various EHR systems can be addressed by using separate block chain systems.

V. Conclusions

The COVID-19 pandemic has exposed significant vulnerabilities within our healthcare sector and the broader cybersecurity landscape. The pandemic has highlighted several critical issues, including data infrastructure fragility, misinformation proliferation, privacy concerns, supply chain failures, and increased cyber threats. Maintaining reliable data sources is essential for effective public health responses, yet we've struggled to achieve this, leading to a slower and less effective pandemic response. Misinformation has thrived in the absence of authoritative sources, hampering efforts to control the virus's spread. Privacy concerns have delayed contact tracing, and erosion of trust in institutions has complicated public support for necessary measures. Supply chain failures, especially for personal protective equipment, have put healthcare workers and patients at risk, emphasizing the need for resilient supply chains.

The healthcare sector's susceptibility to cyberattacks due to budget constraints, outdated systems, and the adoption of the Internet of Things (IoT) necessitates modernization and robust cybersecurity measures. IoT introduces privacy risks, and interoperability challenges arise when sharing healthcare data across borders. To address these challenges, a comprehensive strategy is essential, involving robust encryption, energy-efficient security solutions, interoperability standards, and international cooperation. Federal regulations can establish data security and privacy standards. Vigilance is necessary against Advanced Persistent Threat groups and nation-state actors targeting healthcare services.

Key technologies like tactile edge technology, 5G, and beyond 5G networks have played vital roles in managing the pandemic. They enhance scalability, reduce latency, and ensure data security. Efficient diagnostic methods have been developed, and blockchain technology has secured healthcare data, tracked assets, and improved supply chain visibility.

The COVID-19 pandemic has reshaped our work and brought security risks to the forefront. Mitigating these risks requires robust cybersecurity measures and strict policies. Blockchain technology has proven invaluable in enhancing healthcare security. As we navigate the challenges of the pandemic and beyond, integrating innovative technologies and cybersecurity practices is crucial for a safer and more resilient future. Prioritizing healthcare cybersecurity, infrastructure investment, and regulations are essential for safeguarding public health and well-being.

References

1   Scherer F.M. and David Ross's work (1990), titled "Industrial Market Structure and Economic Performance," published by Houghton Mifflin in Boston, pages 613-660.

2   Case C-418/01, IMS Health, from the year 2004, reported in the European Court Reports (ECR) under volume I, page 5039.

3   Case T-201/04, Microsoft v. Commission, dated [2007] and located in ECR volume II.

4   C-241 and 242/91 P, RTE and ITP v. Commission ('Magill'), detailed in [1995] ECR I-743.

5   World Health Organization Coronavirus disease 2019 (covid- 19) situation report , https://apps.who.int/iris/handle/10665/332 388 7Google Scholar.

6   C. Sohrabi, Z. Alsafi, N. O'Neill, M Khan, A. Kerwan, A. Al-Jabir, C. Iosifidis, R. Agha, World health Organization declares global emergency: a review of the 2019 novel coronavirus (covid-19), Int. J. 76 71-76.

7   Barka, F. Breitinger, K.-K.R. Choo the Role of National Cybersecurity Strategies on the Improvement of Cybersecurity Education Computers and Security (2022), Article 102754.

8   S. Alrabaee, R. Manna Boosting students and teachers' cybersecurity awareness during covid-19 pandemic 2021 IEEE Global Engineering Education Conference (EDUCON), IEEE (2021), pp. 726-731.

9   S. Alrabaee, M. Al-Kfairy, E. Barka Efforts and suggestions for improving cybersecurity education 2022 IEEE Global Engineering Education Conference (EDUCON), IEEE (2022), pp. 1161-1168.

10   M. Jakovljevic, S. Bjedov, N. Jaksic, I. Jakovljevic Covid-19 pandemia and public and global mental health from the perspective of global health security Psychiatry. Danub., 32 (1) (2020), pp. 6-14.

11   J. K. Wagner, Health, housing, and "direct threats" during a pandemic, J. Law Biosci, 2020(1):1.